Structural Concrete

Strut-and-Tie Models for Unified Design

Structural Concrete

Strut-and-Tie Models for Unified Design

Salah El-Din E. El-Metwally
Wai-Fah Chen

CRC Press
Taylor & Francis Group
Boca Raton London New York

CRC Press is an imprint of the
Taylor & Francis Group, an **informa** business

CRC Press
Taylor & Francis Group
6000 Broken Sound Parkway NW, Suite 300
Boca Raton, FL 33487-2742

Printed on acid-free paper

International Standard Book Number-13: 978-1-4987-8384-2 (Hardback)

Library of Congress Cataloging-in-Publication Data

Names: El-Metwally, Salah El-Din E., author. | Chen, Wai-Fah, 1936- author.
Title: Structural concrete : strut-and-tie models for unified design / Salah El-Metwally and Wai-Fah Chen.
Description: Boca Raton : CRC Press, 2017. | Includes bibliographical references.
Identifiers: LCCN 2017015368| ISBN 9781498783842 (hardback : acid-free paper) | ISBN 9781315155500 (e-book)
Subjects: LCSH: Reinforced concrete construction. | Strut-and-tie models. | Structural design. | Structural analysis (Engineering)
Classification: LCC TA683 .E49 2017 | DDC 624.1/8341--dc23
LC record available at https://lccn.loc.gov/2017015368

Visit the Taylor & Francis Web site at
http://www.taylorandfrancis.com

and the CRC Press Web site at
http://www.crcpress.com

Dedication

The strut-and-tie model (STM) is an extension of the limit analysis as applied to continuous media by Drucker, Greenberg, and Prager in 1952, and extended to reinforced concrete members. The concept of using the method of STM to the inelastic-reinforced-concrete analysis was introduced and illustrated for the first time in 1961 by Drucker in his estimate of the load-carrying capacity of a simply supported reinforced concrete beam. The application of the theory of plasticity to the design of members under shear and torsion began in the 1970s, especially by Thürlimann, Nielsen, and others. The efforts of Schlaich, Schäfer and their colleagues, and students at the institute of structural design, University of Stuttgart, in the 1980s and 1990s, have established the method in its current form by the generalization of the original truss model concept proposed to treat shear problems by Ritter (1899) and Mörsch (1902, 1906, and 1909) and utilization of the achievements in the area of the lower bound theorem of limit analysis. They treated all technical problems including model development, material strength of the different components, and solution scope, in order to achieve a unified and consistent treatment of all regions of structural concrete including those with or without web reinforcement and with or without axial force or pre-strains (tension or compression); all are treated in the same manner.

In recognition of their efforts and remarkable contributions to the method of strut-and-tie model, this book is dedicated to:

em. Prof. Dr.-Ing. Jörg Schlaich

em. Prof. Dr.-Ing. Kurt Schäfer

Salah El-Din E. El-Metwally
Wai-Fah Chen

Contents

Preface

One of the most important advances in reinforced concrete in recent years is the extension of lower-bound limit theorem-based design procedures, for example, the strut-and-tie model (*STM*), to shear, torsion, bearing stresses, and the design of structural discontinuities, such as joints, corners, openings, and deep beams. The concept of using the method of strut-and-tie models to the inelastic-reinforced-concrete analysis was introduced and illustrated for the first time in 1961 by Drucker in his estimate of the load-carrying capacity of a simply supported reinforced concrete beam.

The application of the theory of plasticity to the design of reinforced concrete members under shear and torsion began in the 1970s, especially by Thürlimann and Nielsen and their coworkers. This also formed the basis for the method of strut-and-tie models after the work of Schlaich and his coworkers in the 1980s and 1990s. The method has been well-developed worldwide over the past two decades, presented in several texts, and also introduced in many codes of practice, which triggered the acceptance and wide daily use of the method. The development of the method has brought a major breakthrough in design for a consistent theory in the design concept covering both discontinuity- and Bernoulli-regions with similar models. In particular, the method provides a formal design procedure for reinforced concrete detailing.

This book is devoted to the application of the method of strut-and-tie models in the design of structural concrete. In order to put the method into perspective, the theorem of limit analysis with its lower and upper bounds is first presented. The method of *STM*, as a lower bound solution, is further demonstrated with emphasis on model development and optimization and modeling of standard discontinuity-regions. The failure criteria of the model elements are discussed, with attention to the provisions and recommendations of the ACI 318-14 Code. The method is applied to different classes of regions with attention to the detailing. Structural concrete design is treated in a unified manner with consistency in the treatment of both discontinuity- and Bernoulli-regions. The method is also utilized to explain the behavior of concrete elements and regions in response to boundary forces and reinforcement detailing.

This book is addressed to students, researchers, and, in particular, practicing engineers.

We are grateful to the sincere efforts of engineers R. M. El-Garayhi and A. K. Ghoraba who prepared the drawings of the book. The authors especially thank engineer A. K. Ghoraba for his effort in checking the solutions of some book examples.

Salah El-Din E. El-Metwally
Wai-Fah Chen

Authors

Dr. Salah El-Din E. El-Metwally has a broad experience in structural design of many educational and industrial structures, bridges, and large-scale roof structures. For more than three decades, he has been active in research in different areas of structural engineering such as stability, behavior, and design of concrete structures, conceptual design, and application of numerical methods in structural engineering, in addition to solar energy utilization. For many years, he has been an active member in the Standing Committee and Subcommittees of the *Egyptian Code for the Design and Construction of Concrete Structures*. His academic positions in Egypt include head of the Structural Engineering Department at Mansoura University and Tanta University. He was a fellow of Alexander von Humboldt at the Institute of Structural Design, University of Stuttgart, the Department of Concrete and Masonry Structures, the Technical University of Munich, and a visiting professor at the University of Hawaii at Manoa.

S. E. El-Metwally, PhD, Purdue University, MSc, George Washington University, honors include the state prize in structural engineering and the encouraging state prize in engineering science, both granted by the Egyptian Academy of Science and Technology. Honors also include the award of distinction of first class granted by the President of Egypt. He is currently a professor of concrete structures at Mansoura University.

Dr. Wai-Fah Chen, a well-respected leader in the field of plasticity, structural stability, and structural steel design over the past half-century, he has made major contributions to introduce the mathematical theory of plasticity to civil engineering practice, especially in the application of limit analysis methods to the geotechnical engineering field. Having headed the engineering departments at the University of Hawaii and Purdue University, Chen is a widely cited author and the recipient of several national engineering awards, including the 1990 Shortridge Hardesty Award from the American Society of Civil Engineers and the 2003 Lifetime Achievement Award from the American Institute of Steel Construction.

W. F. Chen, PhD, Brown University, Member of U.S. National Academy of Engineering and Member of Taiwan's National Academy of Science (Academia Sinica), Honorary Member of American Society of Civil Engineers, formerly Distinguished professor of civil engineering at Purdue University and dean of engineering at the University of Hawaii.

1 Principle of the Strut-and-Tie Model

1.1 INTRODUCTION

This chapter traces the method of strut-and-tie model (*STM*) back to its roots, the limit analysis. Then, the method as a lower-bound solution is introduced by first explaining why the method is a necessity in those regions where the bending theory cannot be applied. The method has come as an appropriate approach to treat all structural concrete elements and components consistently and in a unified manner.

The chapter covers the limit theorems with both the upper- and lower-bound solutions, the principle of the method of *STM*, the approaches to develop a *STM*, and the constituent elements of the model; struts, ties and nodes. The regions where the bending theory is applicable, B-regions, are distinguished from those disturbed regions, D-regions, where that theory cannot be applied. The chapter also gives a historical sketch on (1) the development of the truss model which is the basis for the design of B-regions; (2) the start of the *STM* from the limit analysis; and (3) the major development of the *STM* method with connection to the truss model for unified and consistent design of structural concrete.

1.2 LIMIT THEOREMS OF PERFECT PLASTICITY

1.2.1 Introduction

In order to carry out plastic analysis and design effectively in the real world of engineering, we shall deal with idealizations of idealizations. As observed from tests, once the material is well into the plastic range, it exhibits relatively low additional resistance to increasing load. This feature can be simply captured by ignoring the small resistance in the plastic range and idealizing the material as perfectly plastic. The consequence of such an idealization and the selection of proper flow strength depend on the problem to be solved as on the material itself. For example, for moderate plastic strain range in most structural engineering design, the flow strength to be chosen for the perfect plasticity idealization should be the average strength of the applicable range of the strains of the nonlinear stress–strain behavior as represented by the horizontal solid lines in Figure 1.1 (Chen and El-Metwally, 2011).

1.2.2 Why Limit Analysis

In order to obtain a valid solution in continuum mechanics, three conditions should be satisfied: equilibrium, compatibility, and constitutive relations. For some cases,

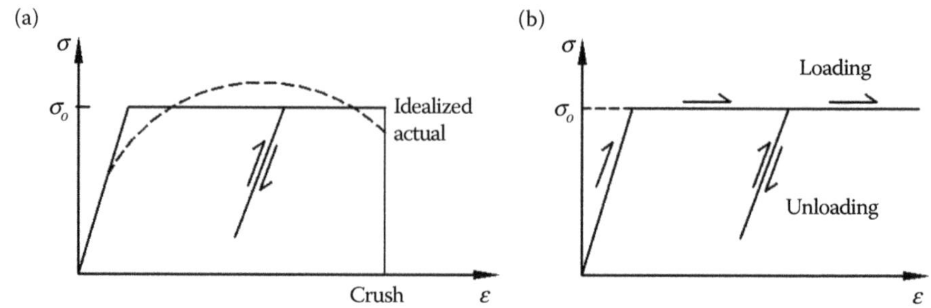

FIGURE 1.1 Uniaxial stress–strain relationship of an elastic-perfectly plastic material: (a) material with limited plastic strain, and (b) highly ductile material.

it is difficult to satisfy all three conditions. For simplicity, in limit analysis instead, only two of the three conditions are necessary for a simpler solution. Historically, engineers, based on intuition, developed many simple solutions from equilibrium for weak-tension material and from kinematics for ductile materials, which have now been justified by limit analysis. The theorems of limit analysis give us a very powerful tool to estimate lower and upper bounds of the collapse load of structures or structural members without having to go through a very tedious calculation procedure. In both theorems, strain-hardening of material is ignored but its effect can be reflected realistically in the selection of a proper level of flow strength as illustrated in the preceding section, which is acceptable from the practice point of view. This further idealization of perfect plasticity enables the proof of the powerful limit theorems, which provide an excellent guide for preliminary design as well as analysis of structure. The development of the limit theorems and their illustrative engineering applications are described next.

In case of a lower-bound solution of the collapse load, only equilibrium and yield criterion are satisfied; equilibrium is satisfied for stress or generalized stress. The solution so obtained represents a good safe guidance for the structural engineer and it can be used to verify solutions from other methods quickly. The method is useful for application to different materials, especially tension-weak material, for example, stones or concrete. Hence, the safety of monumental structures such as cathedrals can be checked very well with such a simple equilibrium method following the flow of forces using simple hand calculations.

In an upper-bound solution of the collapse load, only kinematics and yield criterion are satisfied. The method is especially good for ductile material and even applicable to some material with limited ductility but with modifications. The method uses an engineer's physical intuition on failure modes and their corresponding collapse analysis can be made by hand calculations. Thus, it gives the engineer enough clarity of vision to produce a structure that is understandable and works well with the force of nature.

1.2.3 Basic Assumptions

The collapse load obtained from limit analysis is different from the actual plastic collapse load since it is calculated for an ideal structure in which the deformation is assumed to increase without limit while the load is held constant. Of course, this assumption is not expected to happen in real structures but only in idealized structures in which neither work-hardening of the material nor significant changes in geometry of the structure occur. However, the limit load still represents a good estimate of the real collapse load.

The idealization of a structure analyzed using the limit analysis theorems comes from the following two basic assumptions (Chen and Han, 1988; Chen and El-Metwally, 2011):

1. *Perfectly plastic material*, that is, the material of the structure is assumed to be perfectly plastic with the associated flow rule without strain-hardening or softening, Figure 1.1b. In this simplification, many effects are ignored; for instance, effect of time is eliminated from calculations, while effect of residual stresses on initial yielding and effect of local buckling on maximum plastic moment capacity of steel sections are ignored. In addition, the complex states of stresses and strains in reinforced concrete as a result of bond and cracks are very much simplified.
2. *Small deformation of the structure*, that is, the changes in geometry of the body or the structure which may occur at the limit load are negligible; hence, the geometric description of the body or structure remains unchanged during the deformation at the limit load. This assumption allows the use of the principle of virtual work, which is the key to the proof of the limit theorems.

1.2.4 Tresca Yield Criterion

The elastic limit of a material is defined as yielding and it is determined under a combined state of stresses by a yield criterion. For the simple tension test, this limit is the yield stress, σ_o, while in shear test it is the yield shear stress, τ_o. For the general state of stress, this limit can be expressed as

$$f(\sigma_{ij}, k_1, k_2, \ldots) = 0 \tag{1.1}$$

where k_1, k_2, … are material constants to be determined experimentally.

For isotropic material, the orientation of the principal stresses is immaterial, and the values of the three principal stresses suffice to describe the state of stress uniquely. A yield criterion therefore consists in a relation of the form

$$f(\sigma_1, \sigma_2, \sigma_3, k_1, k_2, \ldots) = 0 \tag{1.2}$$

For isotropic material independent of hydrostatic pressure, such as metals, the influence of hydrostatic pressure on yielding for such type of material is not appreciable; therefore, the yield criterion can be simplified as in the von Mises and Tresca yield criteria.

For illustration only, the Tresca yield criterion (1870) is used here. This criterion states that yielding would occur when the maximum shear stress at a point reaches a critical value k. In terms of principal stresses, the maximum shear stress is the maximum value of half the difference between the principal stresses taken in pairs, that is,

$$\max\left(\frac{1}{2}|\sigma_1-\sigma_2|,\frac{1}{2}|\sigma_2-\sigma_3|,\frac{1}{2}|\sigma_3-\sigma_1|\right)=k \tag{1.3}$$

The material constant k can be determined from the simple tension test

$$k=\frac{\sigma_o}{2} \tag{1.4}$$

where σ_o is the material yield stress. In this criterion, yielding is governed by the maximum and minimum principal stresses.

1.2.5 Lower-Bound Theorem

This theorem states that "*if an equilibrium distribution of stress can be found which balances the applied loads, and is everywhere below yield or at yield, the structure will not collapse or will be just at the point of collapse.*"

Hence, the lower-bound theorem requires the justification of only two of the three sets of conditions necessary for solution in continuum mechanics, that is equilibrium and yield condition (material law). This theory therefore expresses the ability of the ideal body to adjust itself to carry the applied loads if at all possible. In practice, the application of the lower-bound theorem has different versions depending on the structural material of the system. For example, in steel frames the method is called the *statical method*, while in concrete there is the strut-and-tie model (*STM*) method.

In order to illustrate the application of the lower-bound theorem, the following two examples are discussed. In the first example, Figure 1.2, a long prismatic bar of a rectangular cross section ($b \times t$) with one hole of diameter d is subjected to an axial force P. If the yield stress of the bar material is σ_o, the simple two discontinuous stress fields shown in the figure can be assumed. In the two stress fields, the bar is divided into strips, the continuous strips have simple tension σ_o, and the discontinuous strips are stress free. Then the lower-bound load of this bar is

$$P^L=\sigma_o(b-d)t \tag{1.5}$$

The second example is a rigid punch indentation into a half-space of perfectly plastic material, Figure 1.3. Assume that the width of the punch in the direction perpendicular to the plane of paper is so large that this punch problem is considered a plane strain problem.

As a first attempt, consider the simple discontinuous stress field shown in Figure 1.3a, which yields a lower bound on the limit load

$$P_1^L=\sigma_o b=2\text{ kb} \tag{1.6}$$

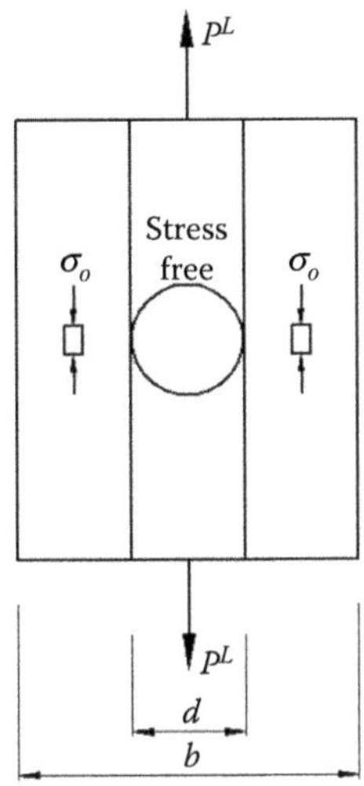

FIGURE 1.2 Lower-bound solution of a bar with a hole.

This is of course not a good lower bound because the load is considered to be carried only by a single vertical strip of material directly beneath the punch. To improve the answer, consider adding a horizontal pressure field as shown in Figure 1.3b. In the overlapping region, the material is subjected to a biaxial compression so that the vertical stress can be increased to $2\sigma_o$ without violating the yield condition. The improved lower bound obtained is

$$P_2^L = 2\sigma_o b = 4 \text{ kb} \tag{1.7}$$

Alternatively, the concept of the truss-action approach can be assumed. If the load P is carried by two inclined truss bars as shown in Figure 1.4a and, further, a vertical leg is added directly below the punch area AA of amount $2k$ to give the stress field shown in Figure 1.4b, in this case the stress discontinuities are admissible. It is noted

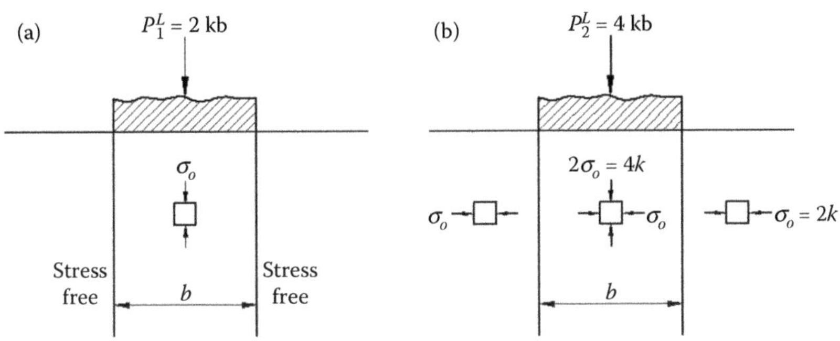

FIGURE 1.3 Stress fields for punch indentation in plane strain: (a) single vertical strip, and (b) two dimensional load transfer.

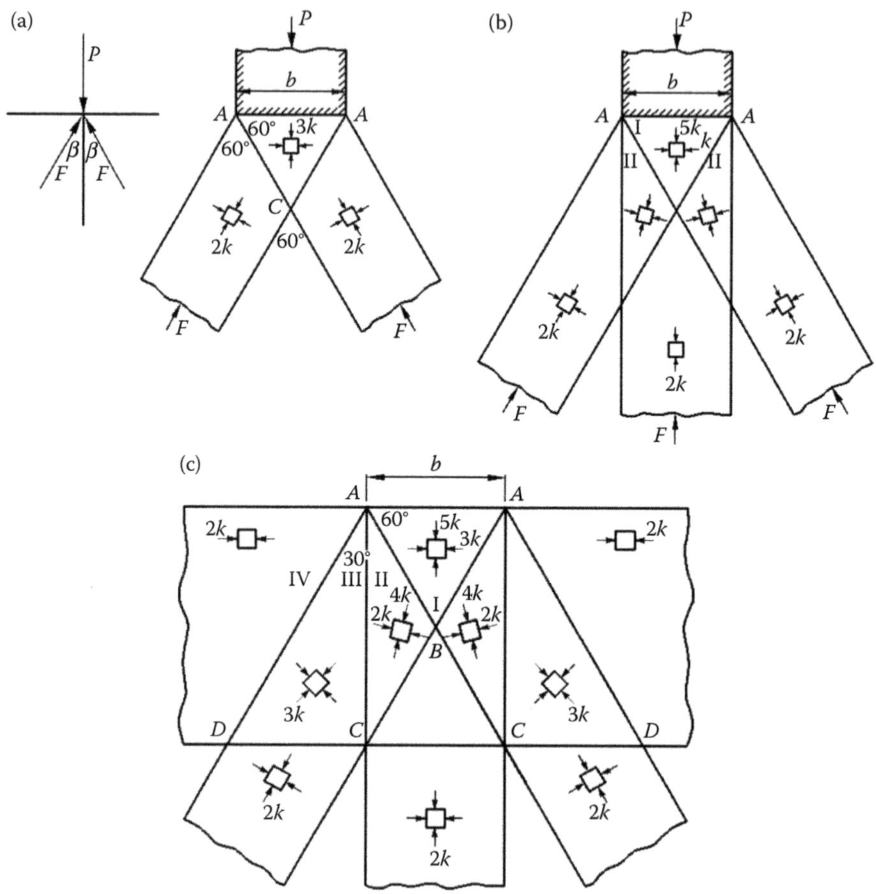

FIGURE 1.4 Load P carried by truss bars: (a) two-leg stress field, (b) three-leg stress field, and (c) combined stress field.

that the yield condition is violated in regions I and II. In region I, for example, the difference between the greatest and the least principal stress is $4k$. This violation can be accommodated by introducing at the free surface a horizontal strip in which there is a horizontal compressive stress $2k$. The width of this strip is as shown in Figure 1.4c. Using this stress field, a better lower-bound solution can be obtained.

$$P_3^L = 5\,\text{kb} \tag{1.8}$$

1.2.6 Upper-Bound Theorem

This theorem states that "*the structure will collapse if there is a compatible pattern of plastic failure mechanism for which the rate of work of the external forces equals or exceeds the rate of internal dissipation.*"

The upper-bound theorem thus requires the justification of only two of the three sets of conditions necessary for solution in continuum mechanics, that is, kinematics and yield criterion. The theory therefore states that if a path of failure exists, the ideal body will not stand up. In practice, the mechanism method of steel beams and frames and the yield line theory of concrete slabs are two different versions of applications of the upper-bound theorem.

In order to illustrate the application of the upper-bound theorem, the following two examples are discussed. The first example is the bar with one hole, Figure 1.2, which was solved using the lower-bound theorem. For this bar, three different compatible discontinuous failure modes are shown in Figure 1.5. In mode 1, Figure 1.5a, the upper and lower parts of the bar move as rigid bodies relative to each other by sliding along the planes AB and CD perpendicular to the face of the bar and making angle α as shown in the figure. If the relative tangential velocity of separation is $\dot{\delta}$, the velocity of separation is $\dot{\delta}\sin\alpha$ and the rate of external work is then $P_1^U\dot{\delta}\sin\alpha$. The rate of energy dissipation over the whole sliding surface is therefore $k\dot{\delta}(b-d)t/\cos\alpha$. Hence,

$$P_1^U = \frac{2k(b-d)t}{\sin 2\alpha}$$

For a minimum value of P_1^U, sin 2α is set equal to 1 ($\alpha = 45°$). This gives

$$P_1^U = 2k(b-d)t \tag{1.9}$$

k is the yield shear stress which is equal to $(\sigma_o/\sqrt{3})$ according to von Mises and $(\sigma_o/2)$ according to Tresca. Based on the von Mises yield criterion

$$P_1^U = \frac{2\sigma_o}{\sqrt{3}}(b-d)t = 1.15P_1^L \tag{1.10}$$

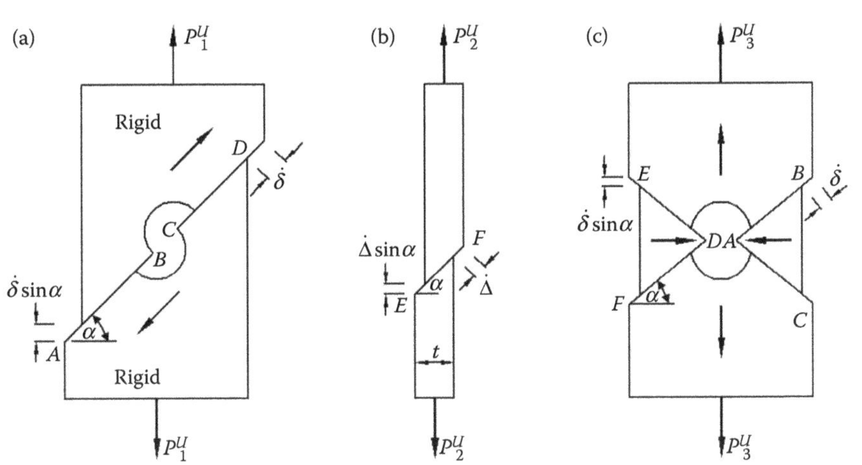

FIGURE 1.5 Kinematically admissible velocity fields of a bar with a hole: (a), (b) and (c) failure modes 1, 2 and 3.

Based on the Tresca yield criterion

$$P_1^U = \sigma_o (b - d)t = P_1^L \tag{1.11}$$

Mode 2, if the bar is assumed to be very thin (i.e., it is an assumed plate), and mode 3 will give the same results as mode 1.

The second example is the rigid punch indentation into a half-space of perfectly plastic material, Figure 1.6, for which a lower-bound solution was derived in the preceding section. Since the punch is assumed to be rigid, the geometric boundary condition requires that when the contact plane moves it always remains plane. Two types of mechanisms, rotational and translational, are discussed next.

The simple rigid-body rotational mechanism about O, Figure 1.6b, is considered geometrically admissible if there are no constraints to hold the punch vertical. The block of material B rotates as a rigid body about O with an angular velocity $\dot{\alpha}$, and there is a semicircular transition layer between the rotating material and the remainder of the body. Since the angular velocity is $\dot{\alpha}$, the rate of work done by the external force P is the downward velocity at the center of the punch, $\dot{\alpha}b/2$, multiplied by P, while the total rate of energy dissipation along the semicircular discontinuity surface is found by multiplying the length of this discontinuity, πb, by the yield stress in pure shear, k, times the velocity across the surface $b\dot{\alpha}$. Equating the rate of external work to the rate of total internal energy dissipation gives

$$P^U \left(\tfrac{1}{2} b\dot{\alpha}\right) = k(b\dot{\alpha})(\pi b)$$

or

$$P^U = 2\pi \text{ kb} = 6.28 \text{ kb} \tag{1.12}$$

It is noted that the upper-bound solution is independent of the magnitude of the angular velocity $\dot{\alpha}$, which means that $\dot{\alpha}$ can be assumed to be sufficiently small not to disturb the overall geometry. In other words, the proofs of the limit theorems can carry through using the initial geometry of the problem.

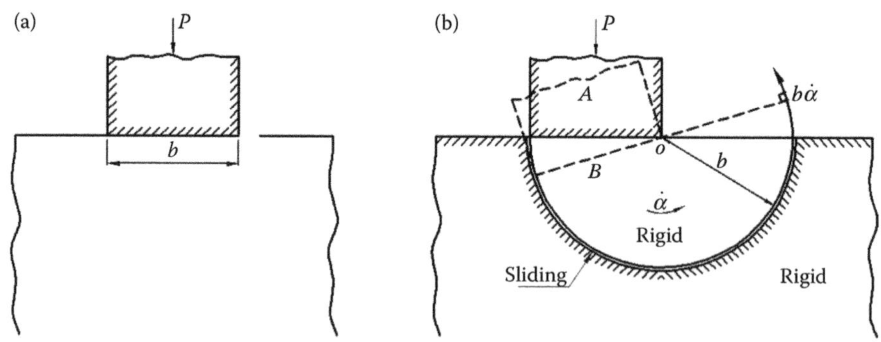

FIGURE 1.6 Punch indentation problem: (a) punch indentation and (b) rotational mechanism.

The rotational mechanism of Figure 1.6b may be generalized by taking the radius and the position of the center of the circle as two independent variables, aiming to obtain a better upper-bound estimate. In this case, a better estimate of the upper-bound solution can be obtained,

$$P^U = 5.53\,\text{kb} \tag{1.13}$$

The mechanism involving only rigid-body translations, shown in Figure 1.7, involves rigid-block sliding separated by plane velocity discontinuities. This mechanism represents a rough punch, which requires that the punch and the triangular block ABC have the same velocity and therefore move together. It is noted that the mechanism is symmetrical about the center line and therefore only the right half of each is examined for kinematics. In this case, the triangular region ABC in Figure 1.7a moves downward with the punch as a rigid body, that is, both have the same velocity, $v_1 = v_o$. The two triangular regions of material BCD and BDE move as rigid bodies in the direction parallel to CD and DE, respectively. The velocity of the triangle BCD is determined by the condition that the relative velocity v_{12} between this triangle and the triangle in contact with the punch must have the direction of BC. The velocity of the third triangle is determined in a similar manner. The information regarding velocities is represented by the velocity diagram (or hodograph) as shown in Figure 1.7b.

From Figure 1.7b, the velocities of the rough punch mechanism of Figure 1.7a are

$$v_2 = v_3 = v_{23} = v_{12}/2 = v_o/\sqrt{3}$$

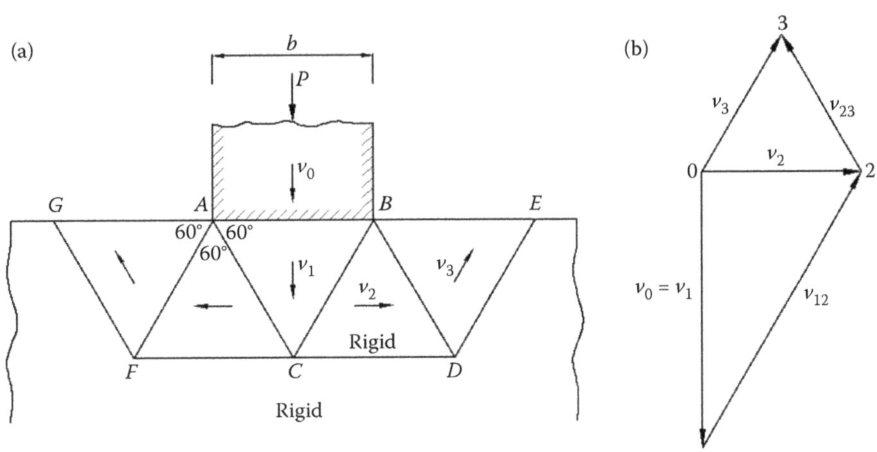

FIGURE 1.7 Simple rigid-block translation and associated velocity diagram for a rough punch: (a) rigid-block translation, and (b) velocity diagram.

and the work equation is

$$P^U v_o = 2k(bv_{12} + bv_2 + bv_{23} + bv_3)$$

Thus,

$$P^U = 5.78\,\text{kb} \tag{1.14}$$

It is noted that the best upper-bound estimate is that given by Equation 1.13, $P^U = 5.53$ kb, while the best lower-bound estimate from the preceding section is $P^L = 5.0$ kb. The correct limit load for this problem is $P = 5.14$ kb.

1.3 THE STRUT-AND-TIE MODEL - A LOWER BOUND SOLUTION

1.3.1 Introduction

The application of stress fields to reinforced concrete design based on the concept of a lower-bound theorem of limit analysis is of more recent development and it represents one of the most important advances in reinforced concrete. *STM* is based on the lower-bound theorem of limit analysis, and hence it provides a *safe solution.* Since the method is based on the equilibrium approach, *only two conditions are justified, equilibrium, and failure criteria.* In this model, the complex stress distribution in the structure is idealized as a truss carrying the imposed loading through the structure to its supports. Similar to a real truss, *STM* consists of compression struts and tension ties interconnected at nodes. Using stress legs similar to those sketched in Figure 1.8, a lower-bound stress field that satisfies equilibrium and does not violate yield criteria at any point can be constructed to provide a safe estimate of capacity of reinforced concrete structures with discontinuities (Chen and Han, 1988; Chen, 1982). As will be illustrated in later examples, these techniques will have the advantage of allowing a designer to follow the forces through a structure with discontinuities which formerly were beyond the scope of engineering practice.

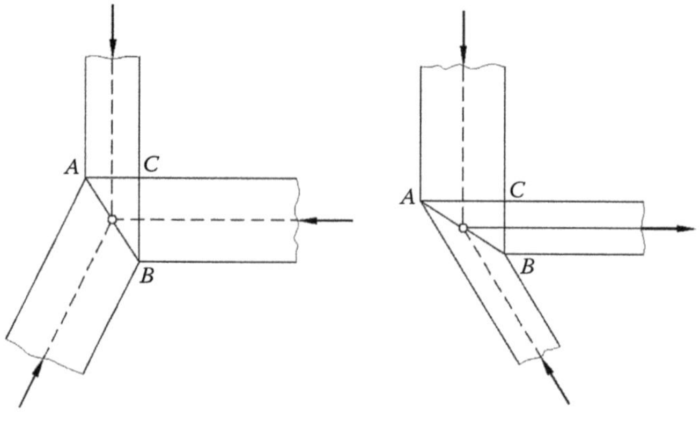

FIGURE 1.8 Using stress-legs as truss members to produce a stress field at a stress joint.

The strut-and-tie model, *STM*, is a logical extension of the truss model and the major difference between the two methods is that *STM* is a set of forces in equilibrium but do not form a stable truss system. Thus, *STM* is a generalization of the truss model. The truss model has been recognized in academia and practice to be the most reliable tool for the treatment of shear and torsion in structural concrete Bernoulli or bending regions (or simply *B-regions*). *STM* is currently recognized as the most reliable tool for the treatment of discontinuity or disturbance regions (or simply *D-regions*).

1.3.2 Concept

STM is an idealization of the stress resultants derived from the flow of forces within a region of structural concrete. The successful model should satisfy two conditions, equilibrium and failure criteria, and the solution so obtained is a safe or lower-bound solution.

STMs are derived from the flow of forces within structural concrete regions, namely, those of high shear stresses, where the Bernoulli hypothesis of flexure, plane sections before bending remain plane after bending, does not apply. Those regions are referred to as discontinuity or disturbance regions (or simply *D-regions*), in contrast to those regions where the Bernoulli hypothesis is valid, and are referred to as Bernoulli or bending regions (or simply *B-regions*). The flow of forces in B-regions can be traced, of course, but in this case the model will yield to as the special case of *STM*.

Discontinuity (which is associated with high shear stresses) is either static (as a result of concentrated loads) or geometric (as a result of abrupt change of geometry) or both. Examples of D-regions are illustrated in Figure 1.9. The dividing sections between B- and D-regions can be assumed to lie approximately at a distance h from the geometric discontinuity or the concentrated load, where h is equal to the thickness of the adjacent B-region, Figure 1.9. This assumption is justified by Saint Venant's principle (1870).

In an *STM*, a strut represents a concrete stress field with prevailing compression in the direction of the strut. On the other hand, a tie represents one or several layers of tension reinforcement. However, concrete ties may exist in models where no reinforcement is available and reliance is on the concrete tensile strength. Examples where tensile stress fields are necessary for equilibrium can be traced in members such as slabs, where no web reinforcement is used, or in bar anchorage with no transverse reinforcement. Meanwhile, compression reinforcement is represented by a strut in case the need arises.

1.3.3 Strut-and-Tie Modeling

Before modeling a D-region, the boundary forces acting from attached B-regions or supports or external forces should be determined, Figure 1.10a. The stress diagrams of all forces applied to the D-region boundaries are subdivided in such a way that the individual stress resultants on opposite sides of the D-region correspond in magnitude and can be connected by streamlines which do not cross each other, Figure 1.10b.

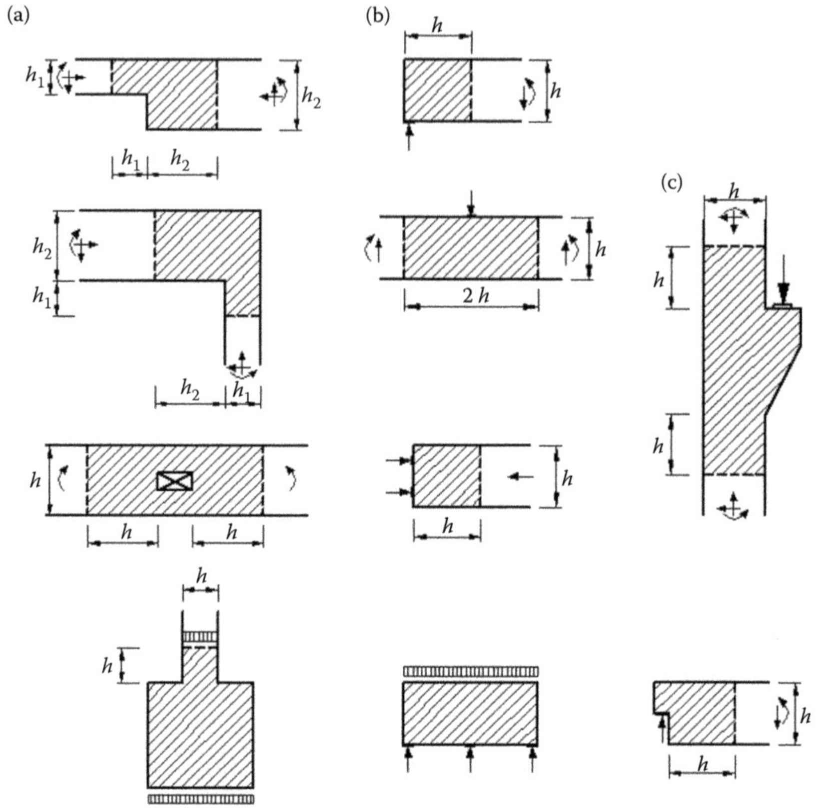

FIGURE 1.9 D-regions (shaded areas) with nonlinear strain distribution due to (Schlaich et al., 1987; Schlaich and Schäfer, 1991, 1993): (a) geometric discontinuity, (b) static discontinuity, and (c) geometric and static discontinuity.

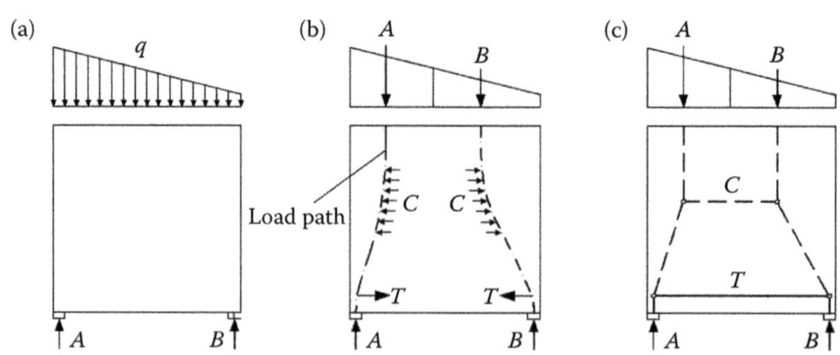

FIGURE 1.10 The load path method (Schlaich and Schäfer, 1991): (a) the region and boundary loads, (b) the load paths through the region, and (c) the corresponding *STM*.

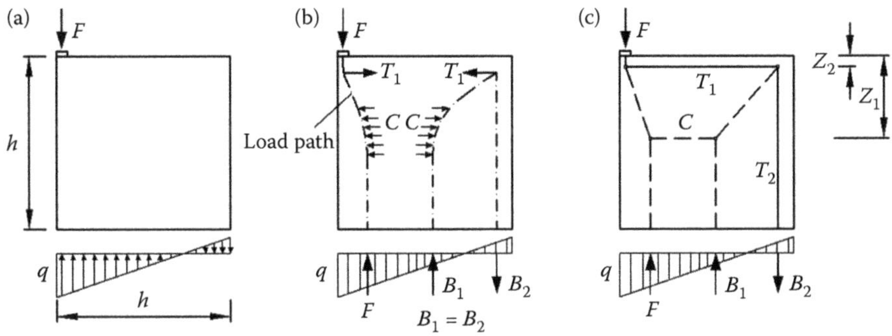

FIGURE 1.11 The load path method including a U-turn (Schlaich and Schäfer, 1991): (a) the region and boundary loads, (b) the load paths through the region, and (c) the corresponding *STM*.

Then the flow of forces through the region can be traced using the *load path* method, Figure 1.10b, which are smoothly curved. Next, the load paths are replaced by polygons as in Figure 1.10c, and additional struts or ties are added for equilibrium, such as the transverse strut and tie in the figure. In some cases, the stress diagrams or forces are not completely balanced with forces on the opposite side; for this, the load path of the remaining forces enters the structure and leaves it on the same side after a U-turn within the region, Figure 1.11.

The development of an *STM* can be simplified if an elastic finite-element analysis is performed to obtain the elastic stresses and principal stress directions (Schlaich and Schäfer, 1991). The location and direction of struts and ties can then be located at the center of stress diagrams, Figure 1.12. The orientation of struts and ties based on results from the theory of elasticity may not be the best choice in some cases where the profile and distribution of stresses may be altered as the load increases from the working load level to the collapse load with the associated nonlinear behavior of structural concrete. However, ductility of structural concrete may account for such a deviation. Also, the ties and hence the reinforcement may be arranged according to practical considerations, that is, the structure adapts itself to the assumed internal structural system. Nevertheless, modeling requires good design experience in order to set up proper design objectives such as safety and economy, and come up with a design which fulfills such objectives.

1.3.4 Elements of *STM*

STM visualizes a truss-like system in the structure, or its components, to transfer the load to the supports. It consists of three types of elements: struts (to resist compression), ties (to resist tension), and the connecting nodes (of the struts and ties) or nodal zones, Figure 1.13. Next, these elements are described in more detail.

Strut: A strut is a compression member in an *STM*, which represents the resultant of a parallel or a fan-shaped compression field. In design, struts are usually idealized as prismatic compression members, as shown by the straight line outlines of the

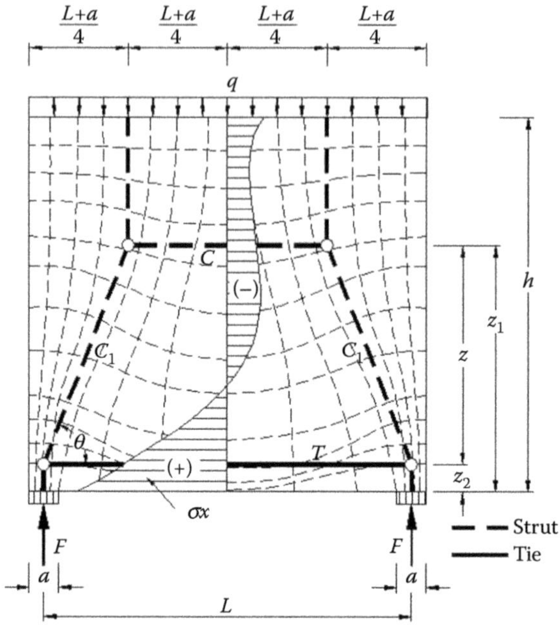

FIGURE 1.12 Elastic stress trajectories, elastic stress distribution, and the corresponding *STM* (Schlaich and Schäfer, 1991).

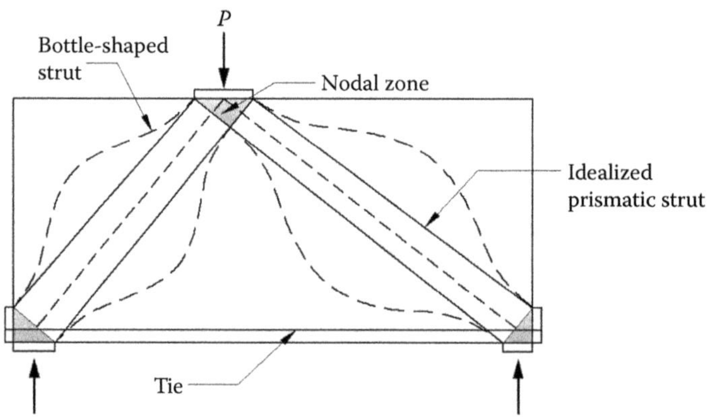

FIGURE 1.13 Description of *STM*.

struts in Figure 1.13. If the effective compression strength (or failure criterion) f_{ce}^{s} differs at the two ends of a strut, due either to different nodal zone strengths at the two ends or to different bearing lengths, the strut is idealized as a uniformly tapered compression member.

Bottle-Shaped Strut: It is a strut that is wider at the mid-length than at its ends and it is located in a part of a member where the width of the compressed concrete at the mid-length of the strut can spread laterally. The curved dashed outlines of the struts

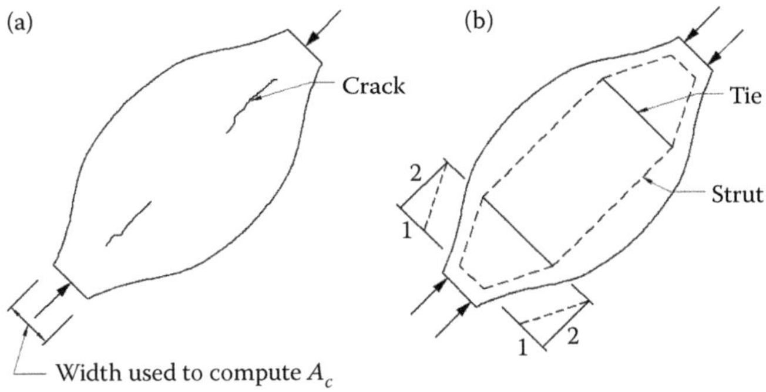

FIGURE 1.14 Bottle-shaped strut: (a) cracking of a strut and (b) *STM* for transverse reinforcement.

in Figure 1.13 and the curved solid outlines in Figure 1.14 approximate the boundaries of bottle-shaped struts. The internal lateral spread of the applied compression force in this stress field is similar to that of a split cylinder test. To simplify design, bottle-shaped struts are idealized either as prismatic or as uniformly tapered, and crack control reinforcement is provided to resist the transverse tension. The amount of confining transverse reinforcement can be computed using the *STM* shown in Figure 1.14 with the struts that represent the spread of the compression force acting at a slope of 1:2 to the axis of the applied compressive force. The strength of a bottle-shaped strut is taken as the smaller of the strengths at the two ends of the strut, Figure 1.14a.

Tie: It is a tension member in an *STM* where the force is resisted by normal reinforcement, prestressing, or concrete tensile strength. The reinforcement may consist of one or more layers and the force is always at the center of these layers.

Node: It is the point in a joint in an *STM* where the axes of the struts, ties, and concentrated forces acting on the joint intersect. For equilibrium, at least three forces should act on a node, as shown in Figure 1.15. Nodes are classified according to the signs of these forces. A $C-C-C$ node resists three compressive forces, while a $C-C-T$ node resists two compressive forces and one tensile force, and so on.

Nodal Zone: The volume of concrete around a node that is assumed to transfer strut-and-tie forces through the node is the nodal zone. Different types of nodal zones are illustrated in Figure 1.16 (ACI 318-14).

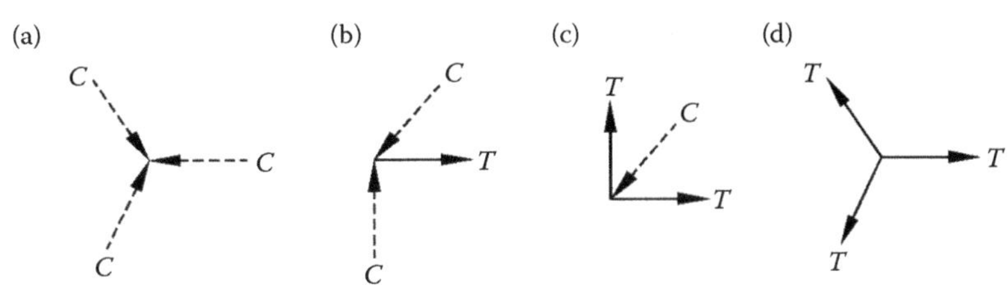

FIGURE 1.15 Nodes classification: (a) $C-C-C$ node, (b) $C-C-T$ node, (c) $C-T-T$ node, and (d) $T-T-T$ node.

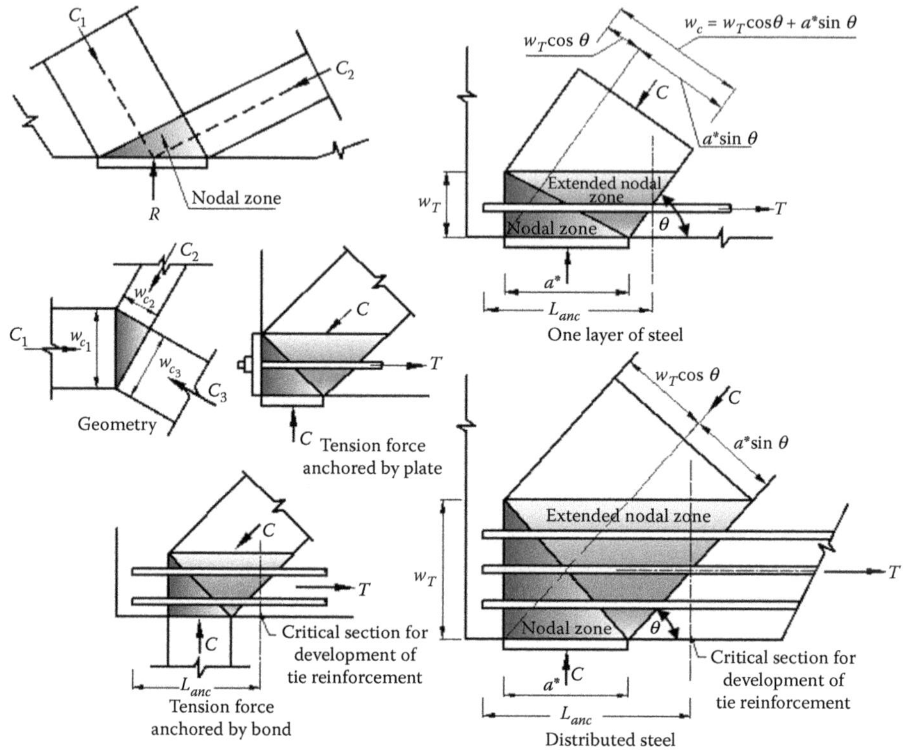

FIGURE 1.16 Different types of nodal zones.

1.4 D-REGIONS VERSUS B-REGIONS

1.4.1 Introduction

A concrete structure can be subdivided into two types of regions based on the strain distribution within a cross section, which is an influential factor in the design approach of these regions. Those regions where the Bernoulli hypothesis of flexure, plane sections before bending remain plane after bending, can be assumed valid, are referred to as Bernoulli or bending regions (or simply *B-regions*). The other regions where the Bernoulli hypothesis does not apply are referred to as discontinuity or disturbance regions (or simply *D-regions*).

B-regions have been successfully treated using the truss model. On the other hand, this truss model has been extended and generalized leading to the *STM* method for the treatment of D-regions. With this, the entire structure is treated in a consistent manner. The validity and success of the method have been proven in academia and in practice.

D-regions are usually the most critical regions in a concrete structure since they are by nature most vulnerable to environmental loading conditions. The *STM* as a transparent and translucent tool represents a rational approach to understanding the behavior of such regions.

1.4.2 B-Regions

B-regions are found in plates and beams where the depth is either constant or changes gradually and loads are continuously distributed. The state of stress at any section of a B-region can be adequately derived from sectional effects (bending, torsion, shear, and normal force).

The solution of uncracked B-regions can be satisfactorily formulated based on the theory of elasticity as in standard mechanics books. On the other hand, if the tensile stresses in B-regions exceed the tensile strength of concrete, the truss model will apply instead of the elasticity-based solutions. In addition to the truss model, codes of practice (ACI 318-14, and Eurocode 2, among others) permit other standard methods that have passed the test of experiment.

1.4.3 D-Regions

In D-regions, the strain distribution is significantly nonlinear as a result of discontinuity which results from the sudden change of geometry (geometric discontinuity) or concentrated loads (static discontinuity). Examples of geometric discontinuity are recesses in beams, frame corners, bends, and openings, Figure 1.9a and c. Examples of static discontinuity are the regions of concentrated loads, reactions, and local pressure (such as prestressing anchorage zones), Figure 1.9b and c. Structures such as deep beams, where the strain distribution is significantly nonlinear, are considered one entire D-region, Figure 1.9b.

Uncracked D-regions can be satisfactorily analyzed based on the theory of elasticity by using, for instance, finite element codes. Nevertheless, this is not the case in most practical applications even under service loads. Once cracks form in a D-region and bond stresses between reinforcement and concrete develop significantly, linear elastic analysis is not applicable any more. On the other hand, a complete nonlinear analysis may turn out to be uneconomical, especially in the early stages of design; besides, it does not help in the development of the right detailing. Moreover, if structure behavior is not precisely simulated, the results may be a cause of poor performance or future failure. With this in mind, the *STM* method represents the rational approach for the treatment of D-regions (Schlaich and Schäfer, 1991, 1993).

In B-regions, the state of stress may be derived from sectional effects, whereas in D-regions this is not the case. Nevertheless, conventional structural analysis is essential, and with the division of structure into B- and D-regions, the boundary forces of D-regions can be identified. These boundary forces come from the effect of attached B-regions and other external forces and reactions, Figure 1.17.

1.4.4 Defining the Boundaries of D-Regions

In contrast to D-regions, the stresses and stress trajectories in B-regions are smooth, Figure 1.18. In D-regions, stress intensities decrease rapidly with the distance from the origin of the stress concentration. Such behavior is the key in the identification of B- and D-regions of structure.

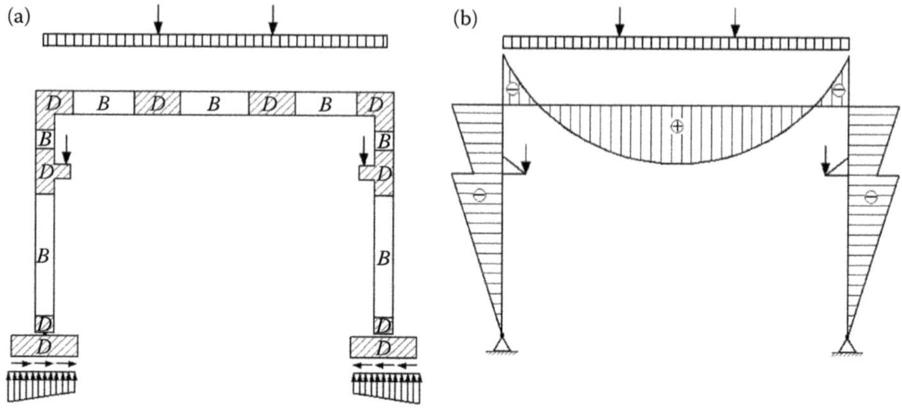

FIGURE 1.17 Frame structure containing both B- and D-regions , (a) frame structure; and (b) bending moment diagram. (Adapted from Schlaich, J., Schäfer, K., and Jennewein, M., *Journal of the Prestressed Concrete Institute*, 32(3), 1987, 74–150.)

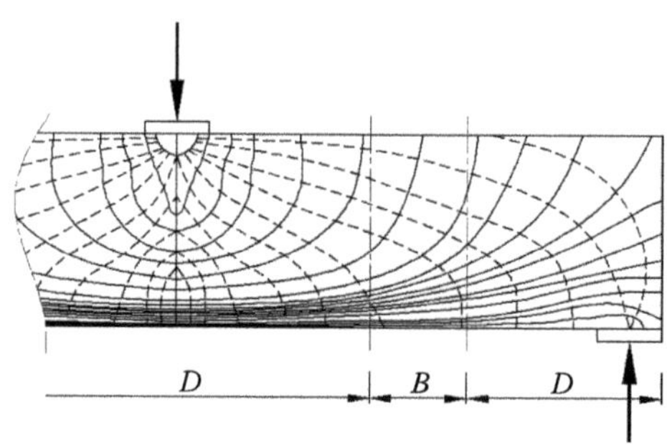

FIGURE 1.18 Stress trajectories in a B-region and near discontinuities (D-regions).

In order to illustrate how the division lines between B- and D-regions are defined, the two illustrative examples shown in Figures 1.19 and 1.20 are considered. The common principle is to subdivide the real structure in Figures 1.19a and 1.20a into the state of stress which satisfies Bernoulli's hypothesis, Figures 1.19b and 1.20b, and the compensating state of stress, Figures 1.19c and 1.20c. Upon applying the principle of Saint Venant (1980), Figure 1.21, it is assumed that the nonlinear stresses in Figures 1.19c and 1.20c are negligible at a distance which is approximately equal to the maximum distance between the equilibrating forces themselves. The distance defines the range of the D-regions, as illustrated in the examples in Figures 1.19d and 1.20d. It should be noted that for most cases of beams, this distance is practically equal to the height of the cross section of adjacent B-regions attached to the D-region.

In cracked concrete members, the stiffness in different directions may be altered as a result of cracking; consequently, the boundaries of D-regions may be altered

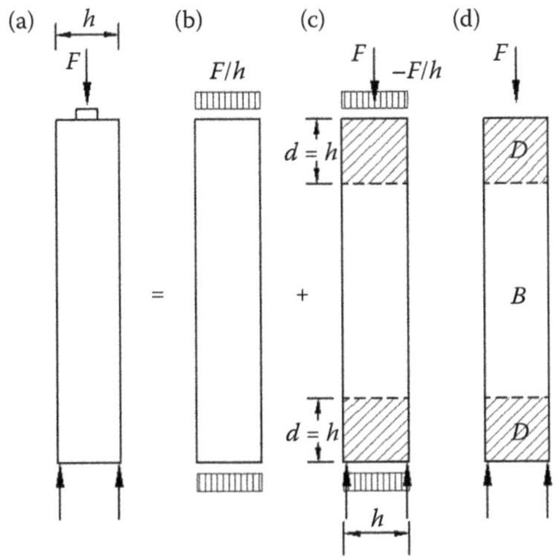

FIGURE 1.19 Example 1 of subdivision of structure into their B- and D-regions using Saint Venant's principle—column or wall with concentrated loads: (a) structure with real load, (b) loads and support reactions applied in accordance with the Bernoulli hypothesis, (c) self-equilibrating state of stress, and (d) real structure with B- and D-regions. (Adapted from Schlaich, J. et al., *Journal of the Prestressed Concrete Institute*, 32(3), 1987, 74–150; Schlaich, J. and Schäfer, K., The design of structural concrete, *IABSE Workshop*, New Delhi, 1993.)

as well. Nevertheless, the preceding approach for the determination of the division lines between B- and D-regions, which was based on elastic material behavior, is still applicable. This is due to the fact that the principle of Saint Venant itself is not precise and the dividing lines between B- and D-regions serve only as a qualitative aid in the development of *STMs*.

1.5 HISTORICAL SKETCH

1.5.1 The Development of the Truss Model for B-Regions Design

The design of the Bernoulli regions, B-regions, where linear strain distribution is assumed valid or applicable, for bending moments and/or axial forces has the same basis in almost all codes leading to minor differences in the design methods for all nations. On the other hand, the design of B-regions for shear and/or torsion still shows wide differences. In the past, there have been two basic approaches used to analyze shear and torsion problems in reinforced concrete, namely, the mechanism method and the truss model method. From a theoretical point of view, the mechanism method cannot satisfy compatibility conditions and, in certain cases, even equilibrium conditions. In contrast, the truss model theory has been conceived by most researchers to provide a more promising way to treat shear and torsion (El-Metwally, 1995).

The original truss model concept was first proposed to treat shear problems by Ritter (1899) and Mörsch (1902, 1906,1909), in which the concrete in the web is separated by diagonal cracks into a series of struts. The compression struts interact

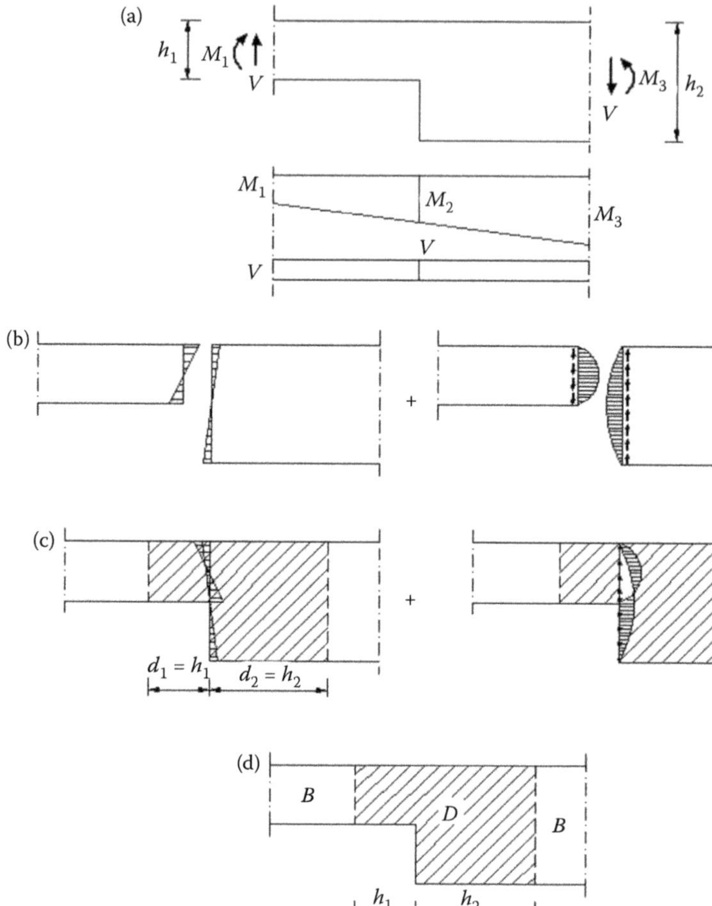

FIGURE 1.20 Example 2 of subdivision of structures into their B- and D-regions using Saint Venant's principle—beam with recess: (a) structure with real load, (b) loads and support reactions applied in accordance with the Bernoulli hypothesis, (c) self-equilibrating state of stress, and (d) real structure with B- and D-regions. (Adapted from Schlaich, J. et al., *Journal of the Prestressed Concrete Institute*, 32(3), 1987, 74–150; Schlaich, J. and Schäfer, K., The design of structural concrete, *IABSE Workshop*, New Delhi, 1993.)

with the stirrups and the longitudinal top and bottom chords (compression and tension chords) to form a plane truss. The stirrups are treated as the vertical members and the concrete struts are considered as the diagonal compression web members. For simplicity, the angle of inclination of the concrete struts is assumed to be 45°, based on the fact that the principal stresses at the neutral axis, from the linear elastic analysis of section subjected to bending, make an angle 45° with the neutral axis. Therefore, this theory has been called the 45° *truss model*. Unfortunately, there is a very wide discrepancy between experiments and the 45° truss model.

In order to improve the predictions of the truss model, the angle of inclination of the concrete struts was assumed by Lampert and Thürlimann (1968, 1969) to

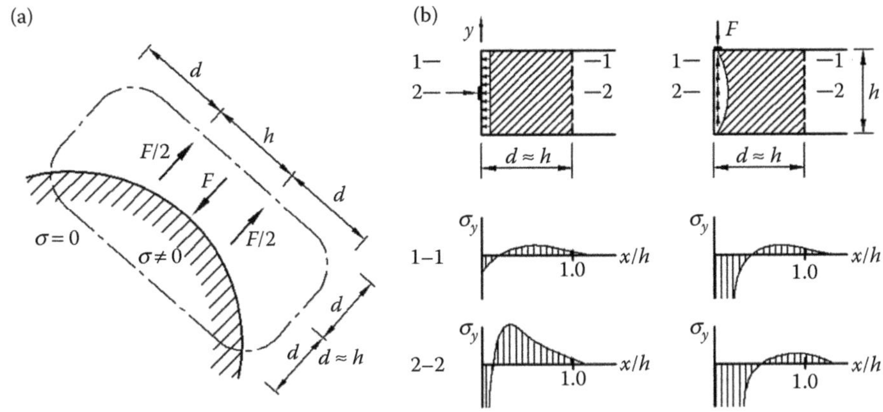

FIGURE 1.21 Principle of Saint Venant: (a) zone of a body affected by self-equilibrating forces at the surface and (b) application to a prismatic bar (beam) loaded at one face.

be variable, and therefore this theory has been known as the *variable-angle truss model*. This theory could also be called the plasticity truss model since plasticity was assumed at failure. The basic equations derived from this model could explain why longitudinal and transverse steel with different percentages can both yield at failure. The angle of inclination of the concrete struts in members subjected to shear was derived by Collins (1978) from compatibility conditions. Since the angle of inclination was assumed by Collins to coincide with the angle of inclination of the principal stress and strain, this theory has therefore been known as the *compression field theory*. The softening of concrete struts was discovered by Robinson and Demorieux (1968, 1972) and quantified by Vecchio and Collins (1986). This quantification of the compressive and tensile strength of concrete subjected to shear has been called the *modified compression-field theory* in which the principal stresses and principal strains were assumed to coincide. The variable-angle truss model with the softening of the struts as predicted by the modified compression field theory has been called by Hsu (1988) the *softened truss model*.

1.5.2 The Start of the Strut-and-Tie Model for D-Regions Design

With the extension of the limit design theorems to continuous media by Drucker, Greenberg, and Prager (1952), applications of the powerful limit analysis techniques were expanded to plates and shells for both metal and reinforced concrete materials as well as soil mechanics. The *yield-line theory* for flexure analysis of reinforced concrete slabs is the most successful application of the upper-bound method of perfect plasticity to concrete structures. Also, for flexure analysis of reinforced concrete beams and frames, the limit analysis has become standard since the 1950s. Considering shear problems, however, very few theoretical advances had been made before the 1970s. The application of the theory of plasticity to the design of members under shear and torsion began in the 1970s, especially by Thürlimann et al. (1975, 1983) and Nielsen and Hoang (1984) and their co-workers. This also formed the basis for the method of *STMs* after the work of Schlaich et al. (1987) and Schlaich and Schäfer (1991, 1993, 2001).

The development of the method of *STMs* has brought a major breakthrough in design for a consistent theory in the design concept covering both D-regions and B-regions with similar models. The method provides a formal design procedure for detailing in design in particular. All this development was brought out in the state-of-the-art report on shear by the ASCE-ACI Committee 445 in 1998. The ACI Committee 318 introduced the method of *STMs* into its 2002 ACI code. Appendix A of the ACI-318-2002 documented this international development in research that formed the basis of other codes around the world. This step was an important milestone in the direction toward the development of a more consistent design concept. It triggered the acceptance and wide use of the method of *STMs* for daily use. The method is now adopted by many international codes; e.g., ACI 318, Eurocode 2, AASHTO, …, etc.

The concept of using the method of *STMs* in the inelastic analysis of reinforced concrete was introduced and illustrated for the first time in 1961 by Drucker in his estimate of the load-carrying-capacity of a simply supported ideal reinforced concrete beam. It took a great physical insight to fully understand the fundamental difference between a tension-weak material such as concrete or soils in its load-carrying capacity through arching compared to that of a ductile material like metal through flexure or bending. Thanks to this revolutionary thinking, the concept of *STMs* was born. The subsequent development, refinement, and expansion resulted in the modern techniques of *STMs* for detailing and design of shear, torsion, joints, and bearing in structural concrete in a consistent manner.

Next, we shall present Drucker's original *simple beam model* (1961) to illustrate his concept of lower- and upper-bound techniques of limit analysis as applied to a reinforced concrete beam in Figure 1.22. For simplicity, he assumed a concrete beam with negligible weight and zero-tensile strength, so it must act as a very flat arch. The outward thrust of the arch is shown in Figure 1.22a as being taken by a steel

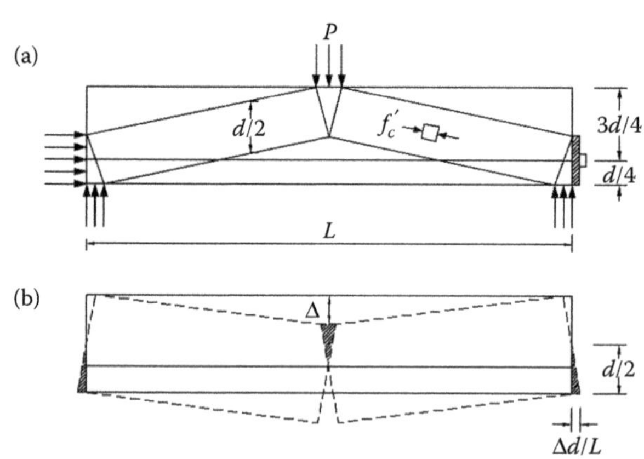

FIGURE 1.22 Drucker's simple beam model (Drucker, 1961): (a) a lower-bound or equilibrium picture of arch action and (b) a kinematical picture of collapse mechanism.

tension tie between two endplates bearing on the concrete. The steel was assumed unbonded. Efficient use of material would seem to dictate at first, at the ultimate or collapse load, that both steel and concrete should be at their yield and failure stresses, respectively.

Since the equilibrium distribution of stress as shown in Figure 1.22a is nowhere tension in the concrete and is everywhere at or below yield in the concrete and steel, the beam would not collapse at this load or would be just at the point of collapse, according to the lower-bound theorem. This approach so far focused on the lower-bound equilibrium technique and thus it might underestimate the strength of the beam.

Figure 1.22b is a kinematical picture associated with an assumed plastic failure mechanism which gives an upper bound on the collapse load. The assumed failure mechanism as drawn shows the stretching or yielding of the steel tie and the crushing plastically of the shaded areas of concrete at the ends as well as in the center. This failure mechanism results in an upper-bound solution which turns out to be equal to the lower-bound solution of Figure 1.22a. Thus, Drucker obtained the correct answer for the idealized beam according to the limit theorem despite the fact that neither the stress field as constructed in Figure 1.22a nor the plastic collapse mechanism as assumed in Figure 1.22b was the real stress distribution or the real failure mechanism, respectively. This simple example clearly illustrates Drucker's basic concept and power of limit analysis as applied to reinforced concrete structures. It also physically shows how the load is carried in a composite structure through arching for tension-weak concrete and stretching for tension-strong steel to its supports or foundations.

1.5.3 *STM* for a Unified and Consistent Design

Professor J. Schlaich and his colleagues, particularly Professor K. Schäfer, and students at the Institute of Structural Design, University of Stuttgart, had worked for decades on the application of the method of *STM* for the treatment to all different D-regions in order to achieve *a unified and consistent treatment of B- and D-regions. Consistency also means that structural concrete with or without web reinforcement and with or without axial force or pre-strains (tension or compression) is treated in the same manner.*

The contributions of Professor J. Schlaich and his colleagues and students have been greatly influential, particularly in exploring and identifying all different D-regions based on geometry and boundary conditions for a unified treatment of these regions. In addition, they illustrated, based on logic and transparency or from mechanics, how D-regions can be modeled. Besides, they set simple, but reliable, failure criteria of *STM* elements. The paper by J. Schlaich, K. Schäfer, and M. Jennewein, entitled "Toward a Consistent Design of Structural Concrete," *Journal of the Prestressed Concrete Institute*, Vol. 32, No. 3, May-June 1987, represents a landmark in this area. The efforts of Professor J. Schlaich and his group from the University of Stuttgart have made the method of *STM* widely recognized in the academia and widely adopted in practice.

In this book, we have used freely the method of *STM* and some of the examples developed by Professor J. Schlaich and his group with proper acknowledgment where it is appropriate.

REFERENCES

AASHTO, *AASHTO LRFD Bridge Specifications*, American Association of State Highway and Transportation Officials, 2nd ed., Washington, DC, 2005.

ACI 318-02, *Building Code Requirements for Structural Concrete and Commentary*, Detroit: American Concrete Institute, USA, 2002.

ACI 318-14, *Building Code Requirements for Structural Concrete and Commentary*, Detroit: American Concrete Institute, USA, 2014, 519 pp.

ASCE-ACI Committee 445 on Shear and Torsion, Recent approaches to shear design of structural concrete, state-of-the-art-report, ASCE, *Journal of Structural Engineering*, 124(12), 1998, 1375–1417.

Beton-Kalender—Teil II, Ernst & Sohn, Berlin, 1993.

Chen, W. F., *Limit Analysis and Soil Plasticity*, Elsevier, Amsterdam, 1975, Reprinted by J. Ross, Florida, 2007.

Chen, W. F., *Plasticity in Reinforced Concrete*, McGraw-Hill, New York, 1982.

Chen, W. F. and El-Metwally, S. E., *Understanding Structural Engineering—From Theory to Practice*, CRC Press, New York, 2011.

Chen, W. F. and Han, D. J., *Plasticity for Structural Engineers*, Springer-Verlag, New York, 1988.

Collins, M. P., Towards a rational theory for RC members in shear, *ASCE, Journal of the Structural Division*, 104(ST4), 1978, 649–666.

Drucker, D. C., On structural concrete and the theorems of limit analysis, *IABSE Publication*, 21, 1961, 49–59.

Drucker, D. C., Prager, W., and Greenberg, H., Extended limit design theorems for continuous media, *Quarterly Applied Mathematics*, 9, 1952, 381–389.

El-Metwally, S. E., For a consistent treatment of shear design of structural concrete B-regions, *Journal of Structural Engineering Review*, 7(4), 1995, 232–335.

Eurocode 2, *Design of Concrete Structures—Part 1: General Rules and Rules for Buildings*, Oct. 1990.

Eurocode 2, *Design of Concrete Structures*, Oct., 2001.

Hsu, T. M., Softened truss model theory for shear and torsion, *ACI Structural Journal*, 85(6), 1988, 624–635.

Lampert, P. and Thürlimann, B., *Torsion Tests of Reinforced Concrete Beams (Torsionsversuche an Stahlbetonbalken), Bericht No. 6506-2*, Institut f. Baustatik, ETH, Zürich, June 1968, 101 pp., and Torsion-Bending Tests on Reinforced Concrete Beams (Torsion-Biege-Versuche an Stahlbetonbalken), Bericht No. 6506-3, Institut f. Baustatik, ETH, Zürich, Jan. 1969, 116 pp.

Mörsch, E., *Der Eisenbetonbau, Seine Anwendung und Theorie*, 1 st Edition, Wayss and freytag, A. G., Im Selbstverlag der Firma, Newstadt a d. Haardt, May 1902, 118 pp.; Der Eisenbetonbau, seine Anwendung und Theorie, 2 nd Edition, Verlag von Konrad Wittmer, Stuttgart, 1906, 252 pp.; 3 rd Edition translated into English by E. P. Goodrich, McGraw-Hill Book Co., New York, 1909, 368 pp.

Nielsen, M. P., *Limit Analysis and Concrete Plasticity*, CRC Press, New York, Washington, DC, 1999.

Nielsen, M. P. and Hoang, L. C., *Limit Analysis and Concrete Plasticity*, CRC Press, Bosa Roca, 1984.

Ritter, W., Die Bauweise Hennebique, *Schweizerishe Bauzeitung (Zürich)*, 33(7), 1899, 59–61.

Robinson, J. R. and Demorieux, J. M., Essais de Traction-Compression sur Modeles d'ame de Pourte en Beton Armae, Part 1, June 1968, 44 pp., and Part 2, Resistance Ultimate Du Beton De L'ame de Pourtes en Double Te en Beton Arme, May 1972, 53 pp.

Schäfer, K. and El-Metwally, S. E., On the role of discontinuity regions detailing in the safety of concrete structures, *Proceedings of the Fifth International Colloquium on Concrete in Developing Countries,* Cairo, Egypt, 2–6 Jan 1994, pp. 43–55.

Schlaich, J. and Schäfer, K., Design and detailing of structural concrete using strut-and-tie models, *Journal of the Structural Engineer,* 69(6), 1991, 113–125.

Schlaich, J. and Schäfer, K., The design of structural concrete, IABSE Workshop, New Delhi, 1993.

Schlaich, J. and Schäfer, K., *Konstruieren in Stahlbetonbau (Detailing of Reinforced Concrete). Betonkalender 90, Tiel II,* Ernst & Sohn Verlag, Berlin, 2001, 311–492.

Schlaich, J., Schäfer, K., and Jennewein, M., Toward a consistent design of structural concrete, *Journal of the Prestressed Concrete Institute*, 32(3), 1987, 74–150.

St. Venant, B. de, Memoire sur l'establissement des equations differentialles des mouvements interieurs operes dans les corps solides ductiles au dela des limites ou l'elasticite pourrait les ramener a leur premier etat, Comptes Rendus de l'Académie des Sciences - Series II, 70, 1870, 473–480.

Thürlimann, B., Grob, J., and Lüchinger, P., *Torsion, Biegung und Schub in Stahlbetonträgern, Fortbildungskurs für Bauingenieure,* Institut für Baustatik und Konstruktion, ETH, Zurich, 1975.

Thürlimann, B., Marti, P., Pralong, J., Ritz, P., and Zimmerli, B., *Anwendung der Plastizitatstheorie auf Stahlbeton, Fortbildungskurs für Bauingenieure,* Institut für Baustatik und Konstruktion, ETH, Zurich, 1983.

Tresca, H., Memoire sur le poinconnage des metaux et des matieres plastiques, *Comptes Rendus de l'Académie des Sciences s - Series IIC* , 70, 1870, 27–31.

Vecchio, F. J. and Collins, M. P., The modified compression-field theory for reinforced concrete elements subjected to shear, *ACI Structural Journal*, 83(2), 1986, 219–231.

2 Developing a Strut-and-Tie Model

2.1 INTRODUCTION

Every individual D-region has different geometry, loading, and boundary conditions. Hence, the design of any D-region generally requires the development of a particular *STM*. In some cases, known models can be adapted to suit specific conditions of D-regions under consideration, as will be illustrated in this chapter. On the other hand, B-regions can be handled by few standard models since they have limited variation in conditions.

Before the development of the *STM* of D-regions in a structure, the boundary conditions (geometric and force) should be fully defined. This can be achieved through the following steps (Schlaich and Schäfer, 1993; Schäfer and Schlaich, 1998):

1. Define the geometry, loading, and support conditions of the entire structure.
2. Subdivide the three-dimensional (3D) structure into different planes to facilitate individual analyses by means of plane *STMs*.
3. Determine the support reactions by means of idealized statical systems, for example, frames, continuous beams, etc.
4. Subdivide the structure into B- and D-regions.
5. Determine the internal stresses of the B-regions and dimension these regions by either the *STMs* or using the standard methods of codes.
6. Define clearly all the forces acting on the individual D-regions. This also includes the boundary stresses or boundary forces in the sections between the D- and B-regions which are to be taken from the B-regions for equilibrium.
7. Check the individual D-regions for equilibrium.

Due to its particularity in geometry, loading and boundary conditions the design of any D-region generally requires the development of a particular strut-and-tie model (*STM*). Nevertheless, in some cases, known models can be adapted to suit specific conditions of D-regions into consideration. In order to develop a *STM*, designer has to rely on one of two approaches; either the load path method or the elastic finite element analysis, to determine the flow of forces for best serviceability. The merit of limit analysis in terms of strut-and-tie model procedures to design lies in the fact that engineers can make practical and safe decisions on the detailing of complex structural discontinuities in reinforced concrete on the basis of relatively simple calculations. The procedure of developing a *STM* is explained in the following sections. It should be noted that such a procedure requires some practice and training; afterwards, it becomes a very simple matter. Accounting for practical detailing in developing an *STM* in addition to model optimization are discussed. The *STM*s of some basic discontinuous stress fields, which are frequently encountered in different versions and combinations even in apparently different structures, are developed. Moreover, practical examples of discontinuous stress fields are modeled. These include local pressure, beam with a dapped end, beam with a recess, deep wall-like column with a recess, walls with openings, deep beam with an eccentric large opening and B-regions with web reinforcement. The chapter also explains how to handle three-dimensional reinforced concrete blocks such as pile caps, with either two-dimensional or three-dimensional modeling.

2.2 THE LOAD PATH METHOD

The development of an *STM* using the load path method is illustrated in the following steps (Schäfer and Schlaich, 1998, and Schlaich and Schäfer, 1993) with reference to the two examples in Figures 2.1 and 2.2.

1. The boundary forces acting on a D-region from the attached B-regions or supports or external forces should be determined, as illustrated for the D-regions in Figures 2.1a and 2.2a.
2. Subdivide the stress diagrams to match with opposite forces. Locate the resultants of the subdivided stress diagrams in their center of gravity, Figures 2.1b and 2.2b.
3. Connect the opposite forces by streamlined load paths which are straight or smoothly curved, Figures 2.1b and 2.2b:
   - Start in the direction of load
   - Avoid crossings with other load paths
   - Choose curvature near discontinuities of load or geometry as large as possible
4. Indicate deviation forces as a basis for transverse struts and ties, Figures 2.1b and 2.2b.

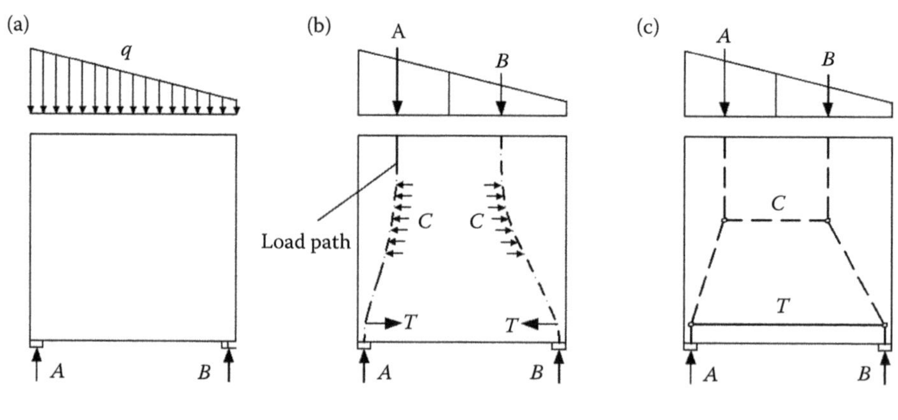

FIGURE 2.1 The load path method (Schlaich and Schäfer, 1991): (a) the region and boundary loads, (b) the load paths through the region, and (c) the corresponding *STM*.

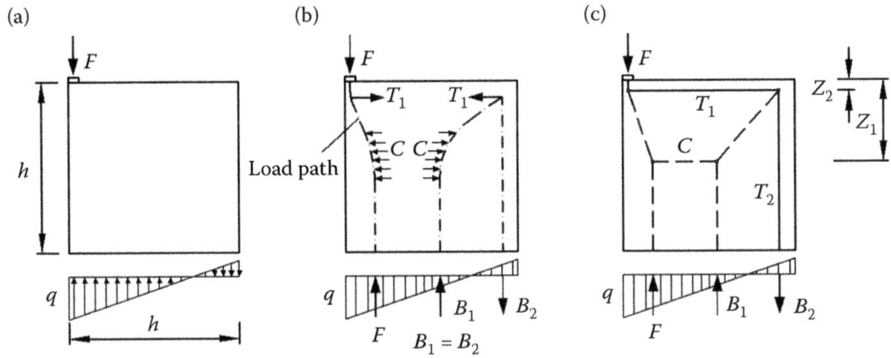

FIGURE 2.2 The load path method including a U-turn (Schlaich and Schäfer, 1991): (a) the region and boundary loads, (b) the load paths through the region, and (c) the corresponding *STM*.

5. Replace the curved load paths by polygons with kinks (bends) at the points of largest curvature or the estimated location of the deviation force resultant, Figures 2.1c and 2.2c. In the model, dashed lines indicate struts and solid lines indicate ties.
6. Connect the remaining stress diagrams by loop-shaped load paths.
7. Extend the ties (reinforcement) preferably straight as far as needed to meet and balance the transverse forces of primary load paths.

It should be noted that additional struts or ties are added for equilibrium, such as the transverse struts and ties in the figures. In some cases, the stress diagrams or forces are not completely balanced with forces on the opposite side; for this, the load path of the remaining forces enters the structure and leaves it on the same side after a U-turn within the region, Figure 2.2.

The previous procedure is explained next in more detail for the previous two examples in Figures 2.1 and 2.2 (Schäfer and Schlaich, 1998).

The D-region in Figure 2.1 is loaded with asymmetric linear stress from the adjoining B-region. The stress diagram is subdivided in such a way that the associated resulting loads in the upper part of the structure find their equivalent counterpart on the opposite side. The pattern that develops when drawing a line to link the opposite forces is the so-called load path (Figure 2.1b).

The load paths begin and end at the center of the corresponding stress diagrams and they have the direction of the applied loads or reactions at their start and end. Streamlines may be straight or smoothly curved in order to connect the opposite forces. Load paths tend to take the shortest possible streamlined way in between. This is based on the fact that a structure will exert the minimum strain energy for load transfer.

The stress fields due to concentrated forces tend to spread out wherever possible in the structure. Considering such behavior, the resulting load paths, originating from the supports or the concentrated forces, propagate inward and show their maximum curvature near such forces.

So far equilibrium in the direction of the applied loads has been considered. Nevertheless, the curvature of the load paths gives rise to deviation forces C, Figure 2.1b, which, for the sake of simplicity, are plotted horizontally. Having in mind that the D-region under consideration is not subjected to horizontal loads, it follows that the deviation forces of the two load path are in equilibrium. From the horizontal equilibrium of the two load paths, the resultants of their deviation forces are in equilibrium. Finally, the load paths are replaced by polygons, the break points of which intersect the resultants of the deviation forces, Figure 2.1c.

The developed model reflects the principal paths of forces and illustrates the load-bearing behavior of the D-region. The tension and compression members are representative of curved 2D or 3D stress fields with the main stream of stress oriented in the direction of their centerlines. Though the nodes of the members should always be in equilibrium, they do not represent real hinges but rather regions where the internal forces (stresses) are diverted or introduced (anchored).

For modeling the D-region in Figure 2.2, the same procedure followed in the first example is applied. In this example, the stress distribution acting at the bottom of the region is determined from equilibrium with the vertical force F, Figure 2.2b. This stress diagram is divided into a stress block of resultant F in addition to two forces B_1 and B_2, where the last two forces are equal in magnitude but of opposite directions and they balance the moment due to the eccentricity of the applied force. The load path of these two forces enters the D-region at B_1, makes a U-turn, and exists at B_2. This U-turn of the load path and its associated deviation forces are necessary to equilibrate the deviation forces of the real load F as it travels through the D-region. In this regard, it is noted that the load path of the force B_2 remains straight since it is tension and for practicality should be carried by straight bars. By replacing the plotted load paths with polygons and linking them by means of the deviation forces, the *STM* is obtained, Figure 2.2c.

2.3 ELASTIC STRESS ANALYSIS

The development of an *STM* can be simplified if an elastic finite-element analysis is performed to obtain the elastic stresses and principal stress directions (Schlaich and Schäfer, 1991). The location and direction of struts and ties can then be located at the center of stress diagrams, Figure 2.3. In addition, the position of the deviation force resultants can be determined from the stress distribution of the individual sections. Thus, for example in Figure 2.3, the distances z and z_1 can be easily determined.

The orientation of struts and ties based on results from the theory of elasticity may not be the best choice in some cases where the profile and distribution of stresses may be altered as the load increases from working load level to collapse load with the associated nonlinear behavior of structural concrete. However, ductility of structural concrete may account for such a deviation. Also, the ties, and hence the reinforcement, may be arranged according to practical considerations, that is, the structure adapts itself to the assumed internal structural system. Generally the strut-and-tie directions should be within $\pm 15°$ from the direction of the compressive and tensile stress trajectories, respectively. Nevertheless, modeling requires good design

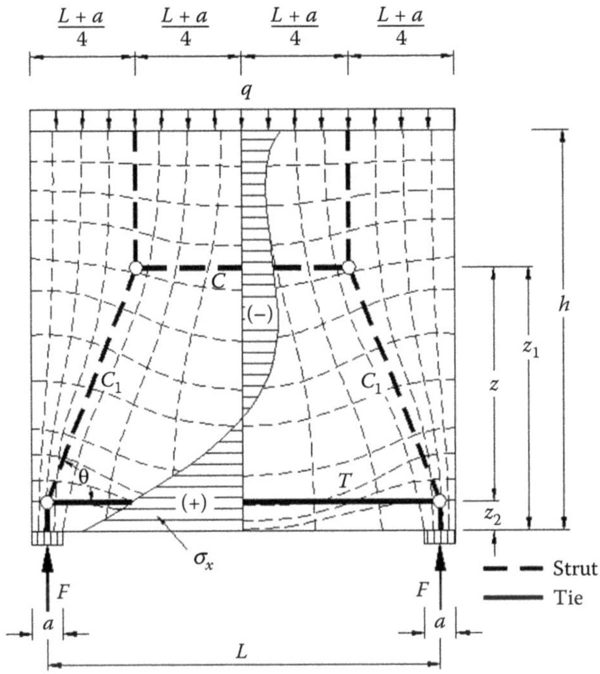

FIGURE 2.3 Elastic stress trajectories, elastic stress distribution, and the corresponding *STM* (Schlaich and Schäfer, 1991).

experience in order to set up proper design objectives, such as safety and economy, and come up with design which fulfills such objectives.

The combination between the load path method and the elastic stress analysis is an efficient approach for developing *STMs* for complicated structures. For the D-region in Figure 2.4, the vertical struts and ties are traced by the load path method as explained before. It is noted that the stress distribution q is obtained from simple statics. The stress diagram is divided into four resultants: C_3 and C_4 which equilibrate the force F, and two equal forces T_2 and C_2 forming a U-turn which balances the moment due to force eccentricity, Figure 2.4c. The curvatures in the streamlines of the compression forces C_2, C_3, and C_4 generate transverse stresses (deviation stresses). The corresponding horizontal struts and ties of these stresses are located at the center of gravity of their diagrams obtained from linear elastic analysis, Figure 2.4b. The diagonal struts are determined subsequently in order to achieve equilibrium at the model nodes, Figure 2.4c.

2.4 MODEL OPTIMIZATION

Knowledge of the stress distribution helps the designer to reduce the great number of possible models in order to ensure good performance of the structure under service loads. It is therefore appropriate to orient the *STM* along the flow of forces from elastic analysis. Nevertheless, in some cases, this approach may seem inconvenient for

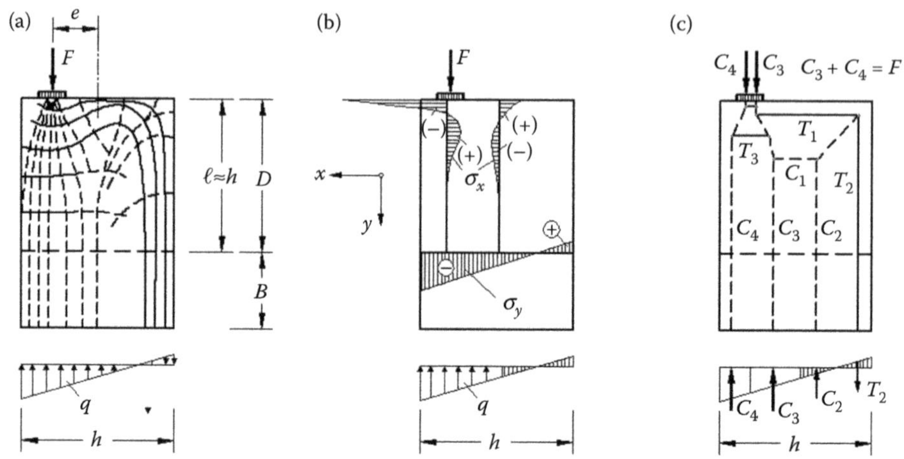

FIGURE 2.4 Application of the load path method, with the location of the deviation forces T_3 and C_1 derived from the elastic stress trajectories (Schlaich and Schäfer, 1993): (a) linear-elastic stress trajectories, (b) linear-elastic distribution, and (c) *STM*.

reinforcement layout. Hence, a model optimization is meant to adjust the mainstream of forces according to elastic stress trajectories and, at the same time, making due allowance for the specific properties and characteristics of structural concrete. This is further illustrated in the following steps (Schäfer and Schlaich, 1998):

1. The arrangement of the reinforcement should satisfy the practical requirements for simplicity of construction. For instance, the designer should use straight bars with a minimum number of bends, laid out in orthogonal arrangement parallel to the edges of the structure, wherever possible.
2. The edges and surfaces of a structure should be fitted with near-surface reinforcement in order to control cracking.
3. In the cracked state of concrete, the reinforcing bars will channel the flow of tensile forces. Therefore, they should be introduced in the model in the form of tensile ties, wherever their position is known in advance. In doing so, several ties, within the same zone of tension, may be resolved into one single tie located in their common axis of gravity.
4. The arrangement of the reinforcement should be so designed that it covers various cases of loadings.
5. The formation of cracks in concrete and the plastic deformations of the structural members will be associated with redistribution of the internal forces, obtained from linear elastic analysis. In selecting the model, it is useful to realize that the structure tends to carry the loads with the least possible strain energy. This can be expressed by the following equation for cracked reinforced concrete:

$$\Sigma F_i \ell_i \varepsilon_i = \text{minimum}$$

where F_i is the force in the strut or tie number i, λ_i is the length of member i, and ε_i is the mean strain of member i. Since reinforced ties are much more deformable than concrete struts, the model with the least and shortest ties will be the best. In case of doubt, the following equation can be used as a simplified criterion for model optimization:

$$\Sigma T_i \ell_i = \text{minimum}$$

where T_i is the tension force in the tie number i.

It should be noted that the *STMs* are quite often kinematic; nevertheless, this does not mean that the structure lacks stability since any slight change of the model geometry associated with any movement of the model nodes evokes diagonal compression forces in the concrete, which stabilize the system. Of course, the designer has the freedom to introduce any diagonal members into the model in order to make the model stable. These additional members will be essentially zero members, or will carry very small forces; hence, they will not affect the flow of forces.

A kinematic model applies to a specific case of loading since it is developed from equilibrium and hence it cannot be employed to other cases of loading. On the other hand, a statically determinate or statically indeterminate truss system is capable of withstanding different cases of loadings. Therefore, it is advisable to deal with *STMs* which are either statically determinate or statically indeterminate. For hand calculations, statically indeterminate models can be developed from a combination of statically determinate models, as illustrated by the example in Figure 2.5.

If the only possible *STM* does not properly match the case of loading under consideration, due reinforcement layout for instance, the structure will adjust its path of forces to the given model. In this case, significant deformations will take place, which can be accommodated with reinforcement; however, the width of cracks under service conditions may increase above the allowed limits. This risk is particularly important when singular cracks issuing from reentrant corners or point loads

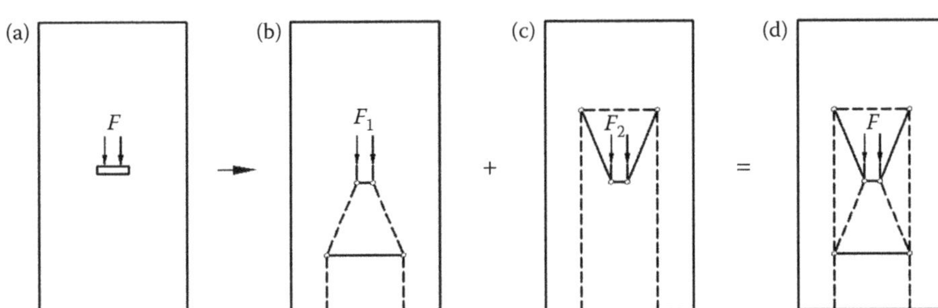

FIGURE 2.5 Statically indeterminate *STM* (Schlaich and Schäfer, 1993): (a) geometry and loading, (b, c) two statically determinate models for different stress flows, and (d) statically indeterminate model from the superimposition of the models of (b) and (c).

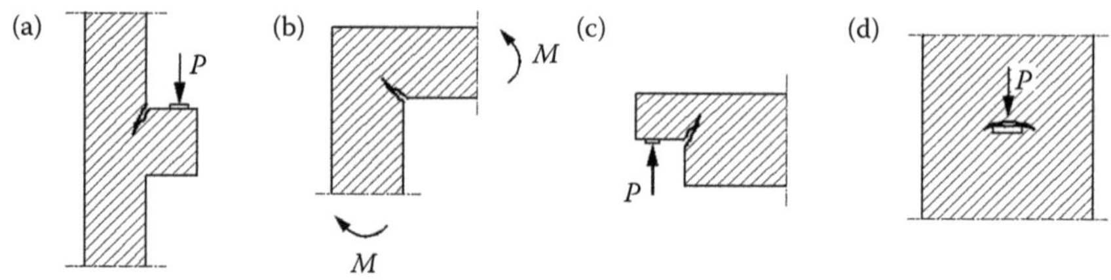

FIGURE 2.6 Risk of unacceptable width of cracks at discontinuities; orientation of appropriate oblique reinforcement (Schlaich and Schäfer, 1993): (a–c) single cracks issuing from reentrant corners and (d) internal transverse crack from concentrated load application within a deep beam (tendon anchorage).

intersect the reinforcement steel at acute angles. Therefore, it is recommended to either choose such a model where reinforcement intersects the cracks at an angle as close to 90° as possible or use additional reinforcement, not included in the model, for crack control (Schlaich and Schäfer, 1993), Figure 2.6.

2.5 BASIC DISCONTINUOUS STRESS FIELDS

2.5.1 Why

In the practical applications of the method of *STM*, some typical models are frequently used in different versions and combinations even in apparently different structures. An example of the application of the model of the deep beam in Figure 2.3, applied to three different structures, is illustrated in Figure 2.7 (Schäfer and Schlaich, 1998): the distribution of cable forces in a bridge deck, Figure 2.7a; a wall with a large opening, Figure 2.7b; and a box girder with anchor loads from prestressing tendons, Figure 2.7c. The reason for this phenomenon is the fact that the number of discontinuities with substantially different stress patterns is very limited. With realizing such feature of structures, the design engineer can trace the apparently different cases back to their common roots and will thus be able to avoid mistakes in modeling. In addition, it is very confusing to give the same phenomenon different names only because it appears in different contexts.

In their modeling, D-regions with their respective boundary conditions can be looked at as isolated discontinuous stress fields. Many of these fields are basic (or standard) stress fields and hence have standard *STM* solutions. These regions are designated by Schlaich and Schäfer (1993) as D_1, D_2 … D_{12}; nevertheless, only D_1 to D_{10} are the important basic D-regions and are illustrated next.

2.5.2 Region D_1

Region D_1 represents a state of concentric local pressure, Figure 2.8. The deviation of compressive stress trajectories, Figure 2.8a, generates transverse tensile, splitting, stresses, σ_x, Figure 2.8b, which is a function of (a/ℓ), where a and ℓ are as shown in

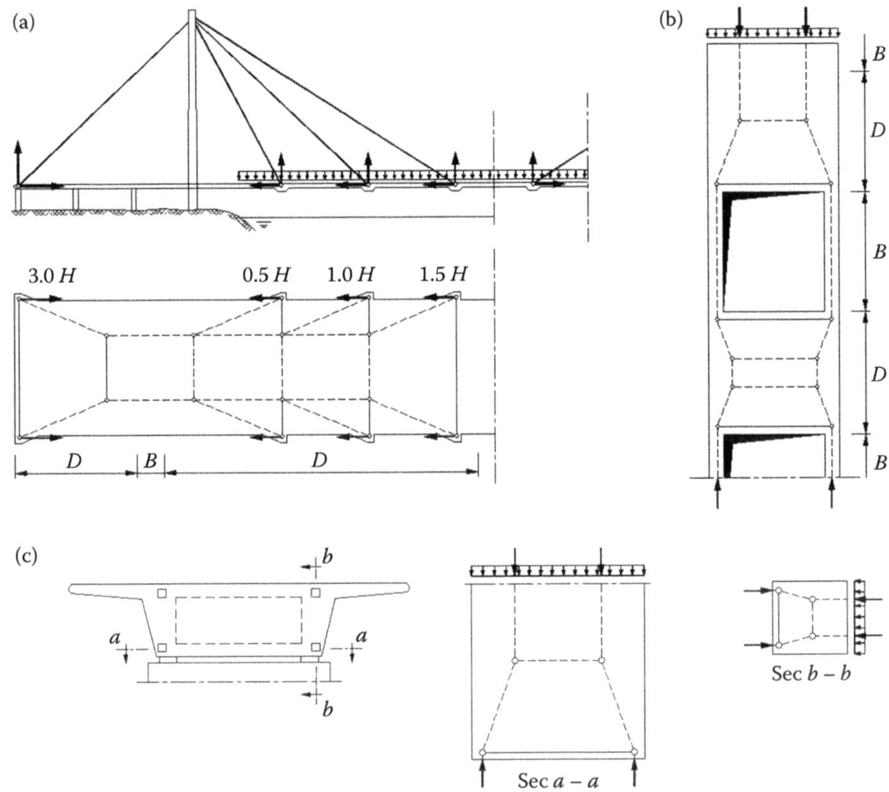

FIGURE 2.7 Three examples for the application of the model of the deep beam of Figure 2.3 (Schäfer and Schlaich, 1998): (a) anchorage of stay cables in a bridge deck, (b) wall with a large opening, and (c) anchorage of tendons in a box girder bridge.

the figure. The resultant of these transverse tensile stresses, T, can be calculated from the respective *STM*, Figure 2.8c, as

$$T = 0.25F\left(1 - \frac{a}{\ell}\right)$$

The value of the lever arm, z, is assumed equal to $\ell/2$ for $h \geq \ell$, based on linear-elastic analysis. If otherwise, $h < \ell$, and the distance z should not exceed $0.8h$. The force T should be resisted by closed stirrups since there is no space for reinforcement anchorage.

2.5.3 Region D_2

Region D_2 represents a case of eccentric local pressure, Figure 2.9. Owing to the applied force eccentricity, the balancing tensile stresses have their resultant, T_2, close

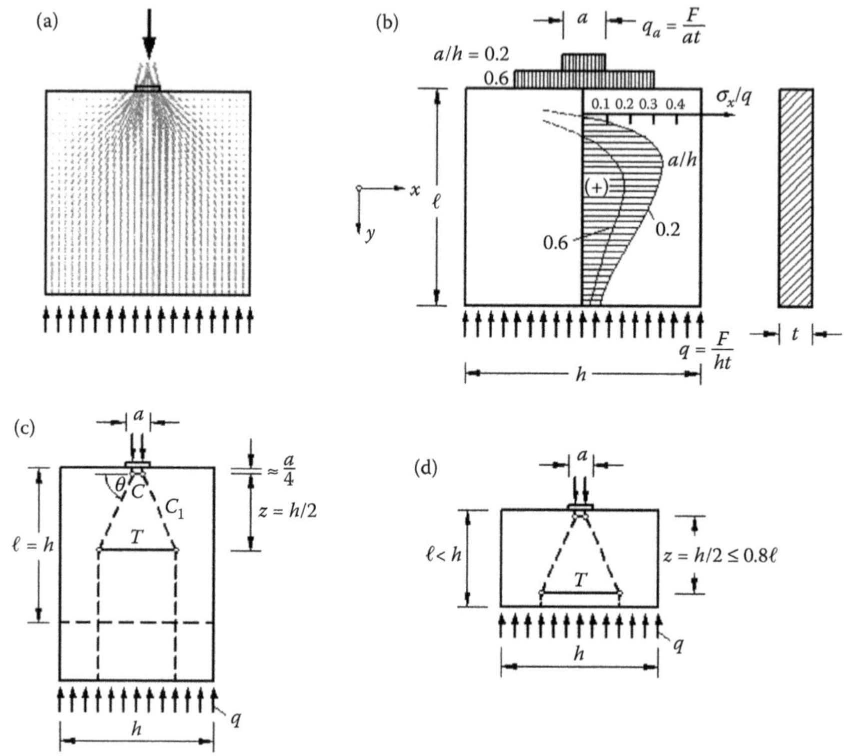

FIGURE 2.8 Region D_1 (Schlaich and Schäfer, 1991, 1993): (a) principal stress pattern from a linear-elastic finite element analysis, (b) corresponding transverse tensile stresses, and (c, d) *STMs*.

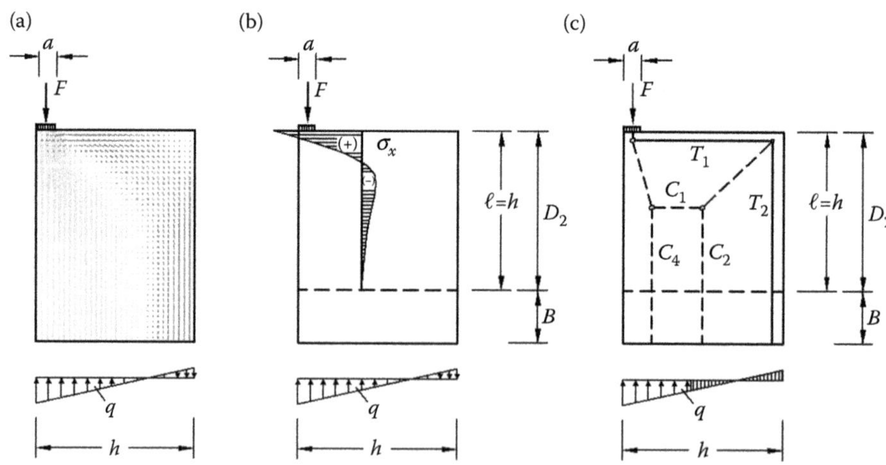

FIGURE 2.9 Region D_2 (Schlaich and Schäfer, 1991, 1993): (a) stress diagram from linear-elastic analysis and (b) *STM*.

to the longitudinal edge. The associated transverse stresses, as shown in the figure, constitute a tensile stress field underneath the load with its resultant T_1 close to the loaded edge, and a compressive stress field with its resultant C_1, as illustrated. The force T_1 may reach a magnitude ($F/3$) with a very limited spread, and hence it should be resisted by closed stirrups located close to the loaded edge in order to avoid the corner to break off. The force T_2 requires longitudinal reinforcement in the form of U-bars in order to overcome the anchorage problem at the right top corner.

2.5.4 Region D_3

Region D_3, Figure 2.10, can be considered as a constituent of two D_1 regions, leading to the *STMs* in Figures 2.10b and c. However, for a very small (h/ℓ) ratio, the refined *STM* in Figure 2.10d is more appropriate.

2.5.5 Region D_4

Region D_4, Figure 2.11, is composed of two D_2 regions, with the stress diagrams shown in Figure 2.11b. Thus, the appropriate *STM* can be generated from that of D_2 as shown in Figure 2.11c and d.

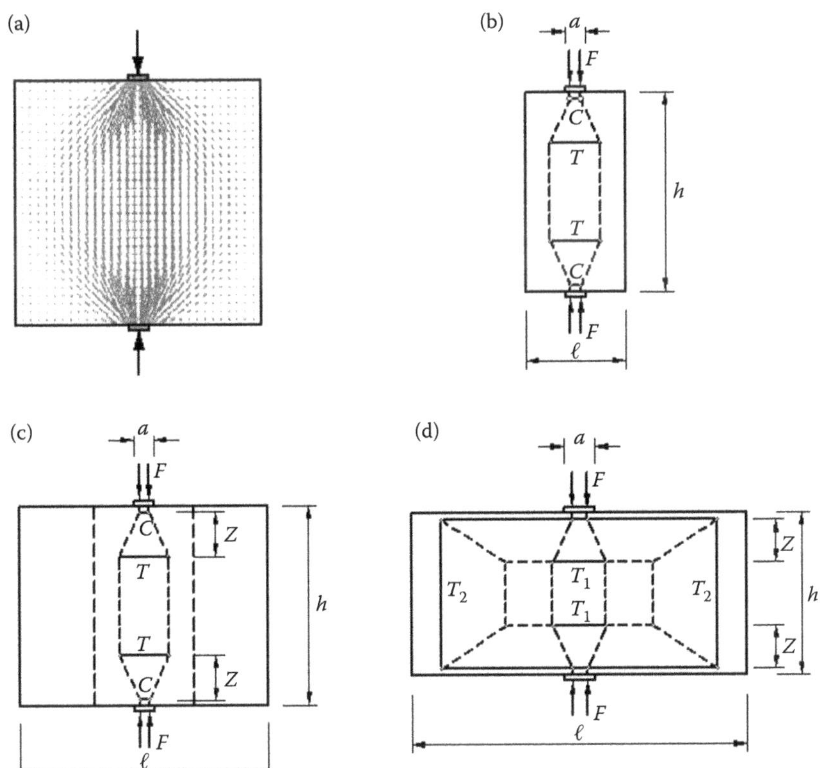

FIGURE 2.10 Region D_3 (Schlaich and Schäfer, 1991, 1993): (a) principal stress pattern from a linear-elastic finite element analysis, (b, c) *STMs*, and (d) refined *STM* for $h/\ell < 1$.

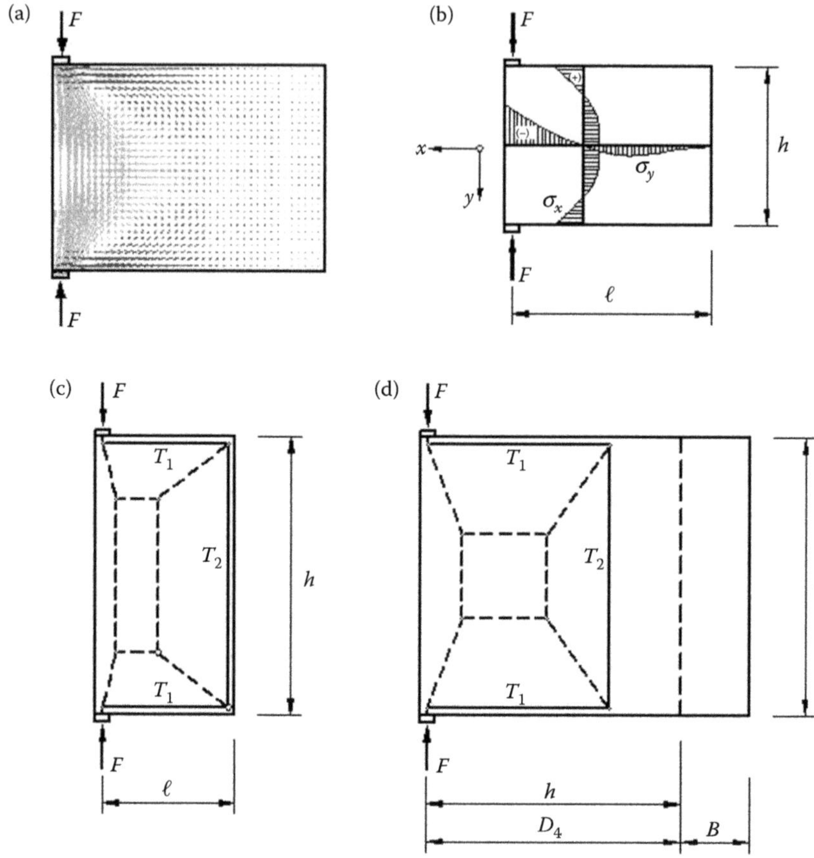

FIGURE 2.11 Region D_4 (Schlaich and Schäfer, 1991, 1993): (a) principal stress pattern from a linear-elastic analysis, (b) corresponding stress diagrams, and (c, d) *STMs*.

2.5.6 Regions D_5 and D_6

Both regions D_5, Figure 2.12, and D_1, Figure 2.8, are the most common patterns. From linear-elastic analysis, the lever arm of region D_6 with a hanging load, Figure 2.13, is the same as that of D_5.

2.5.7 Region D_7

In region D_7, Figure 2.14, for the case of $h/\ell \leq 1$, either the model in Figure 2.14c or the refined model in Figure 2.14d is applicable; however, the refined model directly gives the transverse tension associated with the spread of the compression stress trajectories. For the simple model, the transverse tension can be derived by remodeling separately the individual strut C_2 leading to the same result as the refined model; otherwise, a reduced strength of the concrete strut should be adopted and minimum skin reinforcement, according to the respective code, should be added. If the ratio

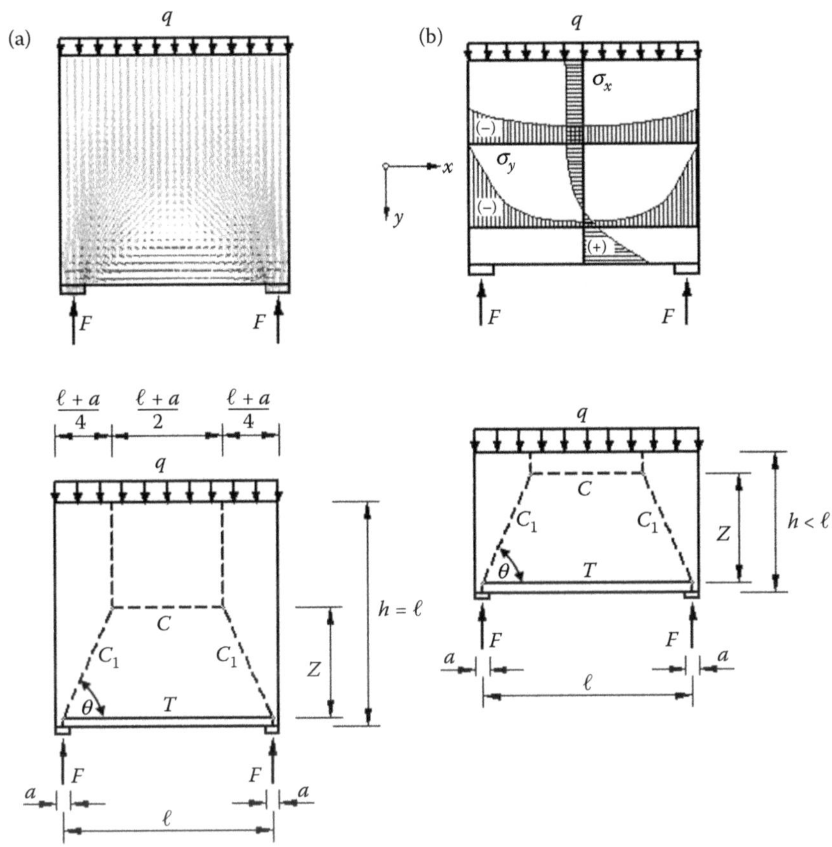

FIGURE 2.12 Region D_5 (Schlaich and Schäfer, 1991, 1993): (a) principal stress pattern from a linear-elastic analysis, (b) corresponding stress diagrams, and (c, d) *STMs*.

h/ℓ is greater than 2, the upper part of region D_7 can be treated as region D_1, Figure 2.8, and the lower part as region D_5, Figure 2.12, as illustrated in Figure 2.14f.

2.5.8 Region D_8

Region D_8, Figure 2.15, can be considered as consisting of two regions D_5, Figure 2.12, for the case of $h/\ell > 2$, and two D_4, Figure 2.11, for the case of $h/\ell < 1/2$.

2.5.9 Regions D_9 and D_{10}

The model of region D_9, Figure 2.16, is a combination of the two models of regions D_1, Figure 2.8c, and D_5, Figure 2.12c. The model of region D_{10}, Figure 2.17, is a combination of the two models of regions D_3, Figure 2.10c, and D_8, Figure 2.15c. For the case of $h/\ell > 2$ in D_{10}, the top and bottom parts of region D_{10} blend into region D_9. It should be noted that the models of D_9 and D_{10}, Figures 2.16b and 2.17, are statically indeterminate.

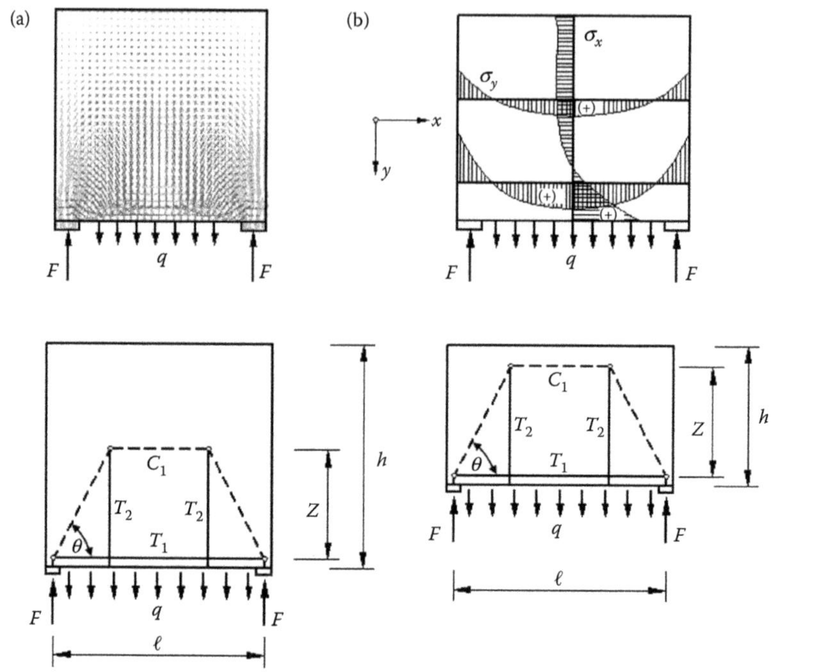

FIGURE 2.13 Region D_6 (Schlaich and Schäfer, 1991, 1993): (a) principal stress pattern from a linear-elastic analysis, (b) corresponding stress diagrams, and (c, d) *STMs*.

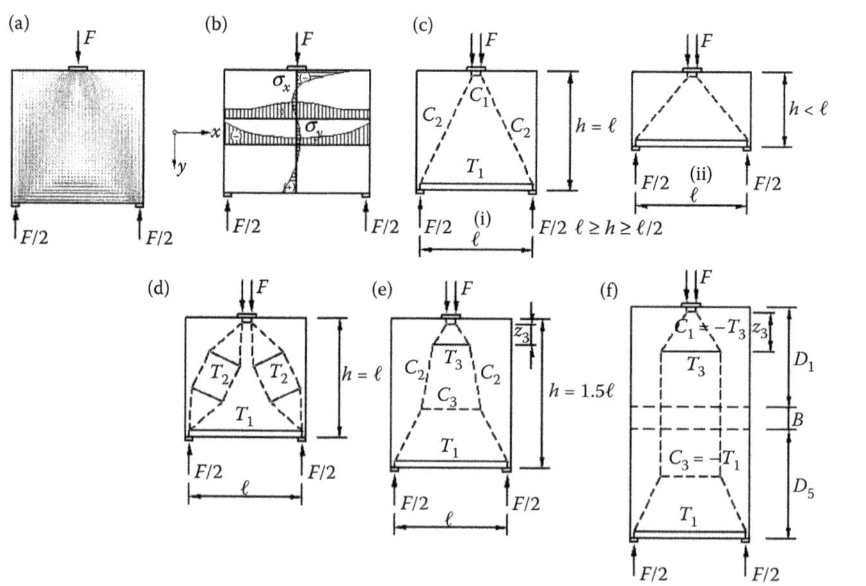

FIGURE 2.14 Region D_7 (Schlaich and Schäfer, 1991, 1993): (a) principal stress pattern from a linear-elastic analysis, (b) corresponding stress diagrams, and (c–f) *STMs*.

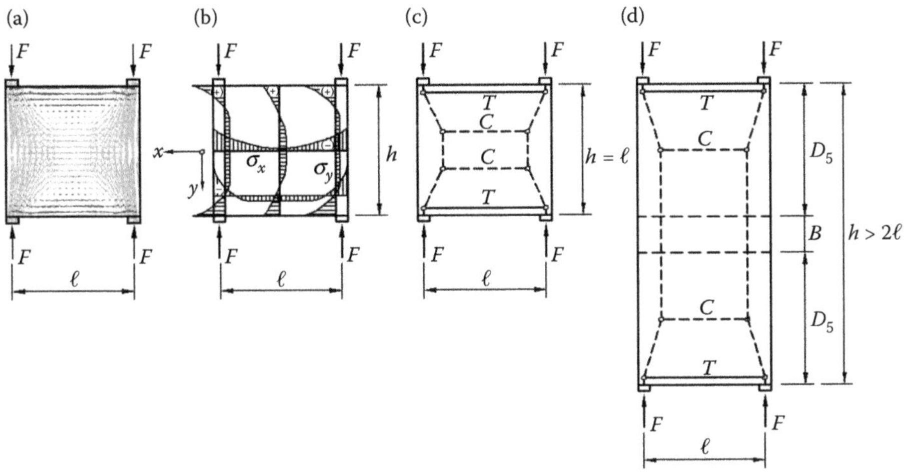

FIGURE 2.15 Region D_8 (Schlaich and Schäfer, 1991, 1993): (a) principal stress pattern from a linear-elastic analysis, (b) corresponding stress diagrams, and (c, d) *STMs*.

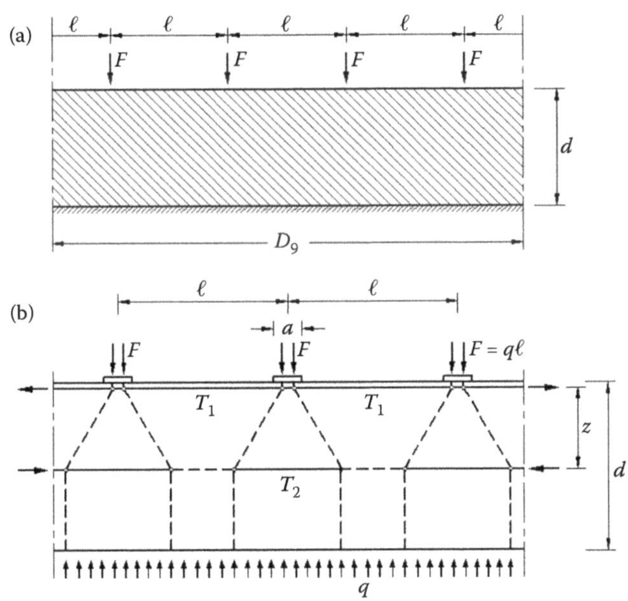

FIGURE 2.16 Region D_9 (Schlaich and Schäfer, 1991, 1993): (a) the region and (b) *STM*.

2.6 EXAMPLES OF DISCONTINUOUS STRESS FIELDS

2.6.1 Local Pressure

The problem of local pressure, Figure 2.18 (Schäfer and El-Metwally, 1994), is simulated as D_1-region for the case of concentric load, Figure 2.8, and D_2-region for the

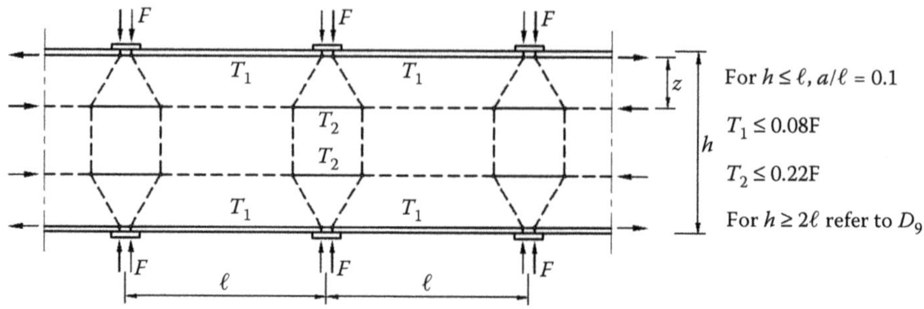

FIGURE 2.17 Region D_{10} (Schlaich and Schäfer, 1991, 1993)—*STM*.

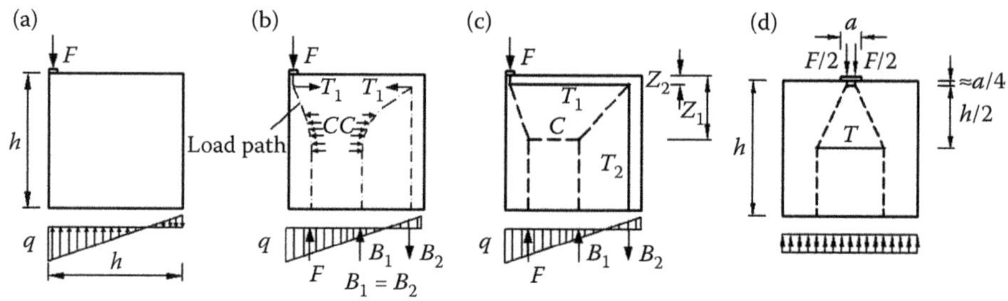

FIGURE 2.18 Local pressure: (a–c) eccentric load and (d) concentric load.

case of eccentric load, Figure 2.9. The amount of necessary transverse reinforcement and its position can be determined from the respective *STM*. It is noted that the two models of local pressure reveal the requirement of different reinforcement in the transverse direction and longitudinal reinforcement for the case eccentric local pressure only. In both cases of local pressure, the transverse reinforcement should be closed stirrups in order to overcome the anchorage problem and in order to avoid the corner to break off in the case of eccentric local pressure. The longitudinal reinforcement in the case of eccentric local pressure should be in the form of U-bars in order to overcome the anchorage problem at the right top corner.

2.6.2 Beam with Dapped End

In order to hang the reaction of a dapped beam, two possible models are shown in Figure 2.19 (Schäfer and El-Metwally, 1994). The first model, Figure 2.19a, illustrates that in addition to the shear reinforcement, T_3, and the reinforcement necessary for hanging the reaction, T_1, additional reinforcement is necessary for a safe transfer of the forces within the D-region, that is, tie T_A and the increase in the magnitude of tie T_2 above the shear requirement.

The second alternative, Figure 2.19b, requires less reinforcement; nevertheless, anchorage of the inclined reinforcement at the upper node may become a problem in case of thick bars. In practice, some reinforcement is always detailed as indicated

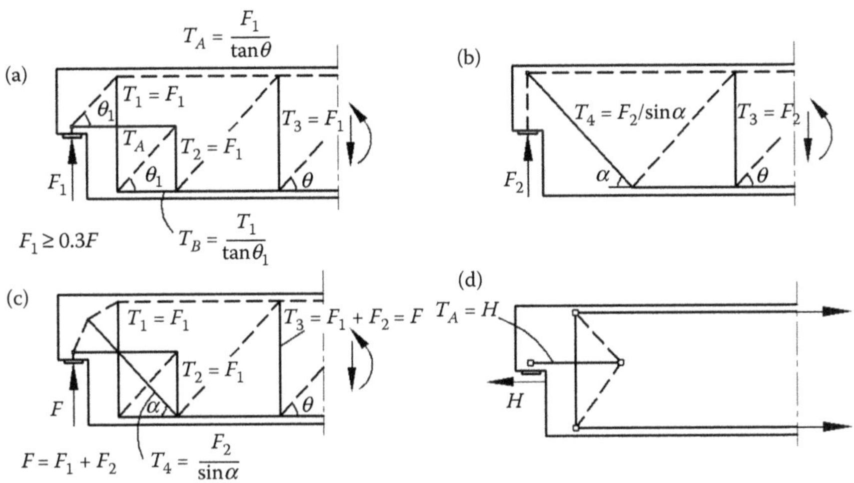

FIGURE 2.19 Dapped beam: (a, b) *STMs* 1 and 2, (c) combined model from 1 and 2, and (d) *STM* for a horizontal reaction H.

by the first model, which is necessary for keeping the integrity of the D-region. Hence, though the two presented models are correct, a more efficient detailing can be achieved by a combination of the two models, Figure 2.19c.

If a horizontal reaction force *H* exists, the model shown in Figure 2.19d can be used for the evaluation of the required additional reinforcement.

2.6.3 Beam with Recess

The beam in Figure 2.20a has the shown shaded D-region as a result of the shown recess. For illustration, the D-region is assumed to be subjected to two cases of constant moment (no shear), positive and negative moment, Figure 2.20b, and the moment lever arm on the left end of the D-region is assumed to be one-half of that on the right end.

Upon the examination of all possible load paths for either case of end moment, the appropriate *STMs* can be derived as shown in Figure 2.20c (Schäfer and El-Metwally, 1994).

2.6.4 Deep Wall-Like Column with Recess

A deep wall-like column, for simplicity referred to as wall-column, with a recess is shown in Figure 2.21a carrying a uniform load and in Figure 2.21b carrying an edge local pressure, with the appropriate *STMs* (Schäfer and El-Metwally, 1994). For the first load case, Figure 2.21a, a balancing moment is required at the bottom section in addition to the axial force which is equal to the pressure resultant. For this case, the *STM* of the D-region reveals that the tension reinforcement resisting the moment tension component, T_1, should be anchored beyond the middle height of the D-region, and half of its anchorage length should start at about the middle

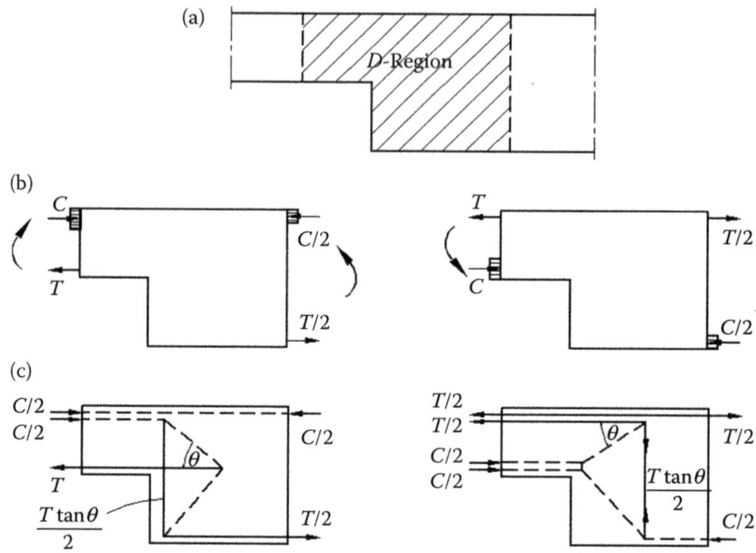

FIGURE 2.20 Beam with recess: (a) D-region, (b) moments applied to the D-region, and (c) *STMs*.

height of the D-region. In addition, stirrup reinforcement should be placed across the D-region (centered about the middle height of the D-region) to carry the tension force T_2. Obviously these stirrups are preferred to have their closing hooks within the compression stress field at the right side of the D-region.

For the case of edge local pressure, Figure 2.21b, the wall-column would have two D-regions, D_2 and D. The D_2-region should be treated in the same manner as in the case of eccentric local pressure. The D-region is similar to the D-region in Figure 21a, noting that a part of the reinforcement of T_1 should be extended to the top of the wall-column to carry T_3 and the rest of the reinforcement can be anchored above the middle height of D_2.

In this example, it is recognized that sectional design would not have revealed the necessity of extending the reinforcement of T_1 a distance within the D-region (beyond sectional requirements) and the stirrups of T_2 and their particular location, for the safety of this structural member. While the reinforcement of T_4 is required for the case of edge pressure, it is not required for the case of uniform pressure, which may not be obvious by merely guessing and also without knowing the distribution of the applied load. The reinforcement of T_4 should be closed stirrups in order to overcome the anchorage problem and in order to avoid the corner to break off.

2.6.5 Walls with Openings

In Figure 2.22, *STMs* of a wall with a rectangular opening are given for two cases of uniform compression and uniform tension applied to two opposite boundaries of the wall (Schlaich and Schäfer, 1991). It is obvious from Figure 2.22a that the tie T_1 would require reinforcement parallel to and near the edge of the opening, which

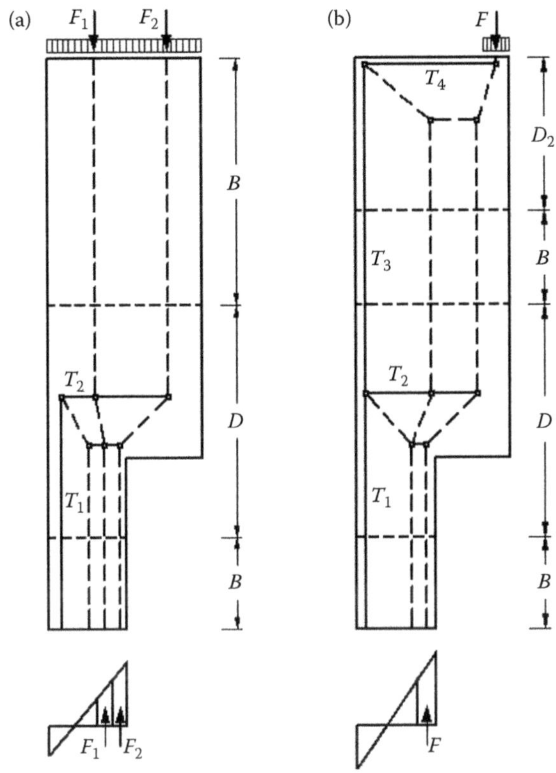

FIGURE 2.21 Deep wall-like column with recess: (a) *STM* for the case of uniform pressure and (b) *STM* for the case of eccentric local pressure.

agrees in principle with normal practice. The quantification of this tie force, and hence the amount of the required reinforcement, is given by the *STM* shown in the figure. On the other hand, for a wall under tension, the model in Figure 2.22b reveals that reinforcement would be required along the edges parallel to the load to carry the tie T_2, which agrees with normal practice too. Nevertheless, the reinforcement of the tie T_1, parallel to the edges and perpendicular to the load direction, has to be placed at a distance from these edges and not along the edges, which may not be satisfied in normal practice. In addition, the anchorage of the reinforcement of ties T_2 and T_3 has to be checked considering the additional length due to the lateral shift of the respective forces as shown by the model.

2.6.6 Deep Beam with Eccentric Large Opening

Owing to the applied concentrated load, the deep beam with an eccentric large opening shown in Figure 2.23a has the stress trajectories shown in Figure 2.23b. Two different *STMs* are combined together for rational representation of the beam behavior (Schlaich et al., 1987). The simple *STM* shown in Figure 2.23c for the right part of the beam can be refined in order to account for the transverse stresses of the strut, as

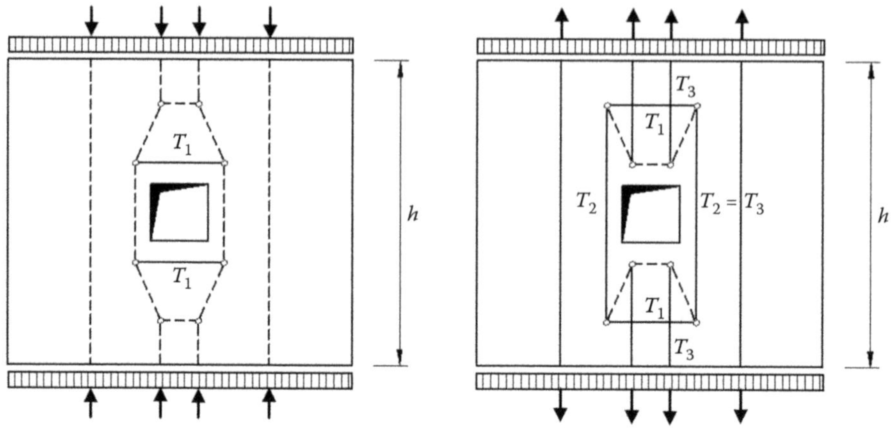

FIGURE 2.22 Walls with openings: (a) wall under uniform compression and (b) wall under uniform tension.

shown by the refined model in Figure 2.23d, which apparently confirms better with the stress trajectories. As for the left part of the beam, Figure 2.23e, either one of the two models shown in Figure 2.23f and g is justifiable by the stress trajectories. In both models, the support reaction A is transferred vertically until a level above the opening by axial action through the B_1-region rather than horizontally by bending action through the B_2-region. This approach is justified by the fact that the axial stiffness of the B_1-region is much greater than the bending stiffness of the B_2-region.

The second model, Figure 2.23g, requires less reinforcement than the first model, Figure 2.23f. Nevertheless, the inclined reinforcement may have an anchorage problem at the upper left node; in addition, some reinforcement has to be detailed according to the first model for keeping the integrity of the concrete material around the opening. A combination of the two models would lead to the most efficient detailing, for example, by assigning 50% of the load to each model. The combined model is shown in Figure 2.23h with the corresponding tension reinforcement layout shown in Figure 2.23i. Of course, web reinforcement and a minimum reinforcement of the B_2-region, below the opening, would still be required.

2.7 MODELING OF B-REGIONS WITH WEB REINFORCEMENT

2.7.1 B-Region with Vertical Web Reinforcement

The beam in Figure 2.24a, apart from the end zones, represents a typical B-region. The appropriate *STM* of such a region is the original truss model as illustrated in the figure, but with the possibility of varying the angle of struts along the beam. The strut angle takes a maximum value at the support node. The mechanism of force transfer within a B-region is illustrated in more detail in this section.

The segment between sections x_1 and x_2 of the beam in Figure 2.24a is illustrated in more detail in Figure 2.24b and c. For section x_2 of this segment, Figure 2.24b,

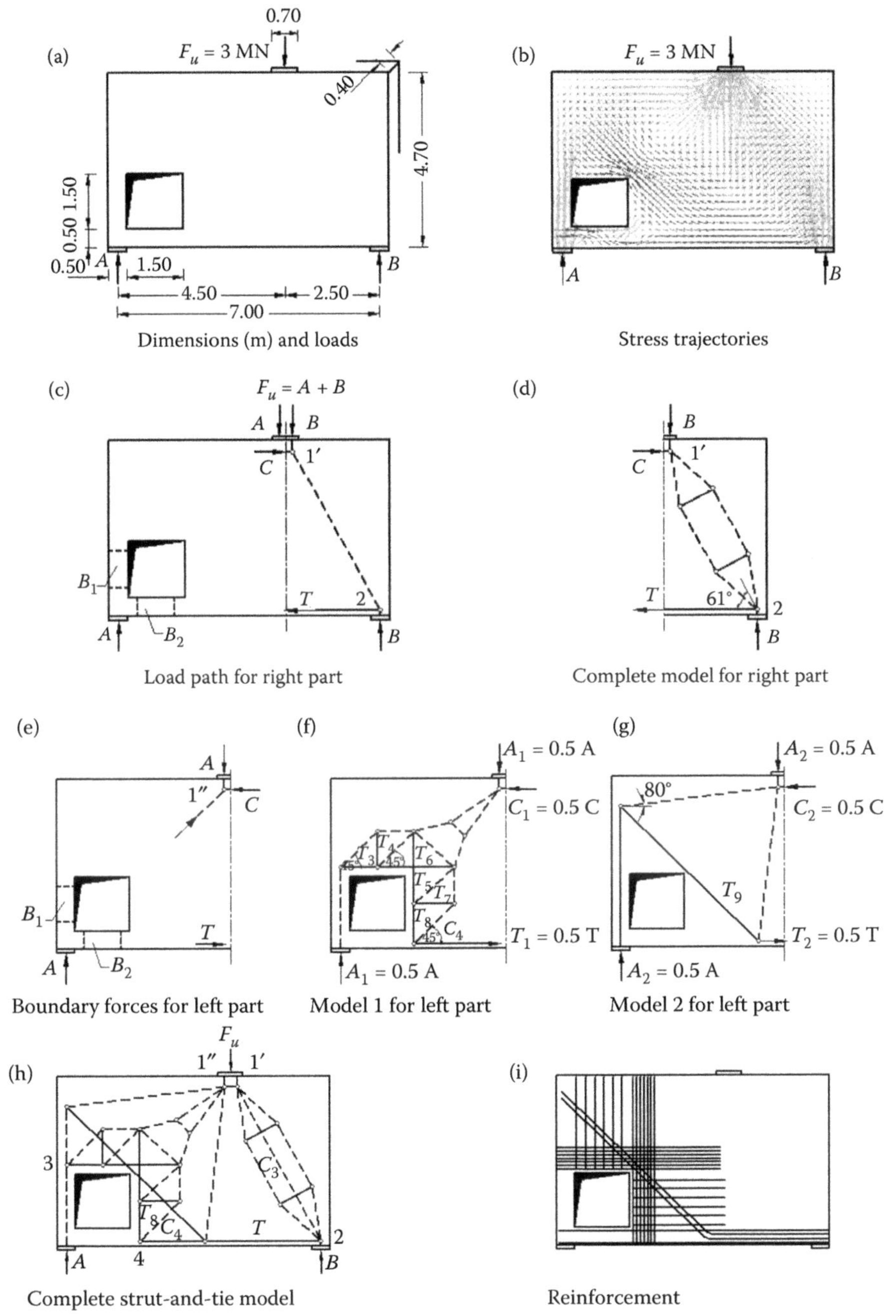

FIGURE 2.23 Deep beam with a large eccentric opening. (From: Schlaich, J. et al., *Journal of the Prestressed Concrete Institute*, 32(3), 1987, 74–150; Schlaich, J., and Schäfer, K., The design of structural concrete, *IABSE Workshop*, New Delhi, 1993.)

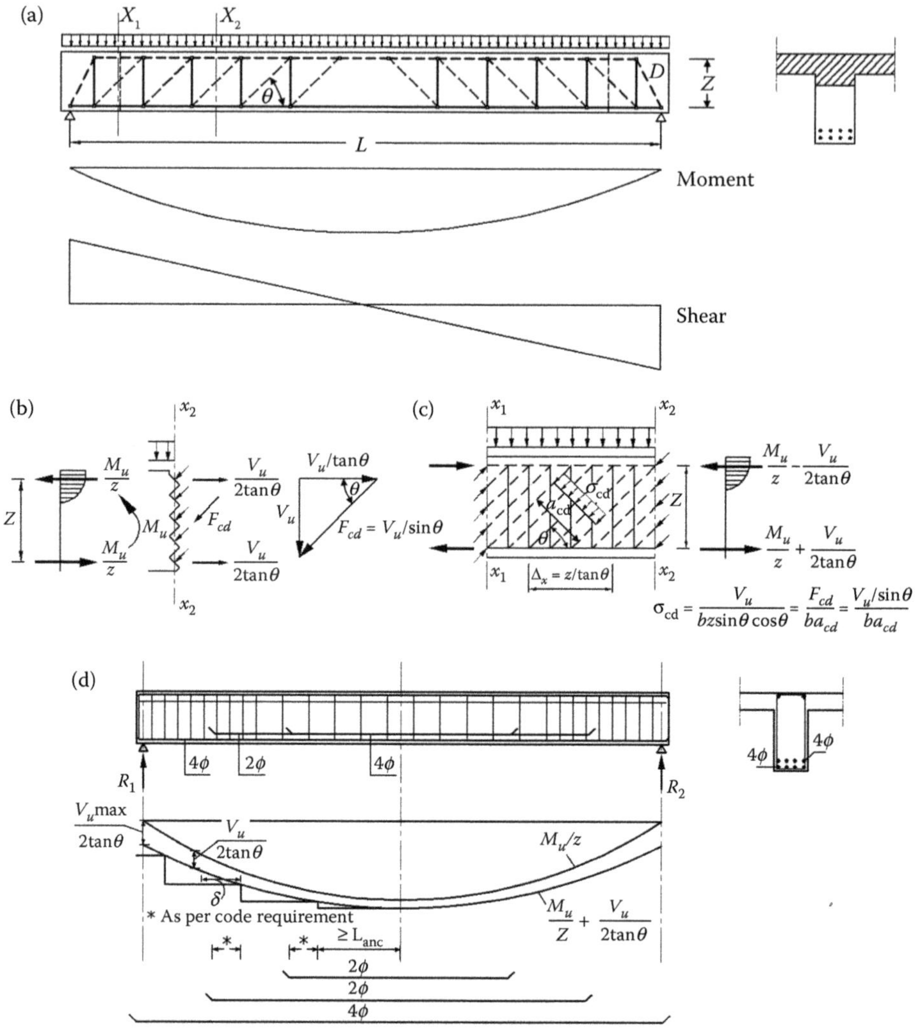

FIGURE 2.24 *STM* of a beam under uniform load: (a) the *STM* (truss model), moment and shear of the beam, (b) force transfer within section x_2 of the segment, (c) force equilibrium of the segment between sections x_1 and x_2, and (d) tension and shifted tension diagrams of the bottom chord.

the moment M_u is resolved into two equal forces, M_u/z (a compression force at the top chord and a tension force at the bottom chord), where z is the lever arm. The shear force V_u is resolved into a diagonal compression force $F_{cd} = V_u/\sin\theta$ and a horizontal tension force $V_u/\tan\theta$, where θ is the angle of inclination of the compression strut (angle of the force F_{cd}). The diagonal compression F_{cd} is carried by the concrete within the web as illustrated in Figure 2.24b and c, causing diagonal compressive stresses, $\sigma_{cd} = (F_{cd}/ba_{cd}) = (V_u/\sin\theta)/ba_{cd}$, where b is the beam breadth and a_{cd} is the strut width, $a_{cd} = z \cos\theta$. Thus, $\sigma_{cd} = V_u/bz \sin\theta \cos\theta = 2v_u/\sin 2\theta$, where v_u is the factored shear stress. The horizontal tension force $V_u/\tan\theta$ is divided equally between the top and bottom chords. This finally leads to a compressive force at the top chord $\left((M_u/z)-(V_u/2\tan\theta)\right)$ and a tensile force at the bottom chord $\left((M_u/z)+(V_u/2\tan\theta)\right)$. Thus, the final force in the tension reinforcement, Figure 2.24d, is obtained by shifting the force diagram due to M_u, a (vertical) amount equal to $\frac{1}{2}V_u\cot\theta$. The additional force in the bottom chord $\frac{1}{2}V_u\cot\theta$ is accounted for in design practice by shifting the design bending moment diagram a (horizontal) distance equal to $\delta = \frac{1}{2}z\cot\theta$, Figure 2.24d.

The force of the vertical tie, V_u, is resisted by vertical stirrups which cover a distance $\Delta x = z\cot\theta$. This indicates the significance of choosing an appropriate value of the angle θ; the smaller value leads to saving in the web reinforcement. On the other hand, a smaller value of the angle leads to higher diagonal compressive stresses, $\sigma_{cd} = 2v_u/\sin 2\theta$, and a larger shift distance of the moment diagram, $\delta = \frac{1}{2}z\cot\theta$.

The value of the angle θ should be chosen between 30° and 60°; however, it could go as small as 26.56° ($\tan^{-1}\frac{1}{2}$). Usually the solution starts with a small value of θ at the zone of low shear stresses and a check of σ_{Cd} will indicate the appropriateness of the selected value. Near the support region, the value of θ should be increased in order to allow for smaller diagonal compressive stresses and better anchorage of the remaining tension reinforcement.

2.7.2 B-Region with Inclined Web Reinforcement

If the need arises for inclined web reinforcement, the appropriate *STM* of the B-region will be as shown in Figure 2.25. In this model, the inclined stirrups make an angle β with the beam axis, where β should be within the range 45–90°. The different parameters of the model are given in Table 2.1.

2.8 2D AND 3D MODELING

In the preceding sections, structural elements have been treated with 2D *STMs*, planar models. In some cases, it may not be adequate to employ 2D models, such as the

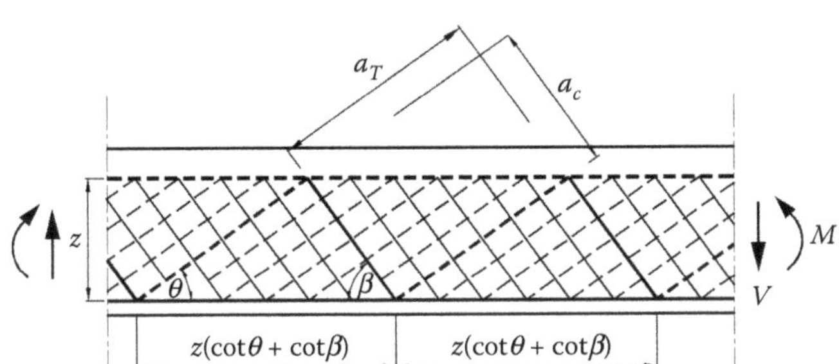

FIGURE 2.25 *STM* of a B-region with inclined web reinforcement.

TABLE 2.1
The *STM* Parameters of a B-Region with Inclined and Vertical Web Reinforcement

Stirrups	**$45° \leq \beta \leq 90°$**	**$\beta = 90°$**
a_T	$z(\cot\theta + \cot\beta)\sin\beta$	$z\cot\theta$
a_C	$z(\cot\theta + \cot\beta)\sin\theta$	$z\cos\theta$
Shift rule δ	$\frac{1}{2}z(\cot\theta - \cot\beta)$	$\frac{1}{2}z\cot\theta$
Compression chord	$C = \left(\frac{M_u}{z}\right) - \frac{1}{2}V_u(\cot\theta - \cot\beta)$	$C = \left(\frac{M_u}{z}\right) - \frac{1}{2}V_u\cot\theta$
Tension chord	$T = \left(\frac{M_u}{z}\right) + \frac{1}{2}V_u(\cot\theta - \cot\beta)$	$T = \left(\frac{M_u}{z}\right) + \frac{1}{2}V_u\cot\theta$
Diagonal compressive stress, σ_{Cd}	$\frac{v_u}{(\cot\theta + \cot\beta)\sin^2\theta}$	$\frac{v_u}{\sin\theta\cos\theta}$
Stirrup force per unit length	$\frac{V_u/z}{(\cot\theta + \cot\beta)\sin^2\beta}$	$\left(\frac{V_u}{z}\right)\tan\theta$

$$v_u = \left(\frac{V_u}{bz}\right)$$

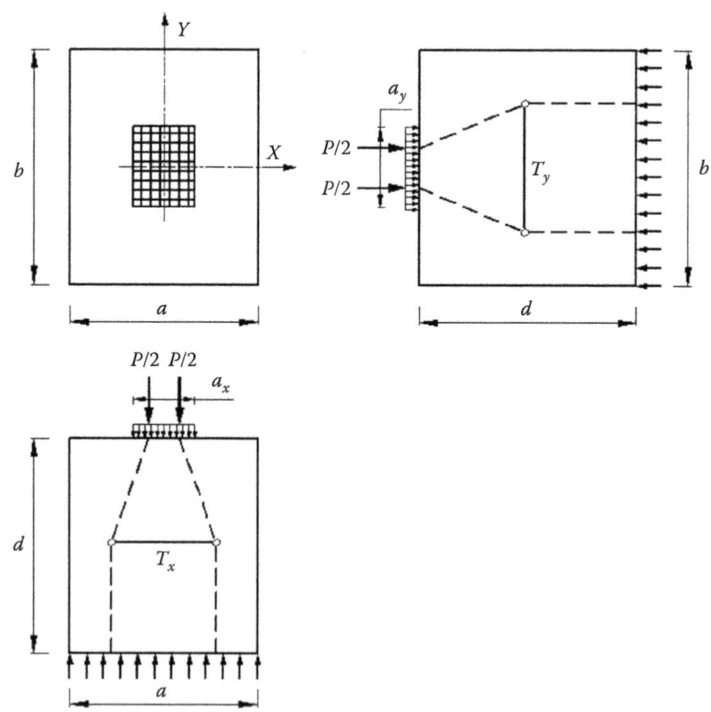

FIGURE 2.26 Applied *STM* in two different planes.

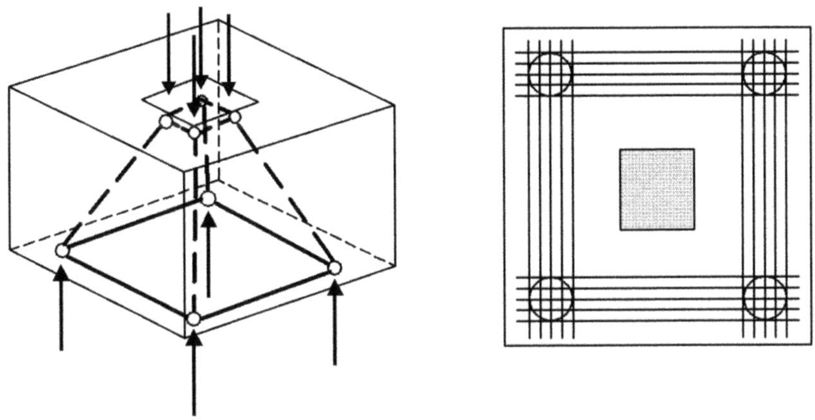

FIGURE 2.27 Example of a 3D *STM* and corresponding reinforcement arrangement for a pile plinth.

design of transfer slabs, pile caps and wind power plant foundations, which are subjected to loads that result in 3D stress fields. If the state of stress is not predominantly plane, *3D STMs* should be used (Schlaich et al., 1987). However, most of the time, details have not been provided on how to deal with 3D *STMs*, for instance concerning the verification of nodes.

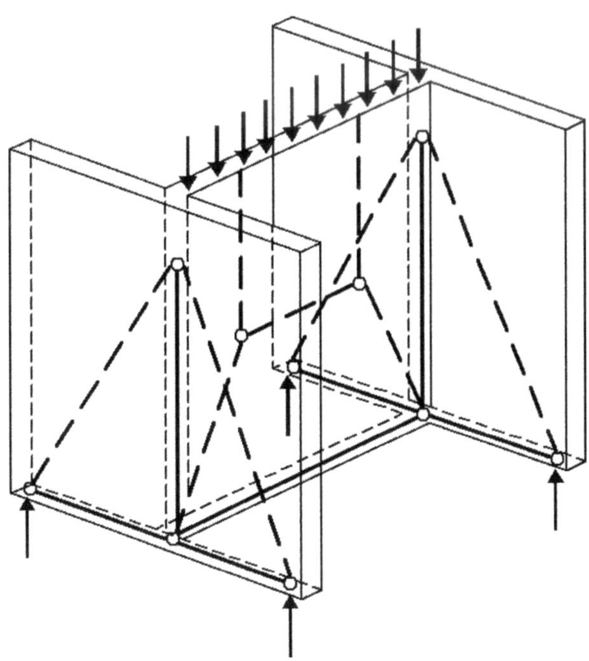

FIGURE 2.28 Combination of 2D *STMs*.

With large 3D reinforced concrete blocks, load paths for the concentrated load may be considered in two directions, as illustrated in Figure 2.26, where the strut-and-tie method is applied in two perpendicular planes. Unlike *2D STMs*, *3D STMs* are required when the structure and loading are considerably spread over all 3Ds, such as pile caps with two or more rows of piles.

There are two different approaches for the construction of a 3D *STM*, by modeling in three dimensions or by combining 2D models. A 3D *STM* for a centric loaded pile cap (Engström, 2011) is shown in Figure 2.27.

Figure 2.28 indicates how 2D *STMs* can be combined to solve spatial structure (Engström, 2011), one model in the plane of the flanges and the other in the plane of the web. For such a model, each *STM* transfers the load in its own plane. The two models are joined with common nodes.

The utilization of *3D STMs* in the design of spatial structures is associated with a very tedious computation procedure. Therefore, the use of such a procedure is restricted to structures where there is no other alternative or for verification purposes. The subject of *3D strut-and-tie modeling* with its challenging issues is discussed in Chapter 8 of pile caps.

REFERENCES

Engström, B., *Design and Analysis of Deep Beams, Plates and Other Discontinuity Regions*, Department of Structural Engineering, Chalmers University of Technology, Göteborg, Sweden, 2011.

Schäfer, K. and El-Metwally, S. E., On the role of discontinuity regions detailing in the safety of concrete structures, *Proceedings of the Fifth International Colloquium on Concrete in Developing Countries*, Cairo, Egypt, 2–6, January 1994, pp. 43–55.

Schäfer, K. and Schlaich, J., *Strut-and-Tie Models for the Design and Detailing of Structural Concrete*, Seminar Notes, Department of Civil Engineering, the University of Calgary, Calgary, Canada, 1998.

Schlaich, J. and Schäfer, K., Design and detailing of structural concrete using strut-and-tie models, *Journal of the Structural Engineer*, 69(6), 1991, 113–125.

Schlaich, J. and Schäfer, K., The design of structural concrete, IABSE Workshop, New Delhi, 1993.

Schlaich, J., Schäfer, K., and Jennewein, M., Toward a consistent design of structural concrete, *Journal of the Prestressed Concrete Institute*, 32(3), 1987, 74–150.

Yun, Y. M., Computer graphics for nonlinear strut-tie model approach, *ASCE, Journal of Computing in Civil Engineering*, 14(2), 2000, 127–133.

3 Failure Criteria

3.1 INTRODUCTION

As explained in Chapter 1 (Section 1.2) and with reference to the illustrative example in Figure 3.1, an *STM* consists of three types of elements:

1. Concrete strut in compression (with or without reinforcement).
2. Ties with or without reinforcement. Ties with reinforcement using reinforcing steel or prestressing steel are dominant. Nevertheless, concrete ties in zones without tension reinforcement are necessary too if equilibrium cannot be achieved unless tensile forces exist; for example, lap splice, slab without stirrups, …, etc.
3. Nodes and nodal zones. Nodes are the points of intersection of struts, ties, and concentrated forces. The volume of concrete around a node that is assumed to transfer *STM* forces through the node is the nodal zone.

The *STM* shown in Figure 3.1 may fail by one of the following modes:

- The tension tie could yield, or anchorage of the ties may fail.
- One of the struts could crush when the stress in the strut exceeds the effective compressive strength of concrete.
- A nodal zone could fail by being stressed greater than the effective compressive strength of concrete.

For safe design, the different elements of an *STM* have to be checked or dimensioned according to the material failure criterion of the element. Dimensioning not only means sizing and reinforcing the individual struts and ties for the forces they carry, but also ensuring the load transfer between them by checking the node regions. There is a close relation between the detailing of the nodes and the strength of the struts bearing on them and the ties anchored in them, since such detailing affects the flow of forces. Therefore, it is necessary to check whether the *STM* initially chosen is still valid after detailing or needs adjustment. Thus, both the modeling and the dimensioning converge to refined solutions through an iterative process.

In the literature, there have been different assessments and different approaches for calculating strength values for elements of *STMs*. However, there is a noticeable variation and inconsistency between reported values. This chapter covers the strength of the different elements of strut-and-tie model (*STM*), struts, nodes (or nodal zones) and ties. For the concrete struts, the behavior of concrete under different loading conditions and the different basic compression stress fields; prismatic, bottle-shaped and fan-shaped, are discussed. The geometry and strength of nodes (nodal zones) under different boundary conditions are explained. The strength of

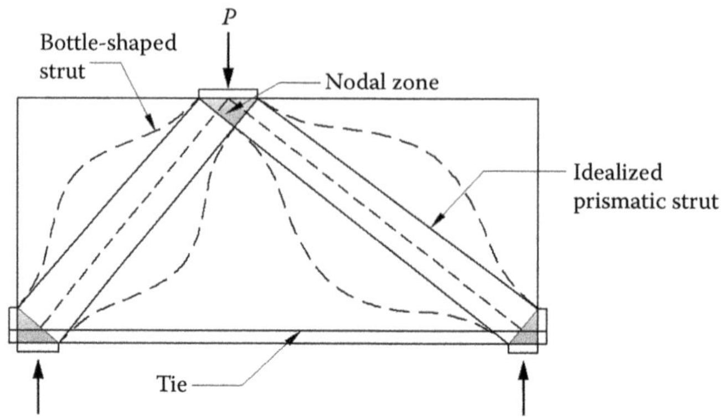

FIGURE 3.1 Components of *STM* of a deep beam.

reinforced ties and the influence of their detailing on the geometry of the connecting nodal zones are examined. The chapter also presents the nominal strength and design strength of struts, nodal zones and ties according to the ACI 318-14 provisions. Finally, this chapter covers the reinforcement anchorage with regard to bond action of straight bars, anchorage length, lap joints and curved reinforcement.

3.2 CONCRETE STRUTS

3.2.1 Behavior and Strength

The shape of a strut is highly dependent upon the force path from which the strut arises and the details of any tie tension reinforcement connected to the strut. Although the reinforced tie, T_s, is essentially linear or a one-dimensional element between two nodes, the concrete strut in compression, C, or the concrete tie in tension, T_c,

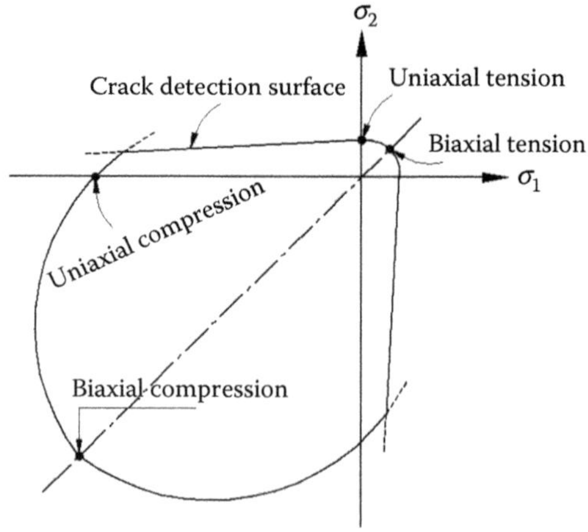

FIGURE 3.2 Kupfer biaxial failure surface of concrete (Kupfer et al., 1969).

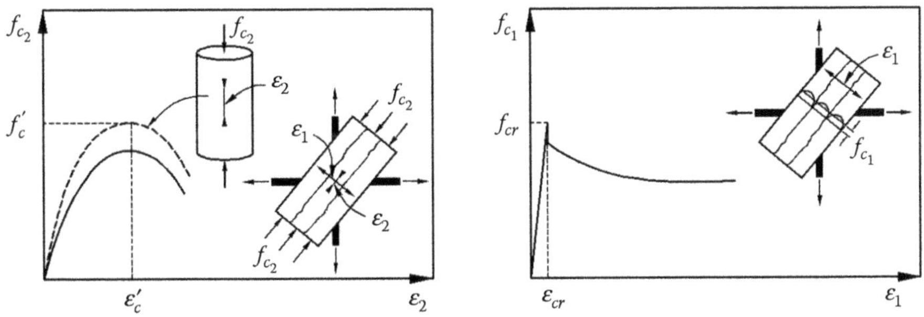

FIGURE 3.3 Stress–strain relationships of diagonally cracked concrete (Vecchio and Collins, 1986).

represents a two-dimensional or three-dimensional stress field that tends to spread in between two end nodes. The associated deviation of the stress trajectories causes transverse tensile and compressive stresses which must be reflected in the design strength of concrete struts, Figures 3.2 and 3.3.

According to Schlaich et al. (1987), there are three major geometric shape classes of struts: *prismatic*, *bottle-shaped*, and *compression fan*, as shown in Figures 3.4 and 3.5.

Prismatic struts are the most basic type of struts, and they are typically used to model the compressive stress block of a beam element as shown in Figures 3.4a and 3.5a.

Bottle-shaped struts are formed when the geometric conditions at the end of the struts are well defined, but the rest of the strut is not confined to a specific portion of the structural element. The geometric conditions at the ends of bottle-shaped struts are typically determined by the details of bearing pads and/or the reinforcement details of any adjoined steel. The best way to visualize a bottle-shaped strut is to imagine forces dispersing as they move away from the ends of the strut as shown in Figures 3.4b and 3.5b. The bulging stress trajectories cause transverse tensile stresses to form in the strut

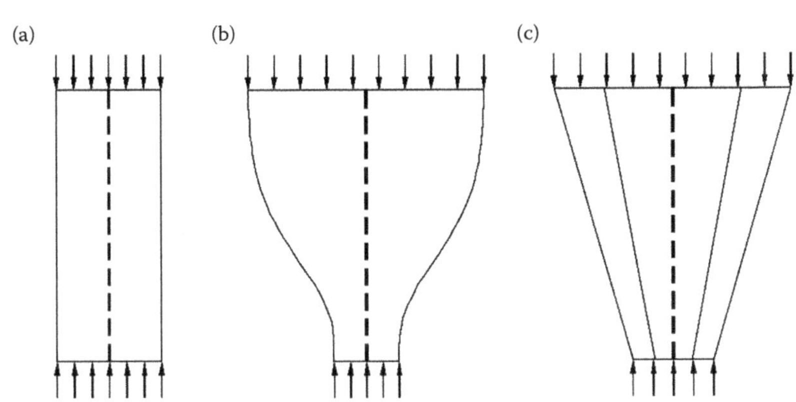

FIGURE 3.4 Basic compression stress fields: (a) the prism, (b) the bottle, and (c) the fan.

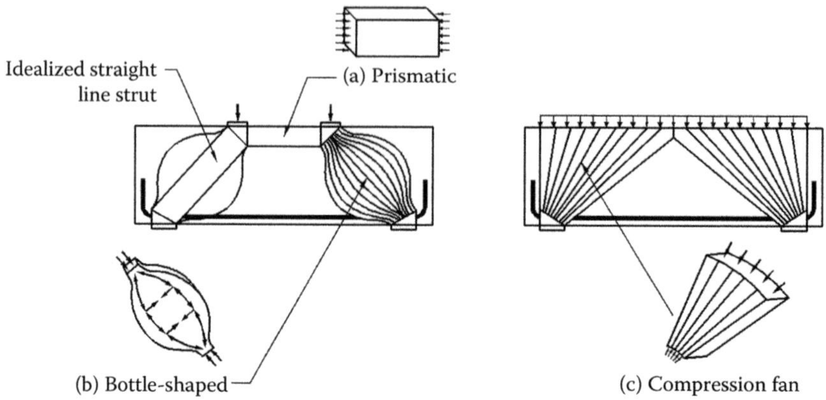

FIGURE 3.5 Geometric shapes of struts.

which can lead to longitudinal cracking of the strut; therefore, appropriate crack control reinforcement should always be placed across bottle-shaped struts to avoid premature failure. For this reason, most design specifications require minimum amounts of crack control reinforcement in regions designed with *STMs*.

Compression fan struts are formed when stresses flow from a large area to a much smaller area. Compression fans are assumed to have negligible curvature, and therefore they do not develop transverse tensile stresses. The simplest example of a compression fan is a strut that carries a uniformly distributed load to a support reaction in a deep beam as shown in Figures 3.4c and 3.5c.

The strength of the concrete in compression stress fields depends to a great extent on the multiaxial state of stress and on the disturbances from cracks and reinforcement. The effective compressive strength of the concrete in a strut f_{ce}^{s} may be obtained from:

$$f_{ce}^{s} = 0.85 f_c' \beta_s \text{ (or } 0.67 f_{cu} \beta_s) \tag{3.1}$$

where f_c' is the concrete cylinder strength, f_{cu} is the concrete cube strength, and β_s is an effectiveness factor of concrete strut, which takes into account the stress conditions, strut geometry, and the angle of cracking surrounding the strut.

Schlaich et al. (1987) proposed the following values of the effectiveness factor for different types of concrete struts.

$\beta_s = 1.0$ for an undisturbed and uniaxial state of compressive stress that may exist in a prismatic strut;

$\beta_s = 0.8$ if tensile strains and/or tensile reinforcement perpendicular to the axis of the strut may cause cracking parallel to the strut with a normal crack width;

$\beta_s = 0.6$ if tensile strains and/or tensile reinforcement cause cracks at skew angles to the strut axis;

$\beta_s = 0.4$ for struts with skew cracks with an extraordinary crack width. Skew cracks would be expected if modeling of the strut departs significantly from the theory of elasticity flow of internal forces.

The values of β_s according to the ACI 318-14 Code are adopted in this book.

The *nominal compressive strength of a strut* without longitudinal reinforcement, F_{ns}, shall be calculated at the two strut end, as follows, and the smaller value is used;

$$F_{ns} = f_{ce}^{s} A_{cs} \tag{3.2}$$

where f_{ce}^{s} is the smaller of:

- The effective compressive strength of the concrete in the strut.
- The effective compressive strength of the concrete in the nodal zone.

and A_{cs} is the cross-sectional area at one end of the strut. In calculating A_{cs}, the strut width is measured perpendicular to the strut axis at its end, Figure 3.6. f_{ce}^{s} shall not exceed the failure criterion of the node at the strut end under consideration. The smaller value of F_{ns} at the two ends of the strut will control the design.

The design of struts shall be based on

$$\phi F_{ns} \geq F_{us} \tag{3.3}$$

In another form

$$\phi(0.85 f_c' \beta_s) A_{cs} \geq F_{us} \tag{3.4}$$

where F_{us} is the largest factored force acting in a strut and obtained from the applicable load combinations and the ϕ factor is the material strength reduction factor which is equal to 0.75 for ties, struts, and nodes according to the ACI 318-14 Code.

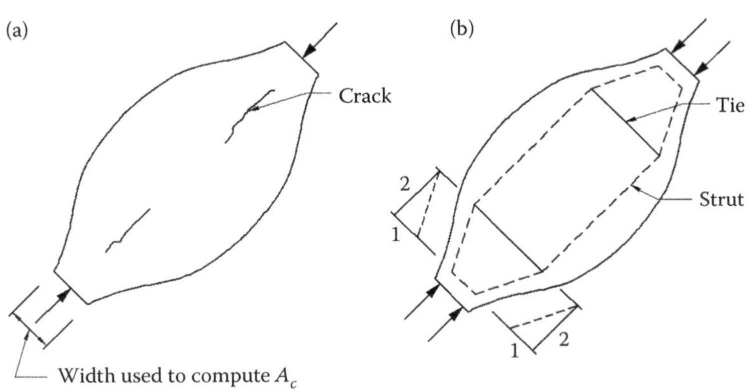

FIGURE 3.6 Bottle-shaped strut: (a) cracking of a strut; and (b) *STM* for transverse reinforcement.

3.2.2 ACI 318-14 Effectiveness Factor for Struts

The values of an effectiveness factor for struts, β_s, are recommended by ACI 318-14 as follows:

$\beta_s = 1.00$ for a strut of uniform cross-sectional area over its length; i.e., prismatic strut.

$\beta_s = 0.75$ for a bottle-shaped strut when providing transverse reinforcement to resist the lateral tension according to the model in Figure 3.6, or if $f_c' \leq 44.0\,\text{MPa}$ and the reinforcement crossing the strut, Figure 3.7, satisfy the following $\sum \frac{A_{si}}{bs_i} \sin\gamma_i \geq 0.003$, where A_{si} is the total area of reinforcement at spacing s_i in a layer of reinforcement with bars at an angle γ_i to the axis of the strut and the other parameters are as illustrated in the figure.

$\beta_s = 0.60\lambda$ for a bottle-shaped strut when the transverse reinforcement does not satisfy the requirement of the model in Figure 3.6. λ is a modification factor to account for the use of lightweight concrete since lightweight concrete has a lower tensile strength and higher brittleness, which can reduce the strut strength. $\lambda = 0.85$ for sand-lightweight concrete and 0.75 for all-other lightweight concretes. $\lambda = 1.0$ for normal weight concrete.

$\beta_s = 0.40$ for struts in tension members or in the tension flanges of members.

$\beta_s = 0.60$ for all other cases; for example, fan-shaped struts, compression fields in the web of a beam where parallel diagonal cracks are likely to divide the web into struts, and struts are likely to be crossed by cracks at an angle to the struts, Figure 3.8.

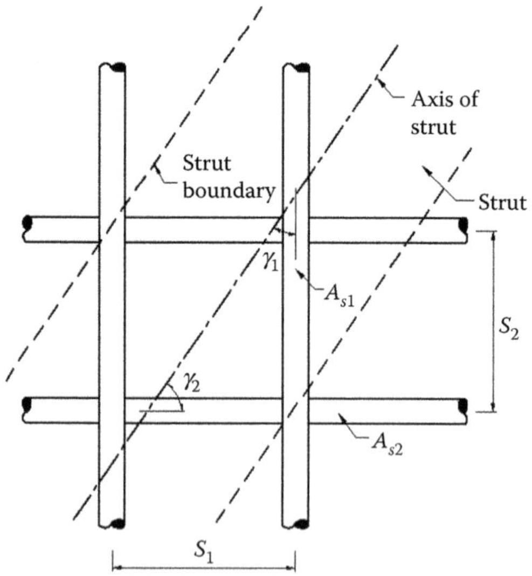

FIGURE 3.7 Reinforcement crossing a strut.

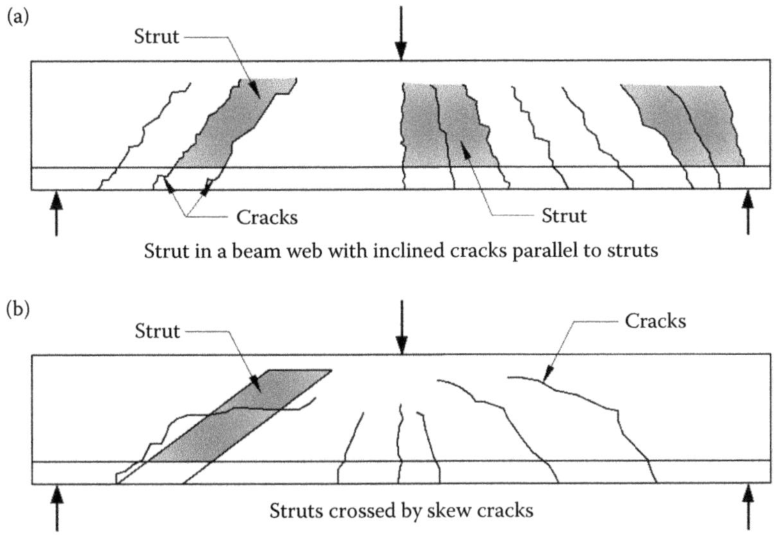

FIGURE 3.8 Types of struts in a beam web.

3.3 NODAL ZONES

3.3.1 Geometry and Strength

The compressive strength of concrete of a nodal zone depends on a number of factors such as the tensile straining from intersecting ties, the confinement provided by compressive reactions, transverse reinforcement, compression struts, anchorage plates of prestressing, and the effects of strain discontinuities within the nodal zone. Different types of nodal zones are illustrated in Figure 3.9.

To distinguish between the different straining and confinement conditions of nodal zones, these zones are identified as follows, Figure 3.10:

- $C - C - C$ nodal zone bounded by compression struts only
- $C - C - T$ nodal zone bounded by compression struts and one tension tie
- $C - T - T$ nodal zone bounded by a compression strut and two tension ties and
- $T - T - T$ nodal zone bounded by tension ties only

As discussed by Brown and Bayrak (2006), nodes can be detailed to be either hydrostatic or non-hydrostatic in theory. For a hydrostatic node, the stress acting on each face of the node is equivalent and perpendicular to the surface of the node, Figure 3.11. Since stresses are perpendicular to the faces of hydrostatic nodes, there are no shear stresses acting on the face of a hydrostatic node. However, achieving hydrostatic nodes for most *STM* geometric configurations is nearly impossible and usually impractical. For this reason, most *STMs* utilize non-hydrostatic nodes. For non-hydrostatic nodes, Schlaich et al. (1987) suggested that the ratio of the maximum stress on a face of a node to the minimum stress on another face of the node

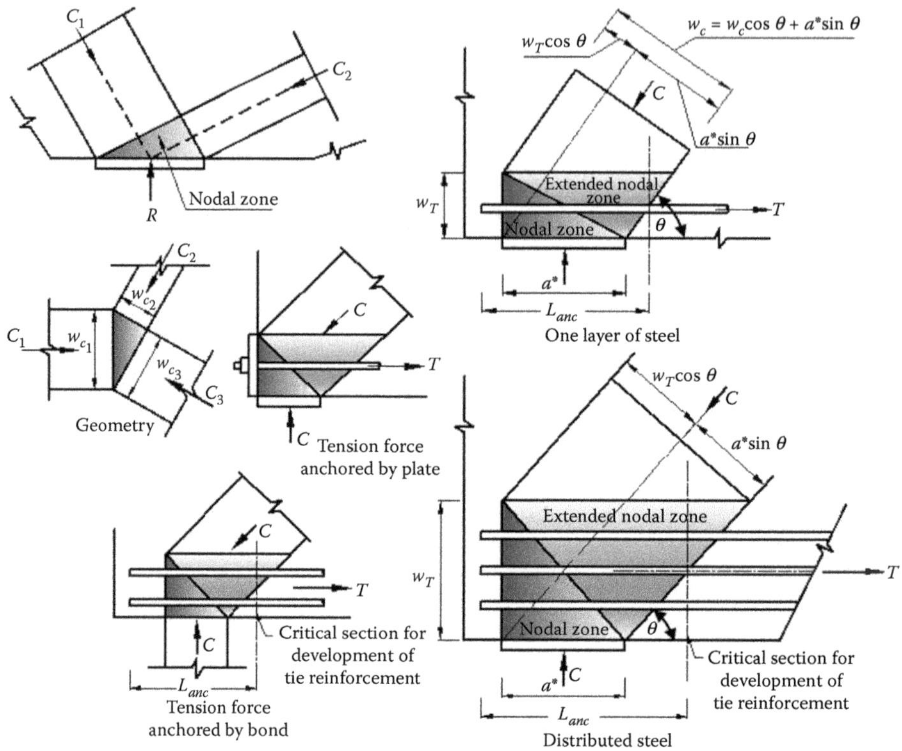

FIGURE 3.9 Different types of nodal zones.

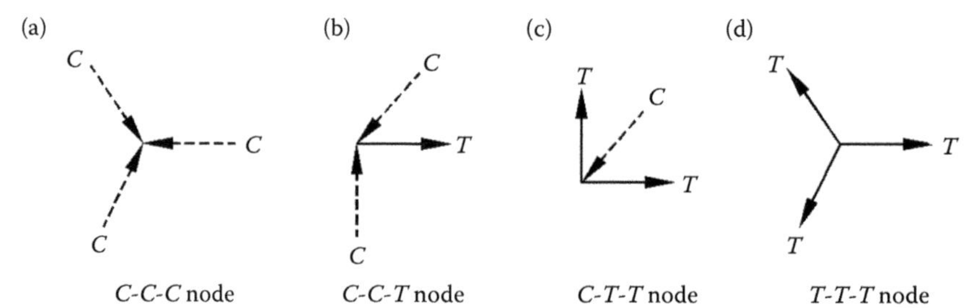

FIGURE 3.10 Classification of nodes.

should be less than 2. The states of stress in both hydrostatic and non-hydrostatic nodes are shown in Figure 3.11.

The effective compressive strength of the concrete in a nodal zone, f_{ce}^n, can be obtained from:

$$f_{ce}^n = 0.85 f_c' \beta_n \text{ (or } 0.67 f_{cu} \beta_n) \tag{3.5}$$

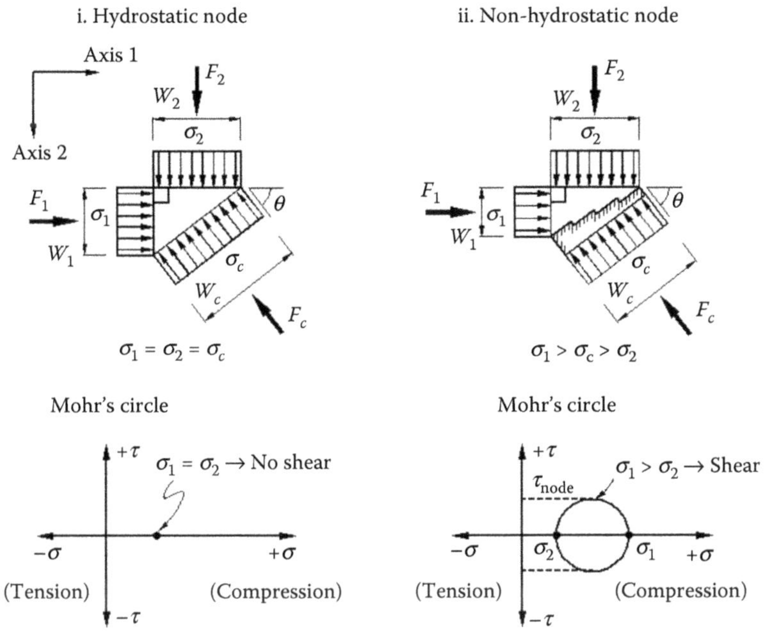

FIGURE 3.11 States of stress in hydrostatic and non-hydrostatic nodes (Brown and Bayrak, 2006).

where β_n is the effectiveness factor of a nodal zone.

Schlaich et al. (1987) proposed the following values of an effectiveness factor for different types of nodes:

$\beta_n = 1.1$ for compression–compression–compression nodes and
$\beta_n = 0.8$ for nodes where reinforcement is anchored in or crossing the node

For safety purposes, for $C - C - C$ nodes, the value of the effectiveness factor of a nodal zone, $\beta_n = 1.1$ suggested by Schlaich et al. (1987) can be reduced to 1.0. Also, in order to appropriately consider the effect of the tensile strains on the concrete compressive strength, the value of 0.8 proposed by Schlaich et al. (1987) for other nodes can be replaced by 0.8 for $C - C - T$ nodes, 0.6 for $C - T - T$ nodes (two or more ties), and 0.4 for $T - T - T$ nodes. The values of β_n recommended by the ACI 318-14 Code are adopted in this book.

The *nominal compressive strength of a nodal zone*, F_{nn}, shall be

$$F_{nn} = f_{ce}^{n} A_{nz} \tag{3.6}$$

where f_{ce}^{n} is the effective compressive strength of the concrete in the nodal zone and A_{nz} is the area of each face of the nodal zone and shall be taken as the smaller of:

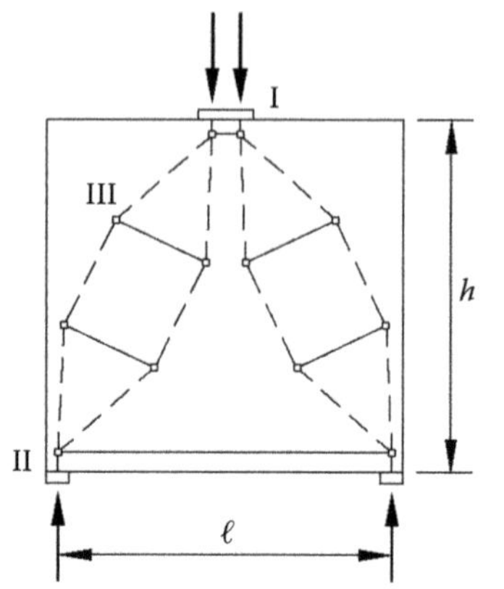

FIGURE 3.12 Types of nodes.

- The area of the face of the nodal zone perpendicular to the line of action of F_{us}, where F_{us} is the largest factored force acting in a strut, reaction, or external force and obtained from the applicable load combinations.
- The area of a section through the nodal zone perpendicular to the line of action of the resultant force on the section.

The nodes of the model are derived from the intersection points of three or more straight struts or ties. A node, as introduced into the model, which implies an abrupt change of direction of forces, is called a singular or concentrated node, for example, nodes I and II in Figure 3.12. On the other hand, for wide concrete stress fields joining each other or with tensile ties, the deviation of forces may be smeared or spread over some length. Therefore, they are called smeared or continuous nodes, for example, node III in Figure 3.12.

In smeared nodes, where the deviation of forces may be smeared or spread over some length, the check of stress is often not critical and the only requirement is to check the anchorage length of the reinforcing bars. On the other hand, singular or concentrated nodes have to be carefully checked.

3.3.2 ACI 318-14 Effectiveness Factorfor Nodal Zones

The ACI 318-14 code recommends the following values of an effectiveness factor for nodal zones, β_n.

$\beta_n = 1.00$ for nodal zones bounded by struts or bearing areas or both, *C–C–C* node

$\beta_n = 0.80$ for nodal zones anchoring one tie, *C–C–T* node

$\beta_n = 0.60$ for nodal zones anchoring two or more ties with the presence of one strut, *C–T–T* node

$\beta_n = 0.40$ for nodal zones anchoring ties only, T–T–T node

3.4 REINFORCED TIES

Reinforced ties are *STM* members that are subjected to tensile forces carried by reinforcement. The tie cross section is constant along its length and is obtained from the tie force and the yield stress of the steel. The *nominal strength of a tie*, F_{nt}, shall be taken as

$$F_{nt} = A_{st} f_y \tag{3.7}$$

where A_{st} is the cross section of area of steel and f_y is the yield stress of steel.

The width of a tie is to be determined to satisfy safety for compressive stresses at nodes. Depending on the distribution of the tie reinforcement, the effective tie width w_t may vary between the following values but with an upper limit given afterward.

- In the case of using one row of bars without sufficient development length beyond the nodal zones (Figure 3.13a):

$$w_t = 0.0 \tag{3.8}$$

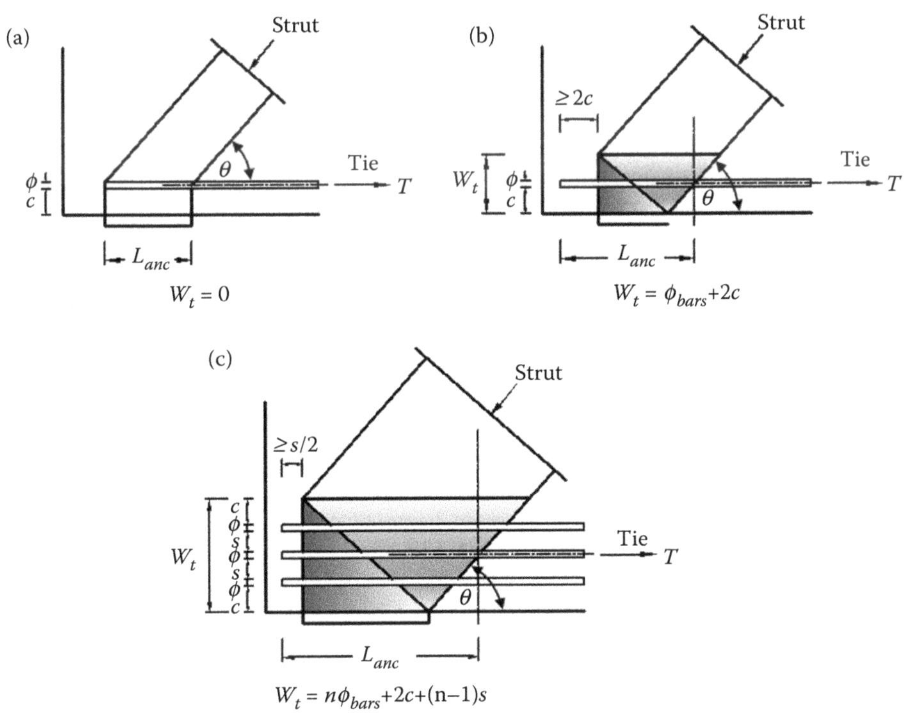

FIGURE 3.13 The width of the tie, w_t, used to determine the dimensions of the nodal zone.

- In case of using one row of bars and providing sufficient development length beyond the nodal zones for a distance not less than $2c$, where c is the concrete cover (Figure 3.13b):

$$w_t = \phi_{bars} + 2c \tag{3.9}$$

where ϕ_{bars} is the diameter of the reinforcing bars.
- In case of using more than one row of bars and providing sufficient development length beyond the nodal zones for a distance not less than $0.5s$, where s is the clear distance between bars (Figure 3.13c):

$$w_t = n\phi_{bars} + 2c + (n-1)s \tag{3.10}$$

where n is the number of layers of reinforcing bars.

In the three cases in Figure 3.13, the development length according to the ACI 318-14, L_{anc}, begins at the intersection of the centroid of the bars in the tie and the extensions of the outlines of either the strut or the bearing area.

In case of multiple layers of reinforcement, the upper limit of the height of the nodal zone, $w_{t,\max}$, is established as the width corresponding to the width in a hydrostatic nodal zone, calculated as

$$w_{t,\max} = \frac{F_{nt}}{\left(f_{ce}^{n} b\right)} \tag{3.11}$$

where f_{ce}^{n} is the applicable effective compressive strength of the nodal zone and is computed from Equation 3.5, and b is the breadth of the beam.

3.5 ANCHORAGE OF REINFORCEMENT

3.5.1 Bond Action of Straight Bars

Inherently, bond is known to form as a result of the reinforcement roughness and adhesion of the surrounding concrete. If the anchorage length is defined as L_{anc}, the average bond stress, τ_b, is then calculated as

$$\tau_b = \frac{T_s}{\pi \phi_{bar} L_{anc}} \tag{3.12}$$

where T_s is the force in the bar to be transferred along the length L_{anc}, and ϕ_{bar} is the bar diameter.

In reality, bond action is developed by compressive and tensile stresses as illustrated by their trajectories in Figure 3.14a and the *STM* in Figure 3.14b. The compressive stresses are carried by concrete which is supported by the bar ribs, whereas the

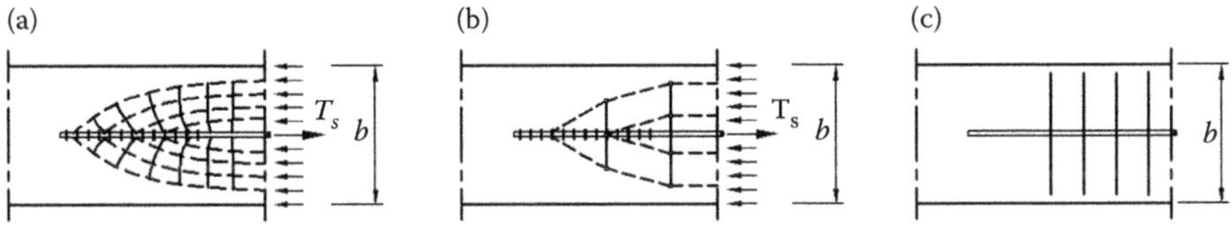

FIGURE 3.14 Symmetrical anchorage of a reinforcing bar in a concrete body: (a) stress trajectories, (b) plane *STM*, and (c) associated transverse reinforcement. (From Schlaich, J., and Schäfer K., *IABSE Workshop*, New Delhi, 1993.)

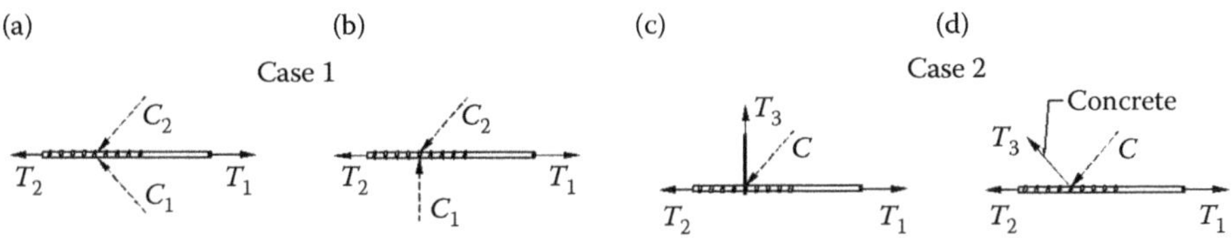

FIGURE 3.15 Typical examples of reinforcement anchorage by bond. Anchor forces are supported by: (a, b) two concrete compression struts, (c) one concrete compression strut and one steel tie, and (d) one concrete compression strut and one concrete tie. (From Schlaich, J., and Schäfer K., *IABSE Workshop*, New Delhi, 1993.)

tensile stresses are either resisted by concrete or by reinforcement, Figure 3.14b and c, hence leading to a higher bond strength.

In the anchorage of reinforcement, there are two possible cases as shown in Figure 3.15. The anchor force ($T_1 - T_2$) is better accommodated in case 1, with a favorable condition in version (a). For this case, version (a) exists where bar anchoring is inside a member, whereas version (b) exists where the bar is anchored above a support or beneath a concentrated load. Case 2 exists where bars are anchored near the surface of a member with version (d) representing anchoring development depending on unreinforced concrete ties.

3.5.2 Anchorage Length

For practical dimensioning of the anchorage length, the respective codes of practice should be referred to. When terminating bar reinforcement, it should be noted that the dispersion of the anchor forces perpendicular to the plane of reinforcement should be considered. Such dispersion of forces generates transverse tensile stresses in the concrete, which may cause splitting of the structure along the anchorage bars, as illustrated in Figure 3.16.

Some of the anchorage may be achieved by not only extending the reinforcement through the nodal zone but also developing it beyond the nodal zone, as shown in Figure 3.13b and c. If the tie is anchored using 90° hooks, the hooks should be confined within reinforcement to avoid cracking along the outside of the hooks in the support region. In deep beams, hairpin bars spliced with the tie reinforcement can be used to anchor the tie forces at exterior supports, provided that the beam

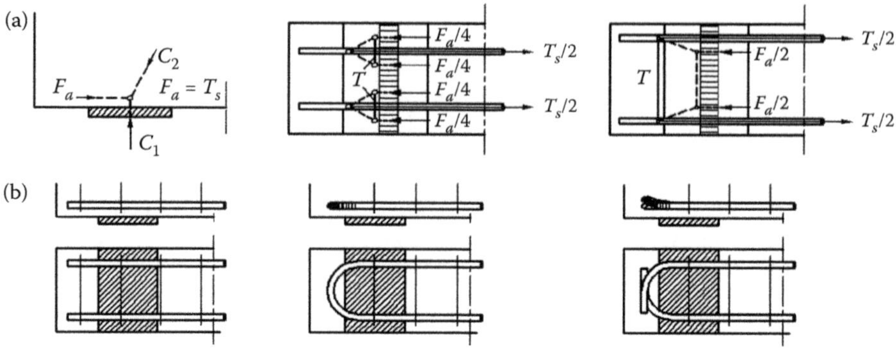

FIGURE 3.16 Distribution of anchor forces perpendicular to the plane of deep beam: (a) *STMs* of two different reinforcement patterns and (b) transverse reinforcement to cover the tensile forces perpendicular to the plane of the deep beam. (From Schlaich, J., and Schäfer K., *IABSE Workshop*, New Delhi, 1993.)

width is large enough to accommodate such bars. The development length of the tie reinforcement can be reduced by using hooks, headed bars, mechanical devices, additional confinement, or by splicing it with layers of smaller bars.

3.5.3 Lap Joints

The common method for force transfer between two steel bars is to place them alongside each other, Figure 3.17a. In this method, the concrete contributes to the force transfer between the two bars via diagonal compression stresses and associated transverse tensile stresses as illustrated by the *STM* in Figure 3.17b. While the compressive stresses are resisted by concrete, the tensile stresses are either resisted

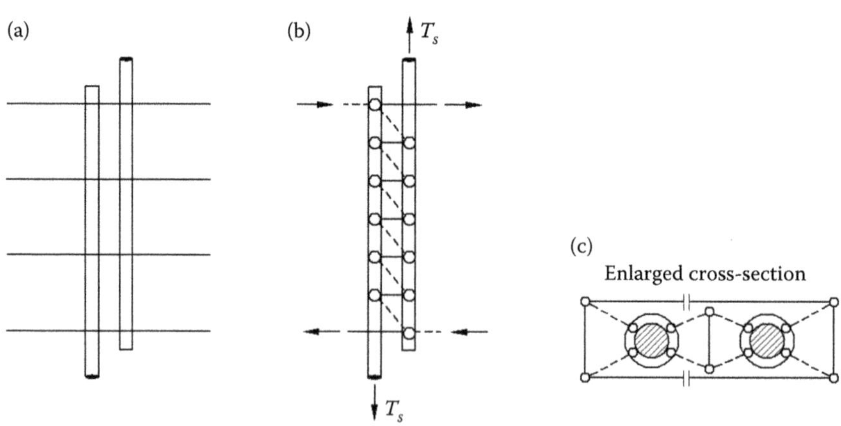

FIGURE 3.17 Lap joint of tension reinforcement: (a) reinforcement and (b, c) models in two planes. (From Schlaich, J., and Schäfer K., *IABSE Workshop*, New Delhi, 1993.)

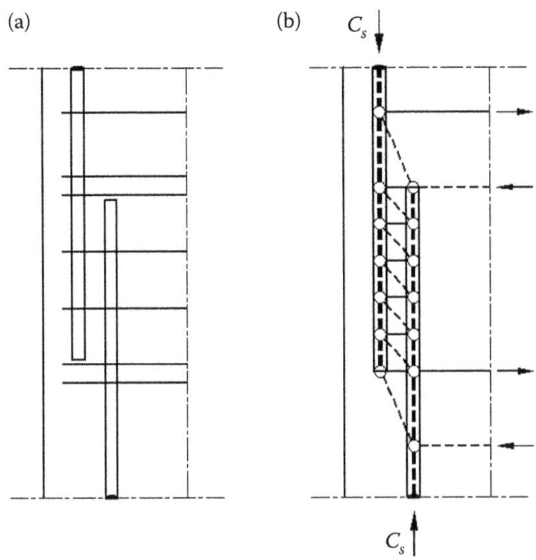

FIGURE 3.18 Lap joint of compression reinforcement: (a) reinforcement and (b) model. (From Schlaich, J., and Schäfer K., *IABSE Workshop*, New Delhi, 1993.)

by concrete alone or by concrete and reinforcement; conservatively, these tensile stresses can be assessed based on a 45° model. With providing such transverse reinforcement the bond strength is significantly improved and hence the required splice length is reduced.

For a compression lap splice, Figure 3.18a, the appropriate *STM* is shown in Figure 3.18b. Hooks are not suited to anchor compression bars as illustrated in Figure 3.19. For a tension lap splice, the model in Figure 3.17b illustrates the

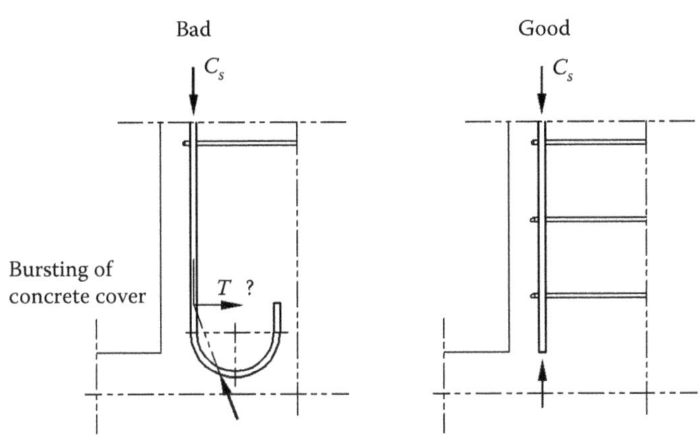

FIGURE 3.19 Hooks are not suited to anchor compression bars. (From Schlaich, J., and Schäfer K., *IABSE Workshop*, New Delhi, 1993.)

necessity of providing transverse reinforcement in the vicinity of the spliced ends. For a compression lap splice, the *STM* in Figure 3.18b illustrates the necessity of extending the transverse reinforcement beyond the splice region, Figure 3.18a. In a lap splice, the transverse reinforcement should be located as close to the concrete surface as possible.

Codes of practice usually give detailed rules on the required lap length and the arrangement of laps and transverse reinforcement in concrete members.

3.5.4 Curved Reinforcement

The deviation forces developed by reinforcement curvature are essentially transferred into the concrete in the form of radial pressure. The dispersion of this pressure in the concrete varies according to the reinforcement arrangement as illustrated by the three cases shown in Figures 3.20 through 3.22.

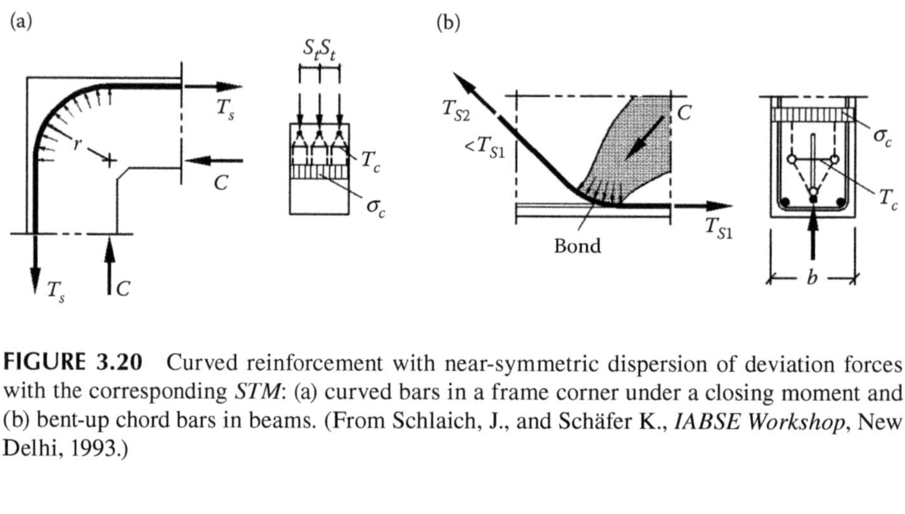

FIGURE 3.20 Curved reinforcement with near-symmetric dispersion of deviation forces with the corresponding *STM*: (a) curved bars in a frame corner under a closing moment and (b) bent-up chord bars in beams. (From Schlaich, J., and Schäfer K., *IABSE Workshop*, New Delhi, 1993.)

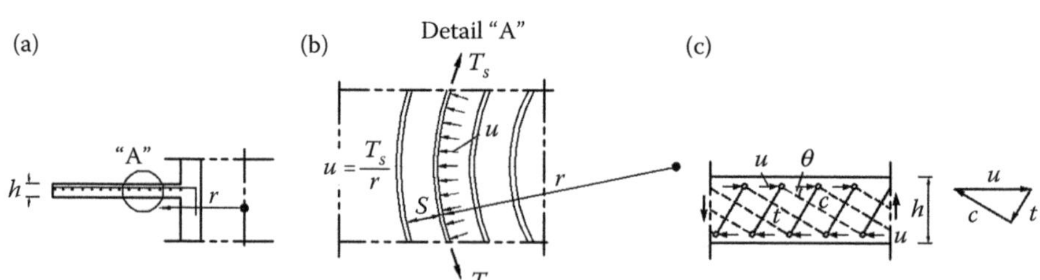

FIGURE 3.21 Hoop reinforcement in an annular plate (deviation forces near the component surface): (a) cross section, (b) plan detail, and (c) *STM* model and forces at the node. (From Schlaich, J., and Schäfer K., *IABSE Workshop*, New Delhi, 1993.)

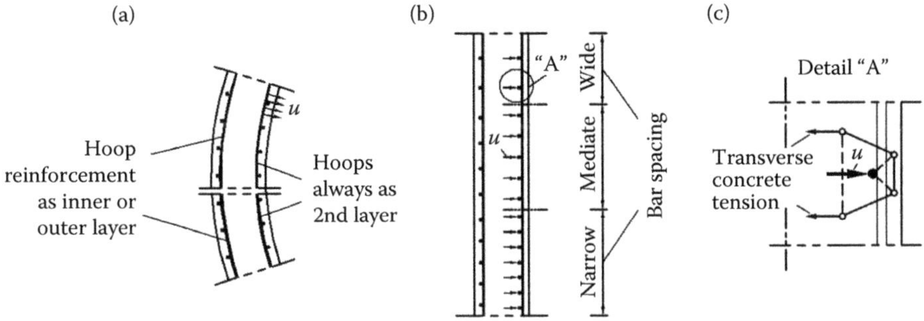

FIGURE 3.22 Reinforcement in a curved wall: (a) plan, (b) vertical section, and (c) *STM* model and forces at the reinforcing bar. (From Schlaich, J., and Schäfer K., *IABSE Workshop*, New Delhi, 1993.)

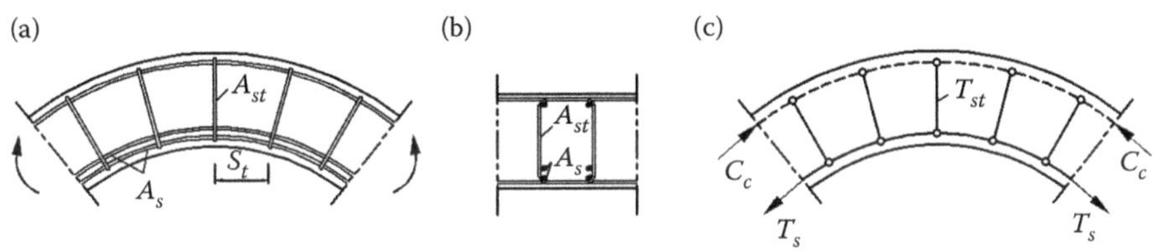

FIGURE 3.23 Narrow stirrups for highly stressed components with a small radius of curvature as a protection against spalling: (a, b) reinforced component and (c) model. (From Schlaich, J., and Schäfer K., *IABSE Workshop*, New Delhi, 1993.)

For the case in Figure 3.20, the admissible bar curvatures should be applied at best such that any transverse tension associated with any force redistribution due to cracking is resisted by transverse reinforcement. The tensile forces, as they develop for the case in Figure 3.21, are equivalent to the ties which are required to carry the shear forces in slabs without stirrups. For the case in Figure 3.22, the magnitude of the transverse tensile stresses associated with hoop forces should not exceed any level that could cause spalling of the concrete cover. It should be noted that the hoops of the inner side should be arranged in the second layer as illustrated in Figure 3.22a.

In case of heavy hoop reinforcement, the associated transverse tensile stresses should be resisted by appropriate transverse reinforcement as illustrated in Figure 3.23.

REFERENCES

ACI 318–14, *Building Code Requirements for Structural Concrete and Commentary*, Detroit: American Concrete Institute, USA, 2014, 519 pp.

Brown, M. D., and Bayrak, O., Minimum transverse reinforcement for bottle-shaped struts, *ACI Structural Journal*, 103(6), 2006, 813–821.

Kupfer, H., Hilsdorf, H. K., and Rüsch, H., Behavior of concrete under biaxial stress, *ACI Journal, Proceedings*, 66(8), 1969, 656–666.

Schlaich, J., and Schäfer, K., The design of structural concrete, IABSE Workshop, New Delhi, 1993.

Schlaich, J., Schäfer, K., and Jennewein, M., Toward a consistent design of structural Concrete, *Journal of the Prestressed Concrete Institute*, 32(3), 1987, 74–150.

Vecchio, F. J., and Collins, M. P.: The modified compression-field theory for reinforced concrete elements subjected to shear, *ACI Structural Journal*, 83(2), 1986, 219–231.

4 Illustrative Design Examples

4.1 INTRODUCTION

The design procedure using the method of *STM* is illustrated with the flow chart in Figure 4.1 (Chantelot and Mathern, 2010). In this chapter, few design examples are given in order to illustrate how the method of *STM* is applied. The examples include (1) a deep beam subjected to a concentrated load; (2) a deep beam with variable depth loaded symmetrically with two concentrated loads; (3) the same aforementioned deep beam with variable depth loaded unsymmetrically with a single concentrated load; (4) a beam with a dapped end; (5) a beam with a recess; (6) local pressure problems; concentric and eccentric; (7) a deep beam with large opening; (8) a high wall with two large openings; and (9) strength assessment of a continuous deep beam with large openings. The failure criteria adopted by the ACI 318-14, as illustrated in Chapter 3, are adopted in in these examples. For most of these examples, a complete design procedure including the development of the *STM*, dimensioning and reinforcement detailing, is presented.

4.2 DEEP BEAM UNDER CONCENTRATED LOAD

It is required to determine the reinforcement for the simply supported transfer girder shown in Figure 4.2a. The single column at the mid-span carries a factored load 2800 kN. The concrete cylinder strength is $f'_c = 32$ MPa and the steel yield stress is $f_y = 420$ MPa. Neglect the beam's own weight. The solution is given in the following steps.

Reactions:
With reference to Figure 4.2a,

$$R_A = R_B = 1400\,\text{kN}$$

Establish an STM:
In this beam, the shear span to depth ratio is less than 2; therefore, the entire beam is considered a D-region, that is, deep beam. The appropriate *STM* is shown in Figure 4.2b, in which the lower nodes are assumed to coincide with the centerlines of the supports. Usually the upper nodes are located approximately at a distance equal to one-quarter the width of the bearing plate; in this example, the distance is 500 mm/4 $\approx$ 120 mm. As for the lower nodes, they are located at the expected centroid of reinforcement; in this example, the distance is assumed equal to 120 mm.

With reference to Figure 4.2b, the length of the diagonal strut = $\sqrt{(1260)^2 + (2000-125)^2} = 2259\,\text{mm}$

The force in strut $C_1 = 1400 \times \dfrac{2259}{1260} = 2510\,\text{kN}$

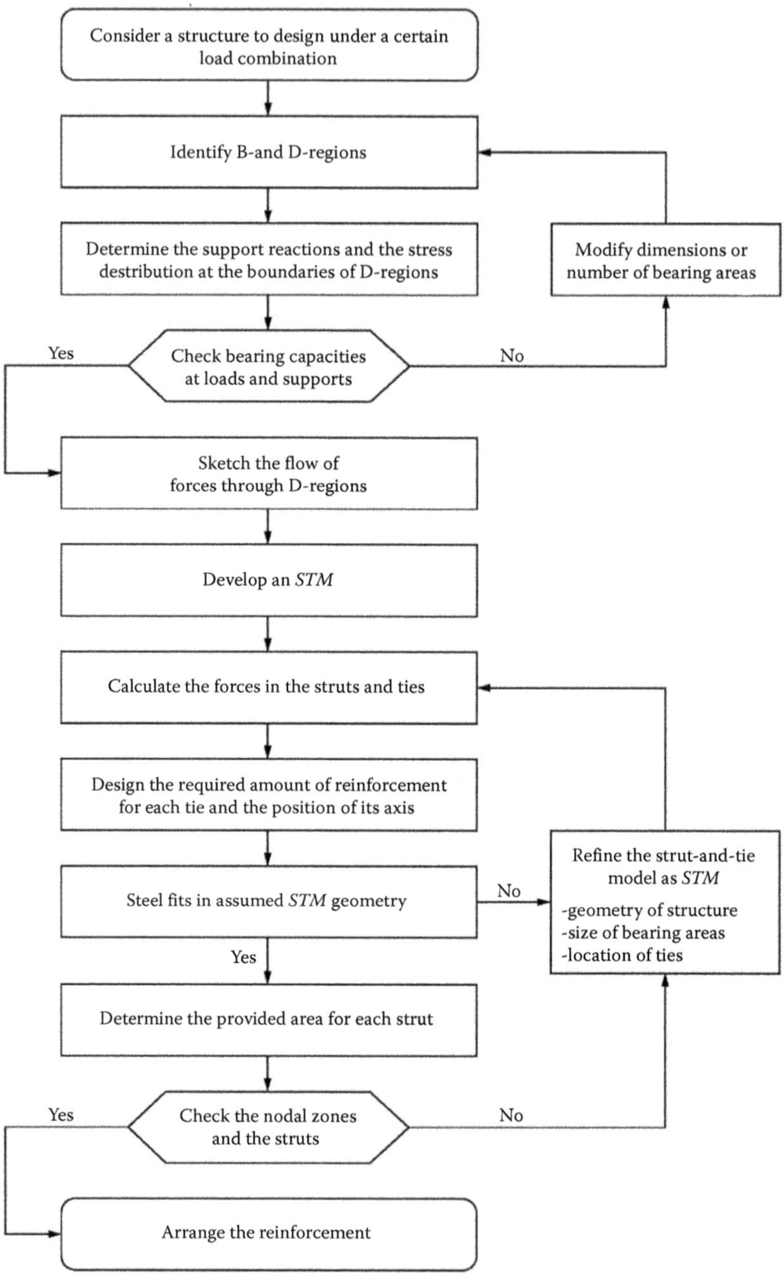

FIGURE 4.1 Flowchart of the design procedure using the *STM*.

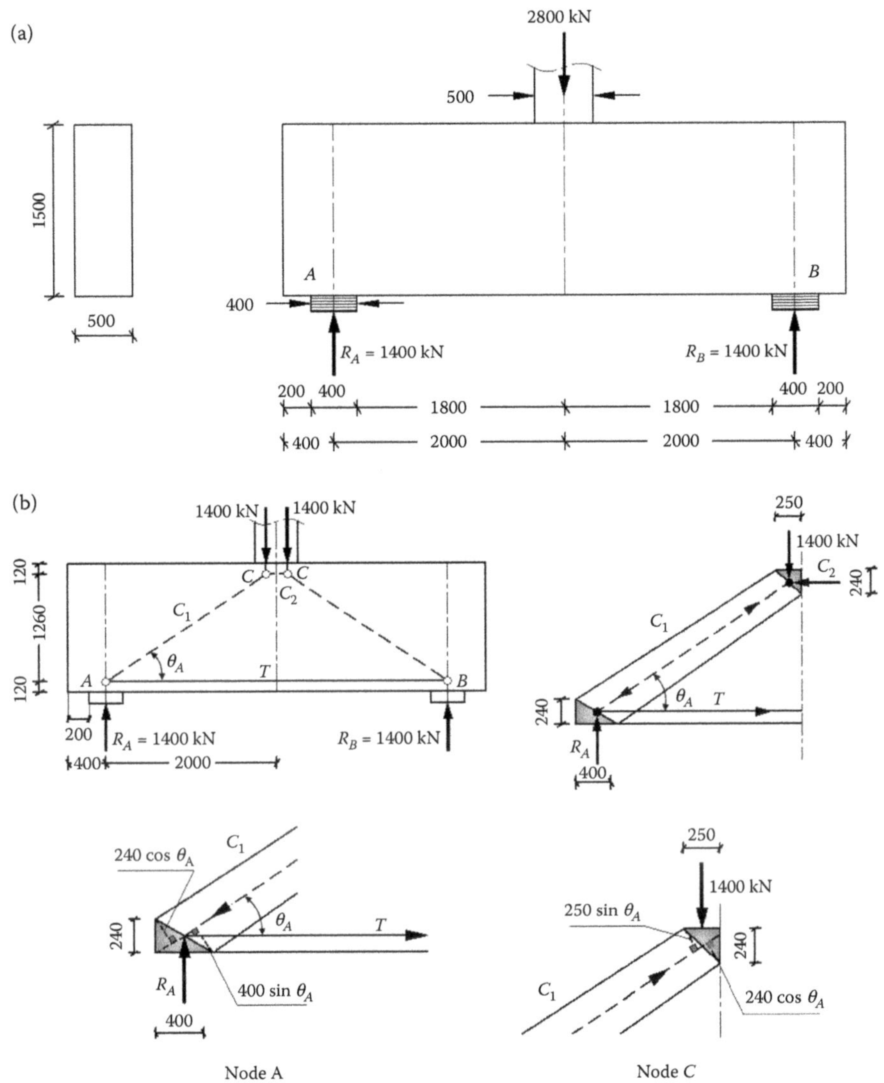

FIGURE 4.2 (a) Deep beam under concentrated load, (b) *STM* of the deep beam under concentrated load.

The force in tie $T = 1400 \times \dfrac{2000 - 125}{1260} = 2083\,\text{kN}$

The force in strut $C_2 = T = 2083$ kN

The angle between C_1 and T, $\theta_A = \tan^{-1}\dfrac{1260}{2000-125} = 33.9° > \tan^{-1}\dfrac{1}{2}$, O.K.

Effective concrete design strength of the struts:

For simplicity in the calculation procedure, the strength reduction factor, $\phi = 0.75$, is implemented in the design strength of the struts and the nodes, as illustrated next.

For strut C_1: $f_{ce}^{s} = \phi(0.85f_c')\beta_s = 0.75 \times 0.85 \times 32 \times 0.75^{*} = 15.30\,\text{MPa}$

* Transverse reinforcement to resist the lateral tension will be provided since strut C_1 is a bottle-shaped stress field.

For strut C_2: $f_{ce}^{s} = \phi(0.85f_c')\beta_s = 0.75 \times 0.85 \times 32 \times 1.0 = 20.40\,\text{MPa}$

Effective concrete design strength of the nodes:
Node A is a $C - C - T$ node; thus, $f_{ce}^{n} = \phi(0.85f_c')\beta_n = 0.75 \times 0.85 \times 32 \times 0.8 = 16.32$ MPa

Node C is a $C - C - C$ node; thus, $f_{ce}^{n} = 0.75 \times 0.85 \times 32 \times 1.0 = 20.40$ MPa

Node C:

The bearing stress is $\dfrac{1400 \times 10^3}{250 \times 500} = 11.20\,\text{MPa} < f_{ce}^{n} = 20.40\,\text{MPa}$, which is O.K.

For the section at the interface between strut C_2 and the node, the design strength should be the smaller of the node strength and the strut strength, both of which are the same in this case, 20.40 MPa.

The required width of strut C_2, $w_{C_2} = \dfrac{2083 \times 10^3}{500 \times 20.40} = 204\,\text{mm}$

The difference between the assumed width (240 mm) and the required width (204 mm) is on the safe side and is not significant. The solution will proceed without modifying the dimensions of the *STM* until strut C_1 is checked.

For the section at the interface between strut C_1 and the node, the design strength should be the smaller of the node strength (20.40 MPa) and the strut strength (15.30 MPa), which is 15.30 MPa in this case. With reference to Figure 4.2b, the width of strut C_1 at node C is

$w_{C_1} = 250\sin\theta_A + 240\cos\theta_A = 339\,\text{mm}$, which corresponds to a stress $\dfrac{2510 \times 10^3}{339 \times 500} = 14.81\,\text{MPa} < 15.30\,\text{MPa}$, which is O.K.

Node A:

The bearing stress is $\dfrac{1400 \times 10^3}{400 \times 500} = 7.00\,\text{MPa} < f_{ce}^{n}(=16.32\,\text{MPa})$, which is O.K.

In the calculations of this node, the tension reinforcement of the tie T is expected to consist of multiple layers and to continue beyond the bearing pad a distance greater than one-half the clear spacing between the reinforcement layers. Hence, an extended nodal zone can be considered.

For the section at the interface between strut C_1 and the node, the design strength should be the smaller of the node strength (16.32 MPa) and the strut strength (15.30 MPa), which is 15.30 MPa in this case. The height of the node, $w_t = 240$ mm, should be checked against a maximum value $w_{t,\max}$, where

$$w_{t,\max} = \frac{T}{f_{ce}^{n} b} = \frac{2083 \times 10^3}{16.32 \times 500} = 255.3\,\text{mm} > w_t, \text{ which is O.K.}$$

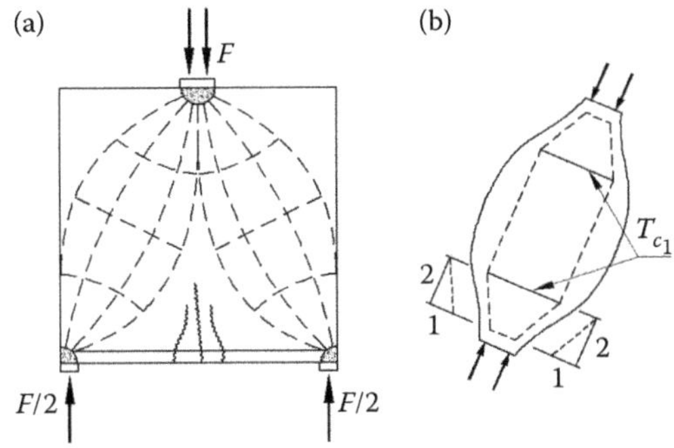

FIGURE 4.3 Stress deviation in a bottle-shaped stress field.

With reference to Figure 4.2b, the width of strut C_1 at node A is $w_{C_1} = 240\cos\theta_A + 400\sin\theta_A = 422\,\text{mm}$, which corresponds to a stress = 2510 × $10^3/422 \times 500 = 11.90\ MPa < 15.30\,\text{MPa}$, which is O.K.

Strut C_1:
The strut strength has been checked at the two ends during the check of nodes A and C. Since the stress field of this strut is bottle-shaped, transverse reinforcement of the strut is required to resist a total force T_{C_1}. From the *STM* of Figure 4.3,

$$T_{C_1} = \left(\frac{1}{2} \times \frac{C_1}{2}\right) \times 2 = \frac{C_1}{2} = \frac{2510}{2} = 1255\,\text{kN}.$$

Thus, the total required reinforcement in perpendicular to the strut is (1255 × 10^3)/(0.75 × 420) = 3984 mm^2. The length of the strut is 2259 mm; hence, the required transverse reinforcement is 1.764 mm^2/mm, in perpendicular to the strut. This can be covered with a skin reinforcement of vertical bars of diameter 16 mm every 200 mm and horizontal bars of diameter 12 mm every 200 mm, on each side. The larger diameter is assigned to the vertical bars since they are more effective in substituting for the inclined reinforcement because the angle θ_A is less than 45°. The selected skin reinforcement is equivalent to a transverse reinforcement along the strut of 1.74 mm^2/mm, which is close enough since the formula used for calculating this reinforcement is slightly conservative. With reference to Figure 3.7, the used transverse steel is equivalent to inclined reinforcement $\sum A_{si} \sin \gamma_i / b s_i = 0.0046 > 0.003$ (the ACI minimum value for $f_c' \leq 44\,\text{MPa}$).

Strut C_2:
The strut strength has been checked at the two ends during the check of node C. Since this strut is prismatic, there are no more checks.

Tie T:
The reinforcement required to resist the force of this tie is (2083 × 10^3)/(0.75 × 420) = 6613 mm^2, which can be covered with 14 bars of diameter 25 mm (14ϕ25). This reinforcement should be extended in the node and beyond, the

anchorage length required to develop the force in the tie with a minimum distance, beyond the bearing plate, equal to one-half the clear spacing between the reinforcement layers.

4.3 SYMMETRICALLY LOADED DEEP BEAM WITH VARIABLE DEPTH

In this section, the design procedure of a beam with an abrupt change of thickness (Schlaich and Schäfer, 1993), Figure 4.4a, is illustrated. The width of the beam $b = 300$ mm and the other dimensions in mm are shown in the figure. The factored design value of the applied force is $F_u = 1200$ kN. The concrete cylinder strength is $f_c' = 30$ MPa and the steel yield stress is $f_y = 460$ MPa.

Reactions and straining actions:
With reference to Figure 4.4a,

$$R_A = R_B = 1200\,\text{kN}$$

The bending moment and shearing force diagrams are shown in Figure 4.4a.

D- and B-regions:
Based on Saint Venant's principle, the boundaries of the D-regions are determined. Figure 4.4b illustrates both the D- and B-regions.

Dimensioning of B-region:
The bending moment of the B-region is a constant value of $M_u = 3600$ kN.m and the shear is zero. If the contribution of the compression reinforcement is neglected,

$$M_u = 3600\,\text{kN.m} = \phi M_n = \phi\left[0.85 f_c' ab\left(d - \frac{a}{2}\right)\right]$$

Since the moment is a constant value, the dividing section between the B- and D-regions requires the use of a strength reduction factor of the *STM*, $\phi = 0.75$ instead of 0.9 (the value used in flexure). Substituting for $b = 300$ mm, $d \approx 1900$ mm and $f_c' = 30$ MPa, the obtained value of $a = 365$ mm. For simplicity in the calculations, assume $a \approx 400$ mm, which leads to a lever arm, $Y_{CT} = d - a/2 = 1700$ mm. From moment equilibrium, Figure 4.4c,

$$C_1 = T_1 = \frac{M_u}{Y_{CT}} = \frac{3600 \times 10^6}{1700} = 2118\,\text{kN}$$

$$A_{s1} = \frac{T_1}{\phi f_y} = \frac{2118 \times 10^3}{0.75 \times 460} = 6138\,\text{mm}^2 = 13\phi 25$$

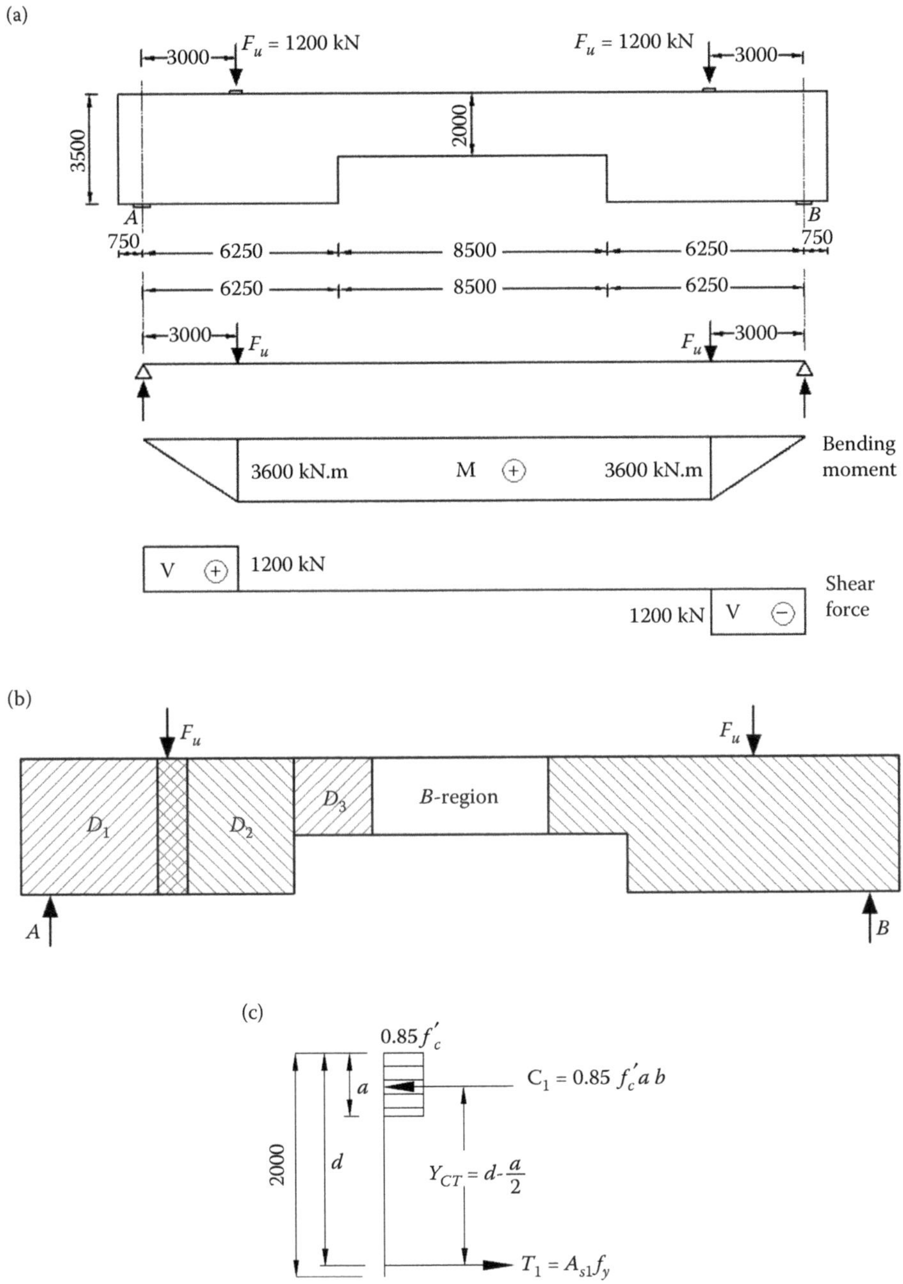

FIGURE 4.4 Symmetrically loaded deep beam with variable depth: (a) beam, dimensions, loads, and straining actions, (b) D- and B-regions, (c) force equilibrium of B-region *(Continued)*

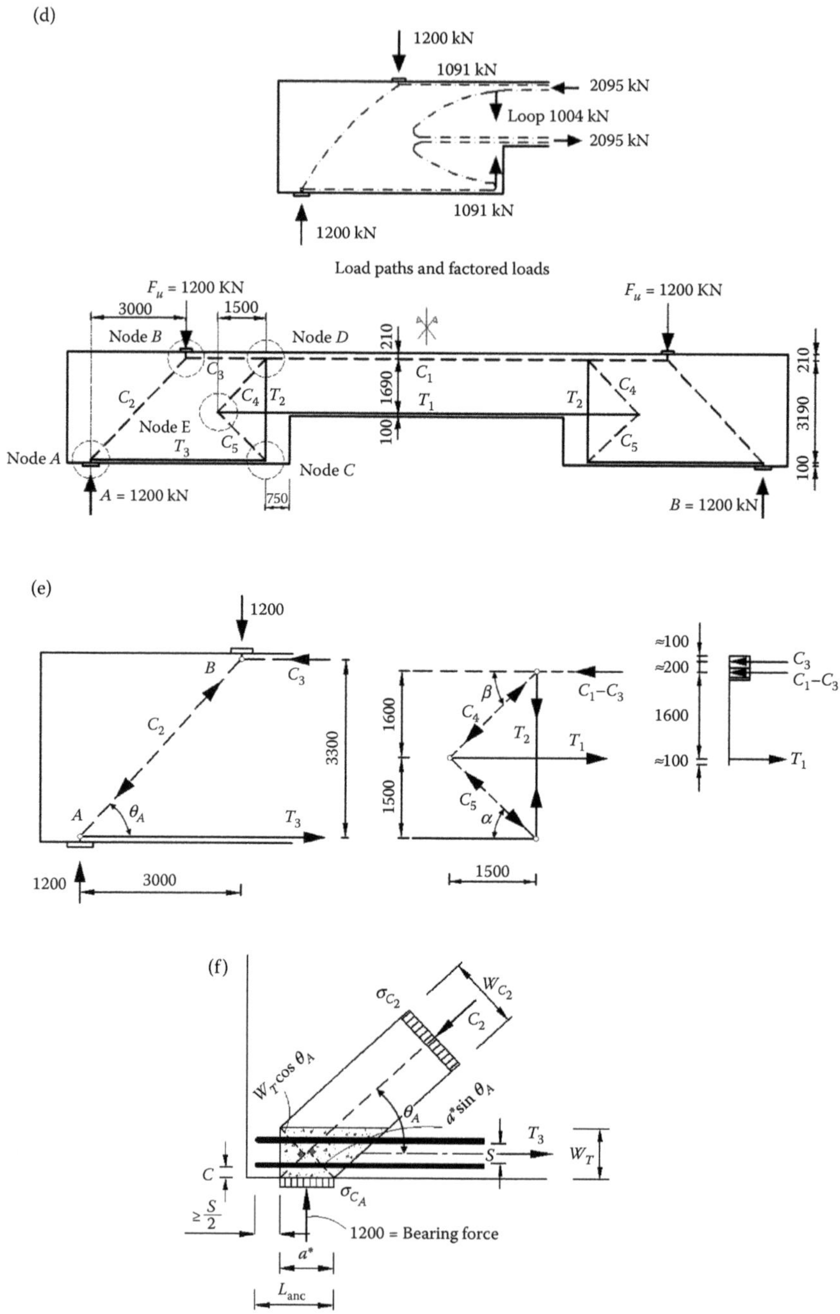

FIGURE 4.4 (Continued) (d) load path and *STM*, (e) geometric relations of the *STM*, (f) node A *(Continued)*

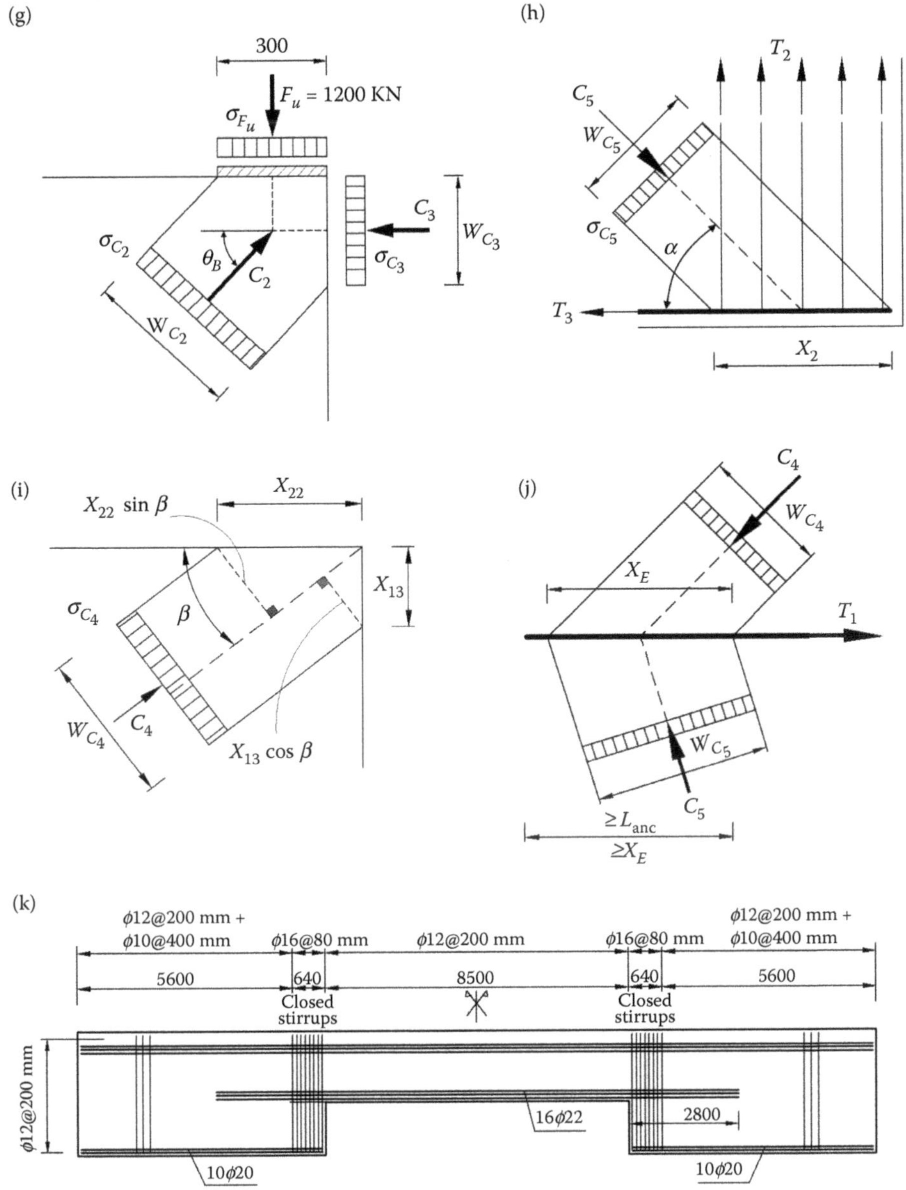

FIGURE 4.4 (Continued) (g) node B, (h) node C, (i) node D, (j) node E, (k) reinforcement layout of the symmetrically loaded deep beam with variable depth.

Establish an STM:

The load path and the corresponding *STM* are shown in Figure 4.4d. The compression block at the dividing section between the B- and D-regions is subdivided into two parts, one to balance the force C_3 and the other to balance the horizontal component of the force C_4. From equilibrium of the D-region, Figure 4.4d, and with the aid of the geometric relations in Figure 4.4e,

$$C_3 = T_3 = 1200 \times \frac{3.0}{3.3} = 1091\,\text{kN}$$

$$C_1 - C_3 = 2118 - 1091 = 1027\,\text{kN}$$

$$C_2 = \sqrt{(1200)^2 + (1091)^2} = 1622\,\text{kN}$$

If the angle α is assumed equal to 45°, then the force T_2 will be

$$T_2 = T_3 = 1091\,\text{kN}$$

which leads to

$$C_5 = \sqrt{T_2^2 + T_3^2} = 1543\,\text{kN}$$

$$C_4 = \sqrt{(C_1 - C_3)^2 + T_2^2} = 1498\,\text{kN}$$

Effective concrete design strength of the struts:
For simplicity in the calculation procedure, the strength reduction factor, ϕ, is implemented in the design strength of the struts and the nodes, as illustrated next.

Struts C_1 and C_3 are prismatic stress fields for which $\beta_s = 1.0$; hence, $f_{ce}^s = \phi(0.85 f_c')\beta_s = 0.75 \times 0.85 \times 30 \times 1.0 = 19.13\,\text{MPa}$.

Struts C_2, C_4, and C_5 are bottle-shaped stress fields for which $\beta_s = 0.75$*; hence, $f_{ce}^s = 14.34\,\text{MPa}$. *Transverse reinforcement to resist the lateral tension will be provided.

Effective concrete design strength of the nodes:
Nodes A, D, and E are $C - C - T$ nodes; therefore, $\beta_n = 0.8$; hence, $f_{ce}^n = \phi(0.85 f_c')\beta_n = 0.75 \times 0.85 \times 30 \times 0.8 = 15.30\,\text{MPa}$.

Node B is a $C - C - C$ node; therefore, $\beta_n = 1.0$; hence, $f_{ce}^n = 19.13\,\text{MPa}$.

Node C is a $C - T - T$ node; therefore, $\beta_n = 0.6$; hence, $f_{ce}^n = 11.48\,\text{MPa}$.

Node A:
With reference to Figure 4.4f, the length of the bearing plate, $a^* = \frac{R_A}{f_{ce}^n b} = \frac{1200 \times 10^3}{15.30 \times 300} = 261\,\text{mm}$

Use a bearing plate 300 × 300 mm.

With reference to Figure 4.4f, if two layers of reinforcement are used, then $w_i \approx 200$ mm. According to ACI 318-14,

$$w_{t,\max} = \frac{T_3}{f_{ce}^n b} = \frac{1091\times10^3}{15.30\times300} = 238\,\text{mm} > w_t, \text{ which is O.K.}$$

From Figure 4.4e, tan $\theta_A = (3.3/3.0) = 1.1$, which gives $\theta_A = 47.7°$. The width of strut C_2 at node A, Figure 4.4f, is

$$w_{C_2} = a^* \sin\theta_A + w_t \cos\theta_A = 357\,\text{mm}$$

The stress in the strut, σ_{C_2}, should be compared with the smaller of the strut strength (14.34 MPa) and node A strength (15.30 MPa), which is 14.34 MPa.

$$\sigma_{C_2} = \frac{C_2}{w_{C_2} b} = \frac{1622\times10^3}{357\times300} = 15.15\,\text{MPa} > 14.34\,\text{MPa}$$

Either increase w_t or use a larger bearing plate. Upon using a bearing plate 400 × 300 mm and redoing the calculations, it is found that $w_{C_2} = 431\text{mm}$ and $\sigma_{C_2} = 12.54\,\text{MPa} < 14.34\,\text{MPa}$; the bearing plate size is adequate.

Node B:
Try a bearing plate 300 × 300 mm, Figure 4.4g. The bearing stress is $(1200\times10^3)/(300\times300) = 13.33\,\text{MPa}$, which is less than $f_{ce}^n (=19.13\,\text{MPa})$. The depth of the compression block of strut C_3 should be computed on the basis of the smaller value of the strut strength and node B strength, both of which have the same value in this case.

The required width of strut C_3 is

$$w_{C_3} = \frac{C_3}{f_{ce}^s b} = \frac{1091\times10^3}{19.13\times300} = 190\,\text{mm}$$

With reference to Figure 4.4g, the width of strut C_2, $w_{C_2} = 300 \sin\theta_B + 190 \cos\theta_B = 350$ mm, where $\theta_B = \theta_A = 47.7°$. The stress of the node at the interface of strut C_2 is

$$\sigma_{C_2} = \frac{C_2}{w_{C_2} b} = \frac{1622\times10^3}{350\times300} = 15.45\,\text{MPa}$$

This stress should be compared with the smaller of the strut strength (14.34 MPa) and node B strength (19.13 MPa), which is (14.34 MPa). Thus, the stress σ_{C_2} is unsafe. Upon increasing the size of the bearing plate to 350 × 300 mm, $w_{C_2} = 387\,\text{mm}$, and $\sigma_{C_2} = 13.97\,\text{MPa}$, which is safe.

Node C:
The width of strut C_5, Figure 4.4h, is based on the smaller value of the strut strength ($f_{ce}^s = 14.34\,\text{MPa}$) and node C strength ($f_{ce}^n = 11.48\,\text{MPa}$), which gives a value of 11.48 MPa in this case.

$$w_{C_5} = \frac{C_5}{f_{ce}^n b} = \frac{1543 \times 10^3}{11.48 \times 300} = 448\,\text{mm} = x_2 \sin\alpha = x_2 \sin 45°$$

which results in: $x_2 = 634$ mm. The reinforcement resisting T_2 should be closed stirrups and placed within a distance at least equal to x_2.

Node D:
The depth of the compression block of ($C_1 - C_3 = 1027$ kN) is x_{13}, Figure 4.4i. This depth is based on the smaller value of the strut strength ($f_{ce}^s = 19.13\,\text{MPa}$) and node D strength ($f_{ce}^n = 15.30\,\text{MPa}$), which gives a value of 15.30 MPa in this case.

$$x_{13} = \frac{C_1 - C_3}{f_{ce}^n b} = \frac{1027 \times 10^3}{15.30 \times 300} = 224\,\text{mm}$$

The width of strut C_4, Figure 4.4i, is based on the smaller value of the strut strength ($f_{ce}^s = 14.34\,\text{MPa}$) and node D strength ($f_{ce}^n = 15.30\,\text{MPa}$), which gives a value of 14.34 MPa in this case.

$$w_{C_4} = \frac{C_4}{f_{ce}^s b} = \frac{1483 \times 10^3}{14.34 \times 300} = 345\,\text{mm} = x_{22} \sin\beta + x_{13} \cos\beta$$

With reference to Figure 4.4e, the angle $\beta \approx \tan^{-1}\dfrac{1600}{1500} = 46.85°$, which gives: $x_{22} = 263$ mm. x_{22} is the minimum distance within which the closed stirrups resisting T_2 should be placed. Since $x_{22} < x_2$, the value of x_2 should be used, which increases the width of strut C_4 to 616 mm.

Node E:
The widths of struts C_4 and C_5 at the node, Figure 4.4j, can be calculated in the same manner as before based on the smaller design strength of the node ($f_{ce}^n = 15.30\,\text{MPa}$) and the struts ($f_{ce}^s = 14.34\,\text{MPa}$), which gives 14.34 MPa. The width of strut C_5 should therefore be $C_5 / f_{ce}^s b = (1543 \times 10^3)/(14.34 \times 300) = 359\,\text{mm}$, which gives a projection in the direction of T_1, $x_E = 507$ mm.

For strut C_4, the width is $C_4 / f_{ce}^s b = (1498 \times 10^3)/(14.34 \times 300) = 348\,\text{mm}$, which gives a projection in the direction of T_1, $x_E = 477$ mm.

Take the larger value of $x_E = 507$ mm in detailing the reinforcement of T_1. The anchorage of the reinforcement resisting T_1 should be extended, from the start of the node, a distance equal to the largest of the anchorage length and x_E, Figure 4.4j.

Struts C_1, C_3 and $C_1 - C_3$:
Since these struts are prismatic stress fields and their safety was checked within the checks of the nodes, there are no more checks for these struts.

Strut C_2:
The strut strength has been checked at the two ends during the check of nodes A and B. Since the stress field of this strut is bottle-shaped, transverse reinforcement of the strut is required to resist a total force T_{C_2}. From the *STM* of Figure 4.3,

$$T_{C_2} = \left(\frac{1}{2} \times \frac{C_2}{2}\right) \times 2 = \frac{C_2}{2} = \frac{1622}{2} = 811 \text{ kN}.$$

Thus, the total required reinforcement in perpendicular to the strut is $(811 \times 10^3)/(0.75 \times 460) = 2351 \text{ mm}^2$. The length of the strut is 4460 mm; hence, the required transverse reinforcement is 0.527 mm^2/mm, in perpendicular to the strut. This can be covered with a skin reinforcement of diameter 12 mm every 200 mm, both vertically and horizontally on each side. With reference to Figure 3.7, the used transverse steel is equivalent to inclined reinforcement $\Sigma A_{si} \sin \gamma_i / bs_i = 0.0053 > 0.003$ (the ACI 318-14 minimum value for $f_c' \leq 44$ MPa).

Strut C_4:
The stress field of this strut is bottle-shaped, and as for strut C_2, the transverse reinforcement of the strut is required to resist a total force $C_4/2 = 1498/2 = 749$ kN, which requires a total reinforcement in perpendicular to the strut $= 2171 \text{ mm}^2$. The length of the strut is 2193 mm; hence, the required transverse reinforcement is 0.99 mm^2/mm, in perpendicular to the strut. This is covered with the predetermined skin reinforcement of diameter 12 mm every 200 mm, both vertically and horizontally on each side. Also, the used transverse steel is equivalent to inclined reinforcement $0.0053 > 0.003$ (the ACI 318-14 minimum value for $f'_c \leq 44$ MPa).

Strut C_5:
The stress field of this strut is bottle-shaped, and as for strut C_2, the transverse reinforcement of the strut is required to resist a total force required to resist a total force $C_5/2 = 1543/2 = 772$ kN, which requires a total reinforcement in perpendicular to the strut $= 2236 \text{ mm}^2$. The length of the strut is 2121 mm; hence, the required transverse reinforcement is 1.05 mm^2/mm, in perpendicular to the strut. This is covered with the predetermined skin reinforcement as stated for strut C_4.

Checking the strength of the B-region and tie T_l:

$$\phi M_n = C_3\left(d - \frac{190}{2}\right) + (C_1 - C_3)\left(d - 190 - \frac{224}{2}\right)$$
$$= 3610 \text{ kN.m} > M_u(3600 \text{ kN.m}), \text{ which is O.K.}$$

The previously selected reinforcement of 13ϕ25 is adequate as a tension reinforcement of T_1.

Tie T_2:
The reinforcement required to resist this force is $T_2/(0.75 \times 460) = 3162\ \text{mm}^2$, which can be covered with two-branches closed stirrups 8ϕ16 to be distributed within a distance $x_2 = 634\ \text{mm} \approx 640\ \text{mm}$.

Tie T_3:
The reinforcement required to resist this force is $T_3/(0.75 \times 460) = 3162\ \text{mm}^2$, which can be covered with 10ϕ20.

Reinforcement:
The used skin reinforcement ϕ12@200 mm on both sides in the vertical and horizontal directions gives a ratio of 0.38% both vertically and horizontally, which is adequate. The final beam reinforcement is illustrated in Figure 4.4k. It should be noted that the ends of the vertical and horizontal web reinforcement should have a U shape.

4.4 UNSYMMETRICALLY LOADED DEEP BEAM WITH VARIABLE DEPTH

In this section, the design procedure of the previous beam with an abrupt change of thickness in Section 4.3 is designed for an unsymmetrical loading case. All the beam data are the same as before except the loading case which consists of a single factored load $F_u = 1200$ kN, Figure 4.5a.

Reactions and straining actions:
With reference to Figure 4.5a

$$R_A = 1028.6\ \text{kN}$$

$$R_B = 171.4\ \text{kN}$$

The bending moment and shearing force diagrams are shown in Figure 4.5a.

D- and B-regions:
The B- and D-regions are the same as in the previous example (Section 4.3).

STM and design:
The appropriate *STM* is shown in Figure 4.5b. The calculation procedure of the forces and the dimensioning of struts, ties, and nodes are the same as followed in the previous example (Section 4.3).

4.5 BEAM WITH DAPPED END

In order to hang the reaction of a dapped beam, two possible models are shown in Figure 4.6 (Schäfer and El-Metwally, 1994). The first model, Figure 4.6a, illustrates

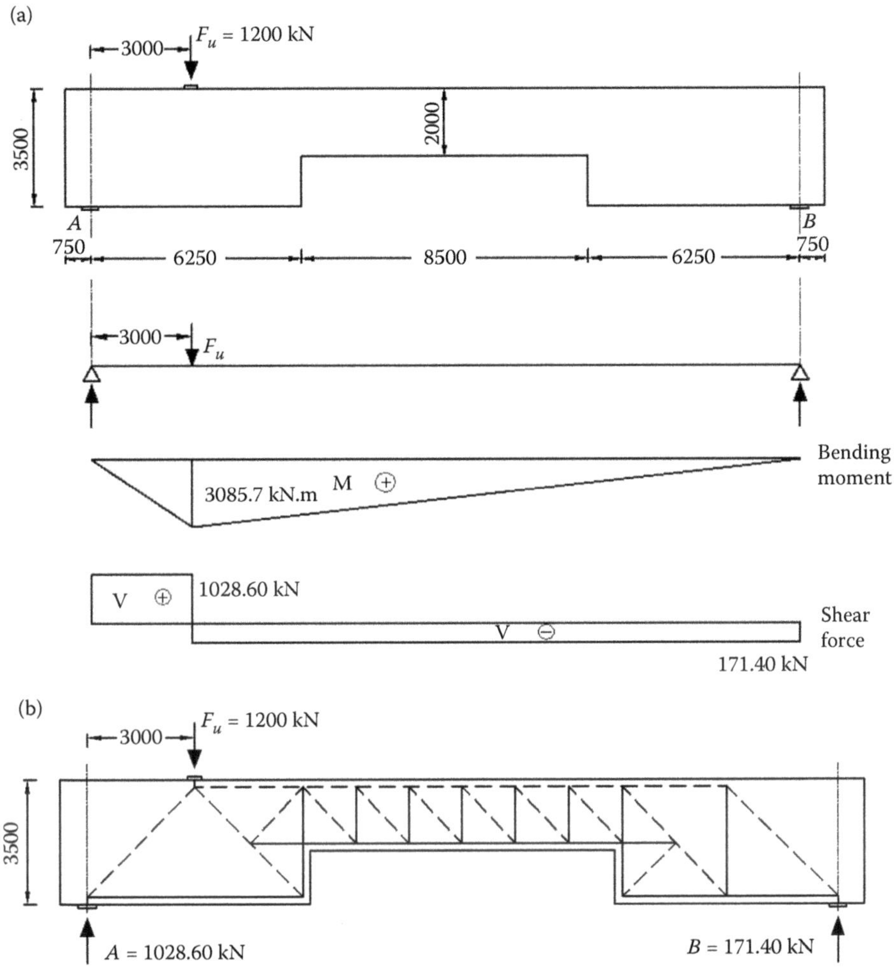

FIGURE 4.5 Unsymmetrically loaded deep beam with variable depth: (a) beam, dimensions, loading, and straining actions, (b) *STM*.

that in addition to the shear reinforcement, T_3, and the reinforcement necessary for hanging the reaction, T_1, additional reinforcement is necessary for a safe transfer of the forces within the D-region, that is, tie T_A and the increase in the magnitude of tie T_2 above the shear requirement.

The second alternative, Figure 4.6b, requires less reinforcement; nevertheless, anchorage of the inclined reinforcement at the upper node may become a problem in case of thick bars. In practice, some reinforcement is always detailed as indicated by the first model, which is necessary for keeping the integrity of the D-region. Hence, though the two models pointed here are correct, a more efficient detailing can be achieved by a combination of the two models. The combined model shown in Figure 4.6c leads to a more efficient detailing with the inclined reinforcement

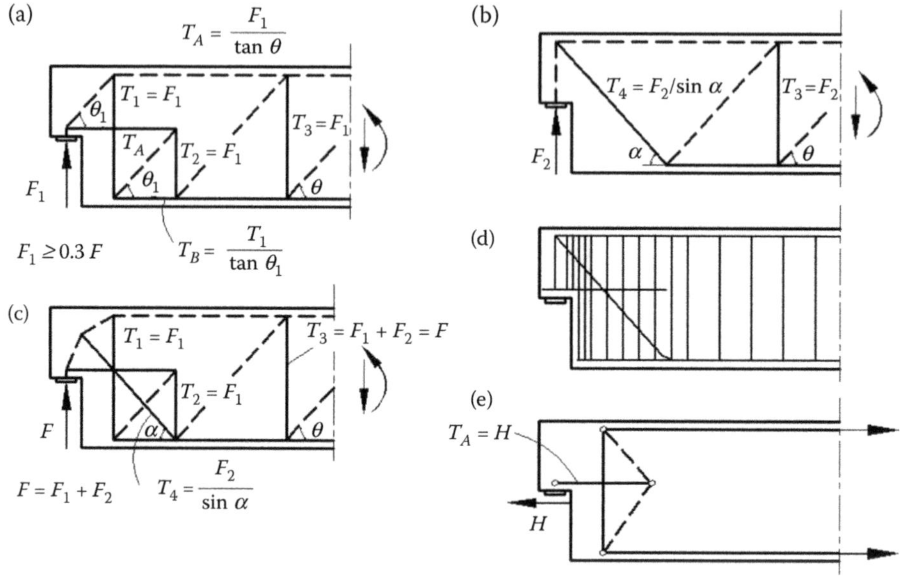

FIGURE 4.6 Dapped beam: (a) *STM*1, (b) *STM*2, (c) combined model from 1 and 2, (d) reinforcement layout, and (e) *STM* for a horizontal reaction H.

assigned at most 70% of the load and the first model assigned at least 30% of the load. While the anchoring of the inclined reinforcement is difficult in the second model, this problem is relieved by the combined model. The reinforcement layout for the combined model is shown in Figure 4.6d. The calculation procedure of the forces and the dimensioning of struts, ties, and nodes are the same as followed in the example of Section 4.3.

If a horizontal reaction force H exists, the model shown in Figure 4.6e can be used for the evaluation of the required additional reinforcement.

4.6 BEAM WITH A RECESS

The beam with a recess in Figure 4.7a has the shown shaded D-region as a result of the recess. For illustration, the D-region is assumed to be subjected to two cases of constant moment (no shear), positive and negative moment, Figure 4.7b, and the moment lever arm on the right end of the D-region is assumed to be double of that on the left end.

Upon the examination of all possible load paths for either case of end moment, the appropriate *STMs* can be derived as shown in Figure 4.7c (Schäfer and El-Metwally, 1994). From the obtained models, it is noted that anchoring the curtailed longitudinal reinforcement should start at a distance beyond what is required by sectional design. In addition, for the safety of the D-region, transverse reinforcement in the shape of closed stirrups is necessary in order to carry the transverse tension (1/2) $T\tan\theta$. The appropriate reinforcement detailing is shown in Figure 4.7d.

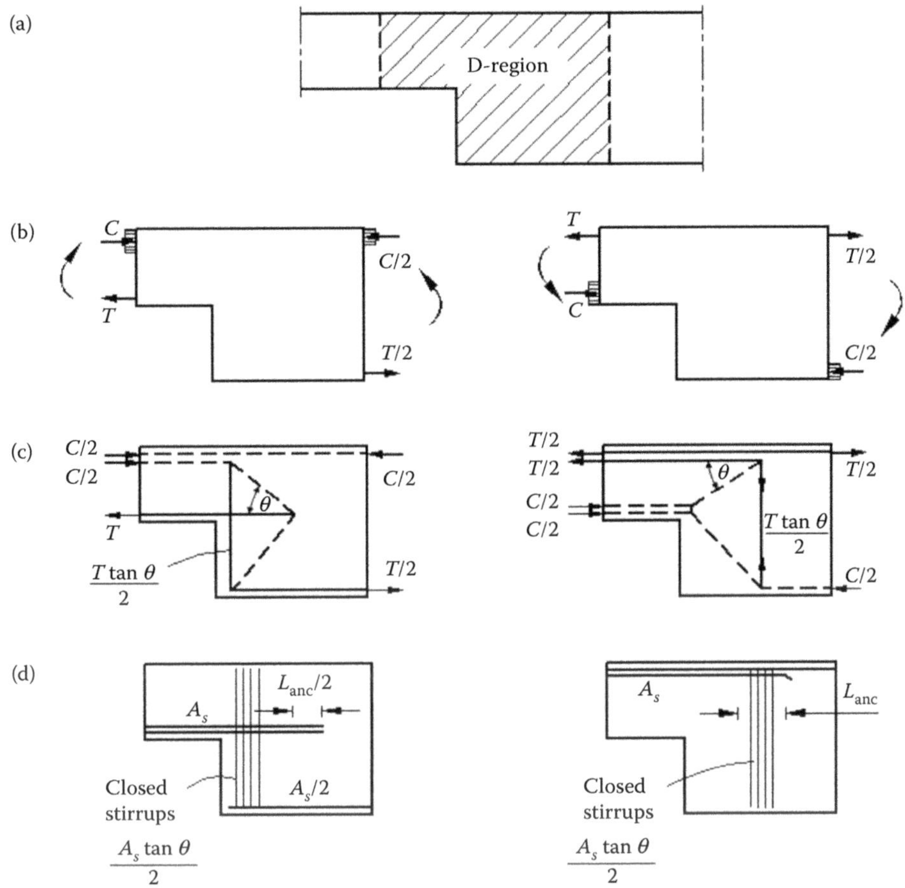

FIGURE 4.7 Beam with a recess: (a) D-region, (b) moments applied to the D-region, (c) *STMs*, and (d) appropriate reinforcement detailing.

4.7 LOCAL PRESSURE

4.7.1 Concentric Local Pressure

The problem of concentric local pressure, Figure 4.8 (Schäfer and El-Metwally, 1994), is simulated as D_1-region, Figure 2.8, of the standard D-regions. The amount of necessary transverse reinforcement and its position can be determined from the respective *STM*. The tension force, *T*, can be calculated from the model as

$$T = \frac{1}{4} F \left(1 - \frac{a}{h}\right)$$

If the width of the bearing plate, *a*, is very small compared to the depth of the D-region, *h*, the tension force, *T*, can be approximated to $T \approx 0.25F$. It is noted that

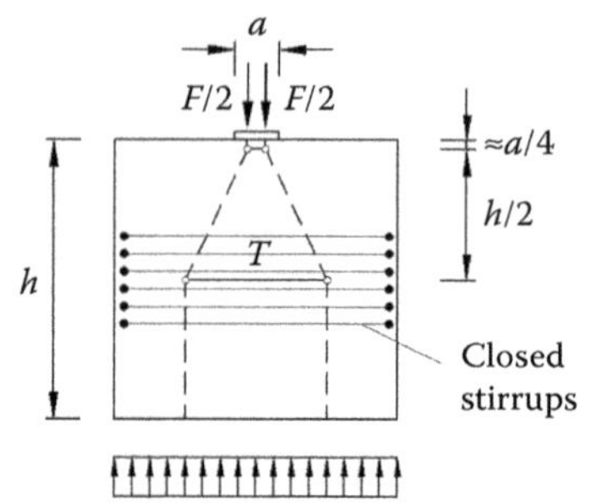

FIGURE 4.8 Concentric local pressure.

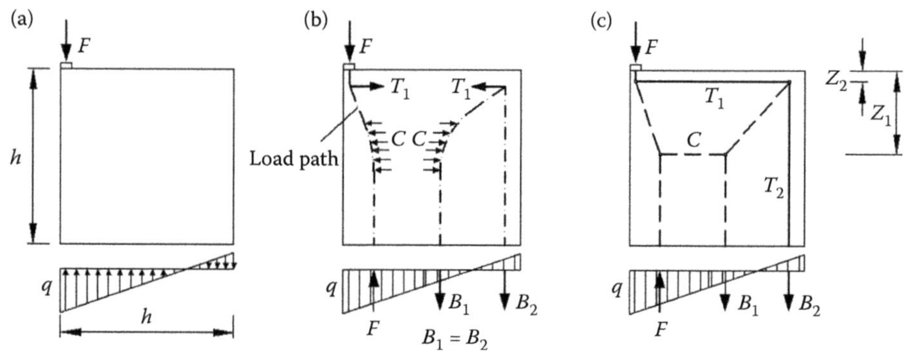

FIGURE 4.9 Eccentric local pressure.

the reinforcement resisting the tension T has to be closed stirrups and its center is located at a distance from the edge equal to $((h/2)+(a/4))$ as shown in the figure.

4.7.2 Eccentric Local Pressure

The problem of eccentric local pressure, Figure 4.9 (Schäfer and El-Metwally, 1994), is simulated as D_2-region, Figure 2.9. For this region, it is required to provide longitudinal reinforcement in addition to the transverse reinforcement, which can be determined from the respective *STM*. The transverse reinforcement of T_1 is very close to the edge and has to be closed stirrups with their center located at a distance z_2 from this edge. This distance can be adjusted in order to properly place this reinforcement, and therefore the geometry of the model is adjusted subsequently. As for the longitudinal reinforcement of T_2, the distance z_2 will not be adequate for anchorage; therefore, it is recommended to use U bars for this case.

4.8 DEEP BEAM WITH LARGE OPENING

The deep beam in Figure 4.10a with an abrupt change of thickness, includes a large opening and carries a concentrated load; and hence, it is entirely considered as a

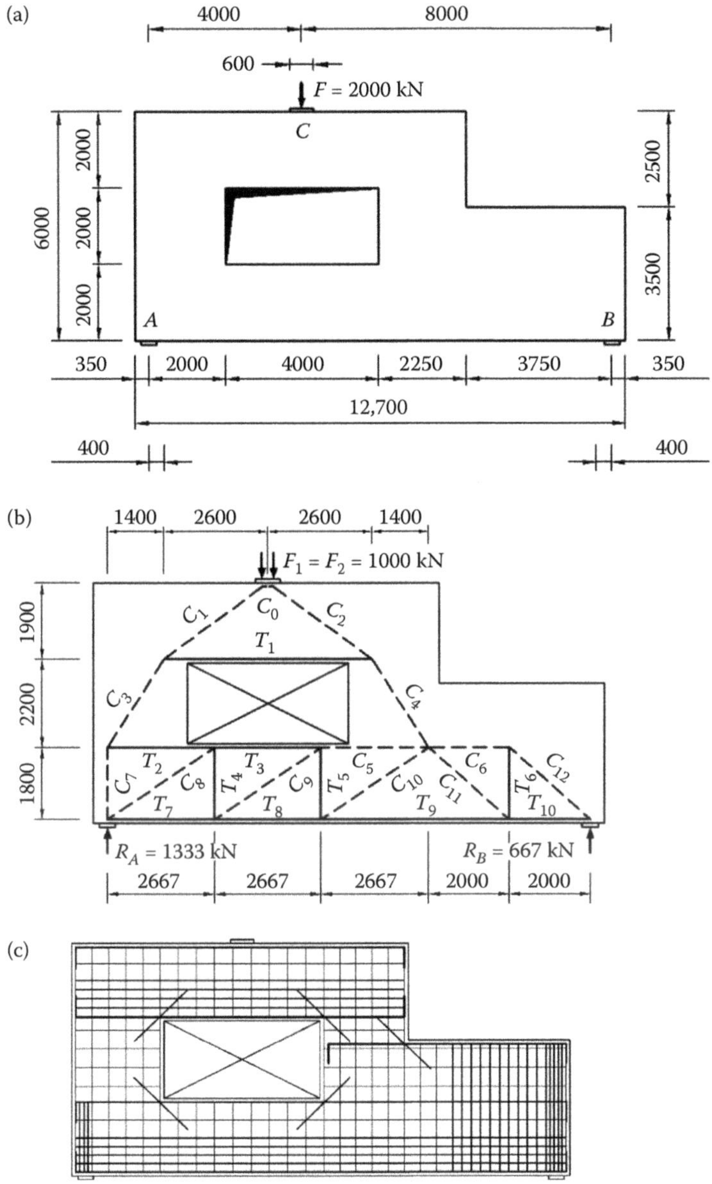

FIGURE 4.10 Example of deep beam with large opening: (a) beam geometry and loading. (b) *STM*. (c) Reinforcement arrangement.

D-region (Novak and Sprenger, 2002). The part of the beam above the opening behaves as a deep beam supported by two inclined struts, C_3 and C_4, the parts to the left and right of the opening. This part can be modeled as illustrated in Figure 4.10b; the model consist of two struts, C_1 and C_2, and one tie, T_1, in addition to the horizontal strut C_o. The part of the beam below the opening supports the two inclined struts

C_3 and C_4, and is supported by the reactions R_A and R_B. With the vertical component of strut C_4 greater than the reaction R_B, the force C_4 has to be transferred to both supports *A* and *B*. As for the force C_3, the vertical component will go directly to support *A* and the horizontal component shall be resisted by a horizontal tie as shown in Figure 4.10b, where the complete *STM* is illustrated. In the model, the horizontal ties are located near the edges where the longitudinal reinforcements are expected to be placed. The forces of the vertical ties are resisted by the vertical web reinforcement. Thus, the web reinforcement should resist both the vertical tie forces and the transverse tension of the struts.

From the *STM* geometry, the forces in the struts and ties can be obtained. The design calculations should start with estimating the dimensions of the bearing plates at *A*, *B* and *C*, followed by the assessment of the ties reinforcement. Afterwards, the struts dimensions can be calculated based on an effectiveness factor $\beta_s = 0.75$ if the skin reinforcement is assumed to cover the transverse tension of struts; otherwise, $\beta_s = 0.6$ if only minimum skin reinforcement is provided. Then, the safety of nodal zones *A*, *B*, and *C* can be verified; the effectiveness factor is $\beta_n = 0.8$ for nodes *A* and *B* and is $\beta_n = 1.0$ for node *C*. In this step, the strength of the nodal zone at any side bounded by a strut is governed by the smaller of the strength of the nodal zone and the strength of that strut.

The reinforcement details at all nodes should ensure adequate anchorage. The design should also justify both minimum reinforcement required by the design code, minimum required for shrinkage and temperature. In addition, the skin reinforcement should justify the code requirements. The reinforcement layout of the beam is shown in Figure 4.10c. It should be noted that the ends of the vertical and horizontal web reinforcement should have a U shape.

4.9 HIGH WALL WITH TWO LARGE OPENINGS

From its geometry, presence of openings and loading, the entire high wall with two large openings in Figure 4.11a is considered a D-region. The wall thickness is 405 mm and each of the left and right bearing support is 1016 × 405 mm. The wall is examined in this section for the following cases of loadings (Barnes, 2002):

1. Two vertical loads of 2000 kN each;
2. Two lateral loads of 760 kN each, acting on the right-hand side of the wall;
3. Combined loads from 1 and 2;
4. Two lateral loads of 760 kN each, acting on the left-hand side of the wall;
5. Combined loads from 1 and 4.

If the wall is modeled as simply supported, the lateral loads resultant has to be resisted by one single support. The choice of the lateral load resisting support is based on the direction of the resultant horizontal force. Upon assuming that the concrete above the support in tension is expected to crack and hence its stiffness is reduced and becomes much less than that of the concrete above the support in compression, this latter support is relied on to transfer the horizontal forces to the foundation by shear-friction (Barnes, 2002).

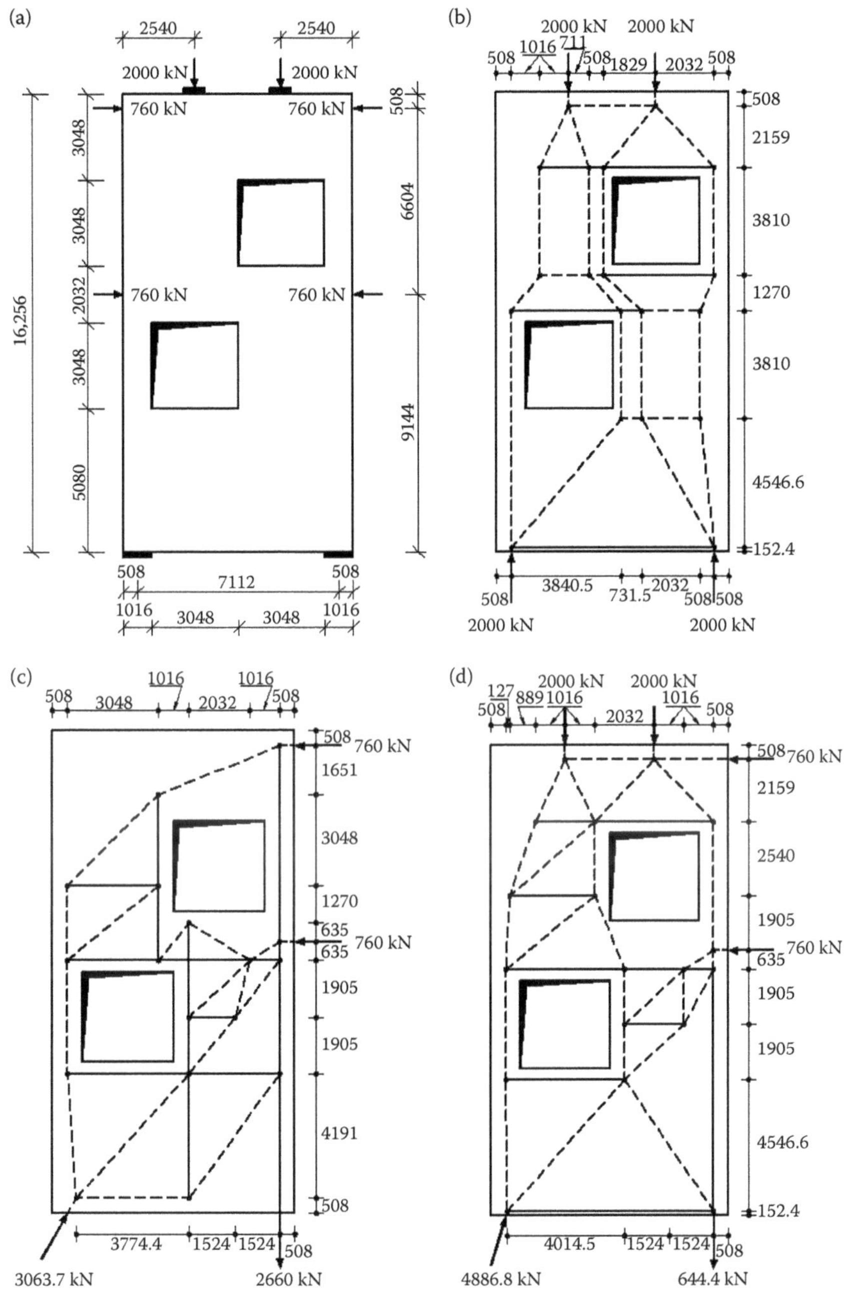

FIGURE 4.11 Example of high wall with two large openings: (a) wall geometry and loading. (b) to (f) *STM* for load cases 1–5, respectively. *(Continued)*

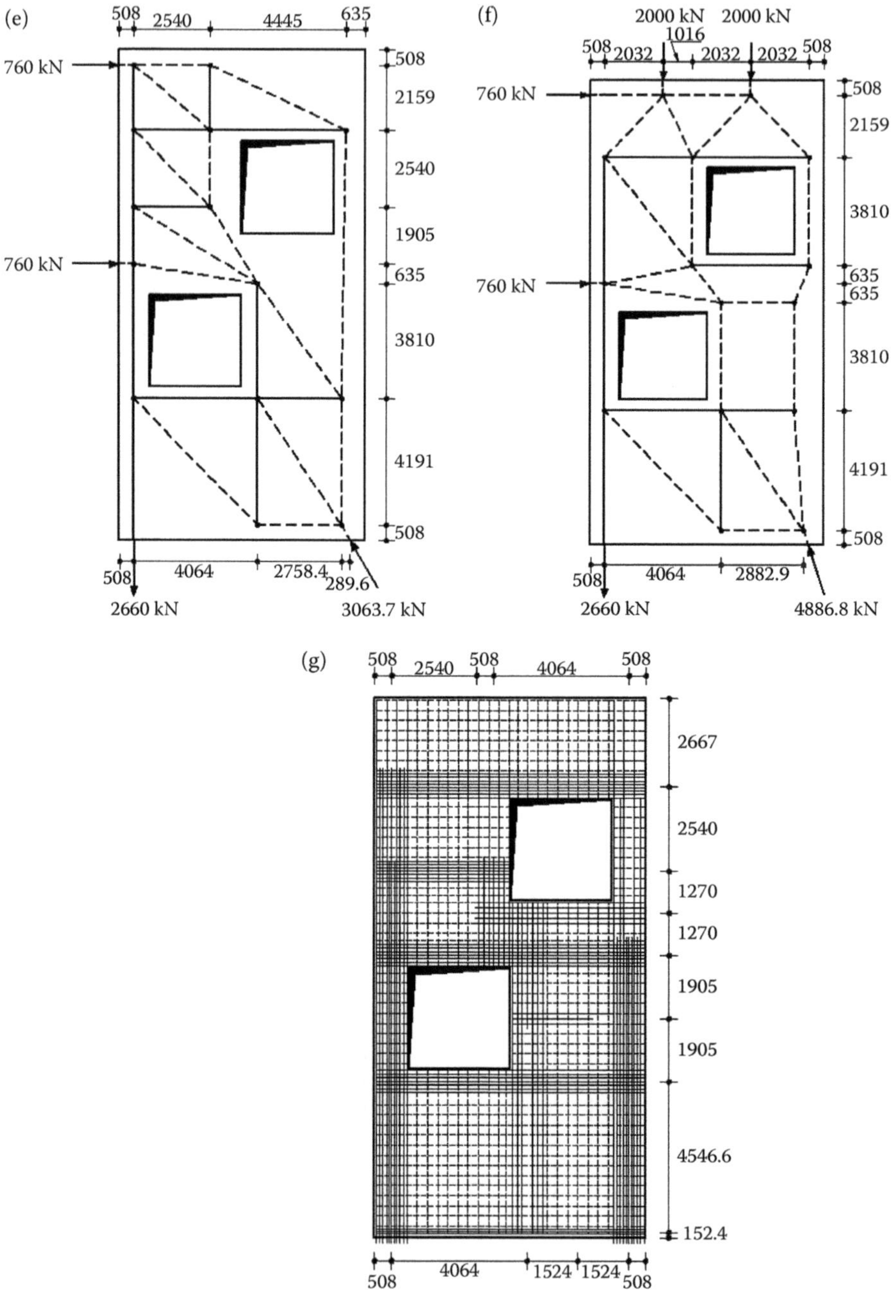

FIGURE 4.11 (Continued) (g) Reinforcement arrangement.

Upon following the load path and placing the tension reinforcements as they are practically arranged, the *STMs* of the wall for the different cases of loading are shown in Figure 4.11b–f. The *STM* of load case 1, Figure 4.11b, is statically indeterminate and in order to estimate the forces in the model, based on engineering judgement, the outer parts of the wall, to the right of the upper opening and to the left of the lower opening, are assigned a larger portion of the load than the passage between the two openings. In addition, a linear elastic finite element analysis may be used to refine the model geometry and hence the forces in the model elements. The significance of this analysis lies in improving serviceability conditions, better crack control, and also the need for load redistribution will be less and hence the required reinforcement will be less. It should be noted that the two forces transferred through the passage between the two openings can be lumped into one single force.

The models for load cases 2–5, Figure 4.11c–f, are also statically indeterminate and a linear elastic finite element analysis may be used to refine the model geometry and the assessment of the forces in the model elements and to improve serviceability conditions. In the shown models, the skin reinforcements are utilized to carry the tension forces of some ties. Also, it is noted that the angle between the struts and the ties connected with these struts satisfy the minimum limit (25°).

For wall design, the solution should start with verifying the bearing stresses at the supports and loading nodes. The appropriate effectiveness factor of struts is $\beta_s = 0.75$ if the skin reinforcement is assumed to cover the transverse tension of struts; otherwise, $\beta_s = 0.6$ if only minimum skin reinforcement is provided. The appropriate effectiveness factor of the nodes is variable based on the boundary conditions of the node. The respective code provisions for reinforcement anchorage, minimum reinforcement requirements and skin reinforcement should be justified. The reinforcement layout of the beam is shown in Figure 4.10g. It should be noted that the ends of the vertical and horizontal web reinforcement should have a U shape.

4.10 EXAMPLE ON STRENGTH ASSESSMENT OF A CONTINUOUS DEEP BEAM WITH LARGE OPENINGS

General Description:

In this example the strength of the continuous deep beam with large openings shown in Figure 4.12a is assessed. The beam was tested by Wu and Li (2009) under equal point loads, *P*, at the middle of each span. The beam consists of two equal spans, 2000 mm length each, with overall height $h = 750$ mm, depth $d = 705$ mm, breadth $b = 180$ mm and the openings and loadings are symmetrically arranged as shown in the figure. Four openings, 400 mm × 200 mm each, with their centers located at the 1/4 and 3/4 points of each span were introduced. The width of bearing plates $b_1 = 150$ mm, and $b_2 = b_3 = 200$ mm. Due to its geometry and also the presence of openings the beam is entirely considered a D-region.

The concrete cylinder strength of the beam was $f_c' = 27.7$ MPa and the steel yield stress was $f_y = 460$ MPa for high strength deformed bars $\phi 10$, $\phi 13$, and $\phi 20$ and $f_y = 250$ MPa for mild steel bars $\phi 10$. The detailed reinforcement layout is shown in Figure 4.12b. The clear cover for the top and bottom reinforcement was 25 mm and it was 20 mm for the transverse reinforcement. $2\phi 20$ were provided for both top and

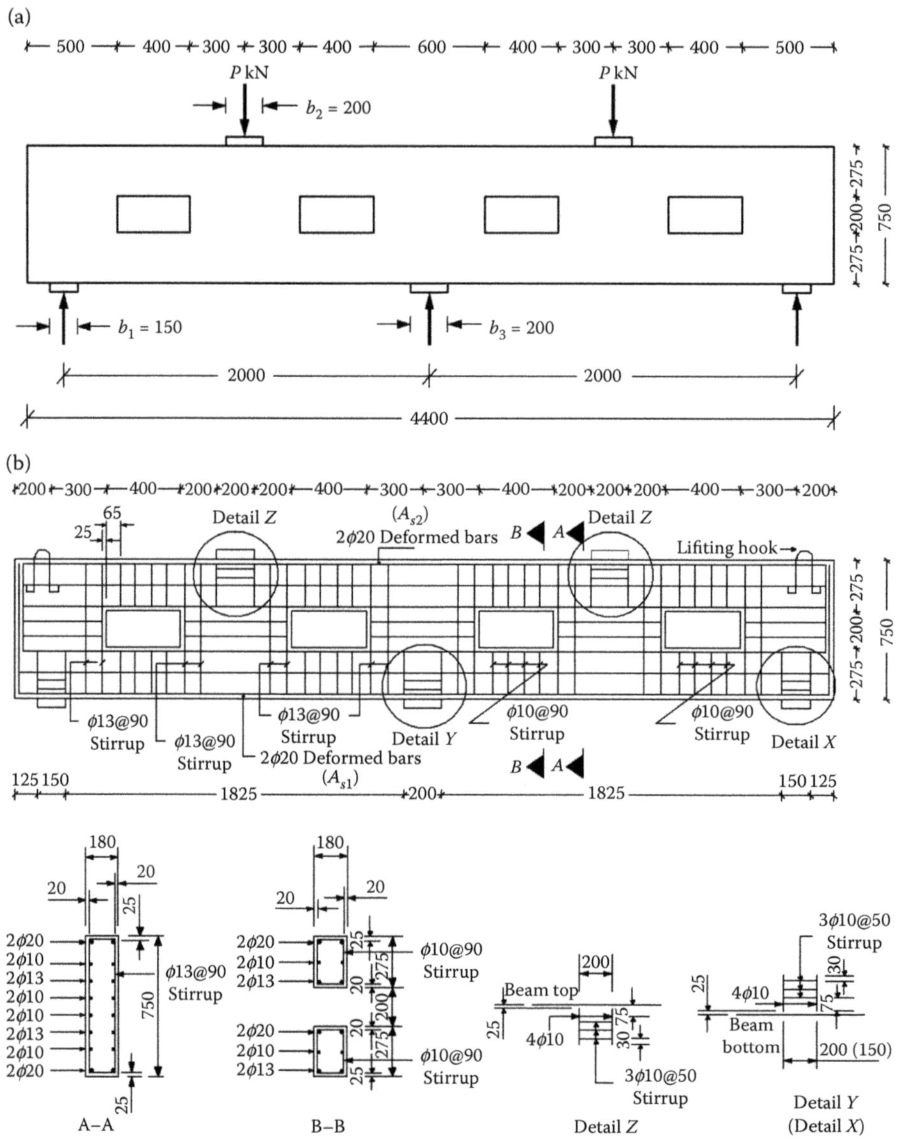

FIGURE 4.12 Example of continuous deep beam with large openings: (a) geometry. (b) Reinforcement details. *(Continued)*

bottom longitudinal reinforcement with corresponding longitudinal reinforcement ratio $\rho_v = 0.5\%$. 2ϕ13 at 90 mm spacing were placed as transverse reinforcement at both sides of the opening. The tension steel, $A_{s1} = A_{s2} = 628.3\ \text{mm}^2$.

In addition to the reinforcements required at the tie positions, confining reinforcements were incorporated to improve the strength of nodes where potentially high compressive stresses were encountered. Confining reinforcements were used under

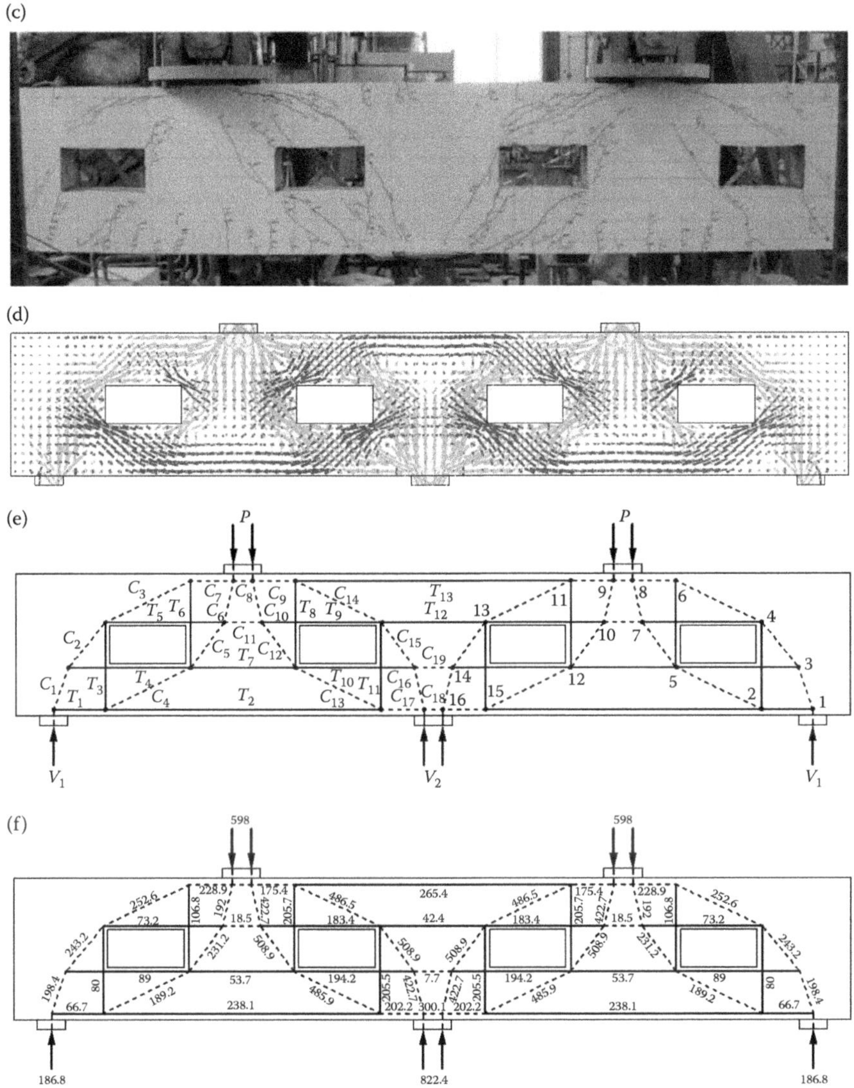

FIGURE 4.12 (Continued) (c) Cracking pattern at ultimate test load (Wu and Li, 2009). (d) Stress trajectories. (e) Proposed *STM*. (f) Computed model forces (kN) assuming yielding of T_{10} reinforcement.

the concentrated loads, above the supports and around the openings as shown in Figure 4.12b.

To eliminate potentially the detrimental effects of apparently inadequate reinforcing details on load carrying capacity, the reinforcing bars passing the nodes were extended beyond the nodal zones and the bottom longitudinal reinforcing bars were extended into the end supports and were anchored at the supports using standard 180-degree hooks.

Experimental Results:
The ultimate test load of the specimen was 555.0 kN. The failure mode was a combination of yielding of the bottom reinforcement and crushing of the strut (ductile failure). The cracking pattern of the specimen at failure is shown in Figure 4.12c.

Establish an STM:
From a nonlinear finite element analysis of the beam, the stress trajectories in Figure 4.12d has been obtained (Ghoraba, 2017). From the cracking pattern in Figure 4.12c and the stress trajectories in Figure 4.12d it is obvious how the struts go around the openings in order to transfer their forces to the supports. In addition, the connecting ties maintaining equilibrium are logical and adhere with the reinforcement details. Thus, developing an *STM* of the beam becomes a straightforward matter. The proposed *STM* of the beam is shown in Figure 4.12e, where the numbering of ties and struts is illustrated on the left part of the model and the numbering of the nodes is illustrated on the right part.

With reference to Figure 4.12e, the reinforcement of each of the ties T_1, T_2, and T_{13} is $2\phi 20$. Since the reinforcement details allow the use of extended nodal zone, the width of each of these ties, height of nodes 1, 2, 11, and 15, is

$$w_{T_1} = w_{T_2} = \cdots = 2(h-d) = 2(750-705) = 90\,\text{mm}$$

or

$$w_{T_1} = w_{T_2} = \cdots = 2(c+\phi_{stirrup}) + \phi_{bars} = 2(25+10)+20 = 90\,\text{mm}$$

Each of the ties T_1, T_2, and T_{13} has a nominal strength, $T_{1n} = T_{2n} = T_{13n} = 460 \times 628.3 = 289$ kN. With reference to Figure 4.12e, each of the ties T_3, T_6, T_8, and T_{11} is represented by two vertical stirrups ϕ_{13}, each stirrup has two branches and the spacing between the stirrups is 90 mm. The width of each of these ties is

$$w_{T_3} = w_{T_6} = \cdots = 2c + n\phi_{bars} + (n-1)s = 2(25) + 2\times(13) + 1\times 90 = 166\,\text{mm}$$

The nominal strength of each of the ties T_3, T_6, T_8, and T_{11} is

$$T_{3n} = T_{6n} = \cdots = 460\times(2\times 2\times 132.7) = 144.4\,\text{kN}.$$

In the same manner, each of the ties T_4, T_5, T_7, T_9, T_{10}, and T_{12} is represented by $2\phi 13 + 2\phi 10$ horizontal bars with width 143 mm and nominal strength 194.2 kN.

Reactions and forces:
From linear elastic analysis the reactions can obtained in terms of the load *P*,

$$V_1 = 0.3125\,P$$

$$V_2 = 1.375\,P$$

To obtain the forces in the model elements, the reinforcement of a selected tie is assumed to reach its yield stress. The choice of the tie to start with is up to the designer; however, some elements come to mind at the first glance; e.g., ties T_2 or T_{13}, Figure 4.12e. Nevertheless, the solution of this problem is initiated with assuming that the reinforcement of tie T_{10} reaches its yield stress, T_{10} = 194.2 kN. Then, from simple truss analysis the model forces are obtained as given in Figure 4.12f.

Effective concrete strength of the struts:
The effective concrete strength of strut, $f_{ce}^{s} = 0.85 f_c' \beta_s$.

For prismatic strut, $f_{ce}^{s} = 0.85 \times 27.7 \times 1.0 = 23.55\,\text{MPa}$.

For bottle-shaped strut with sufficient reinforcement to resist the transverse tension, $f_{ce}^{s} = 0.85 \times 27.7 \times 0.75 = 17.66\,\text{MPa}$. This value will be used throughout the example and at the end there will be a discussion on this point.

Effective concrete strength of the nodes:
The effective concrete strength of a node, $f_{ce}^{n} = 0.85 f_c' \beta_n$

For $C - C - C$ node, $f_{ce}^{n} = 0.85 \times 27.7 \times 1.0 = 23.55\,\text{MPa}$
For $C - C - T$ node, $f_{ce}^{n} = 0.85 \times 27.7 \times 0.8 = 18.84\,\text{MPa}$
For $C - T - T$ node, $f_{ce}^{n} = 0.85 \times 27.7 \times 0.6 = 14.13\,\text{MPa}$

Check the bearing of the nodes:
For node 1, the nominal value of the reaction V_1, V_{1n} = 18.84 × 150 × 180 = 508.6 kN, which is greater than the reaction, V_1 = 186.8 kN.

For nodes 8 and 9, the nominal value of the load P, P_n = 23.55 × 200 × 180 = 847.8 kN, which is greater than the load, P = 598.0 kN.

For node 16, the nominal value of the reaction V_2, V_{2n} = 23.55 × 200 × 180 = 847.8 kN, which is greater than the reaction, V_2 = 822.4 kN.

Check of stresses:
Node 1:
Since the bearing stress has been checked before, there is no need to check it again. For strut C_1, its angle of inclination θ_1 = 70.3°, which gives a width of the strut.

$w_{C_1}^{1} = 150 \sin\theta_1 + 90 \cos\theta_1 = 171.5\,\text{mm}$. Then, the nominal strength of the strut.

C_{1n} = 17.66# × 171.5 × 180 = 545.1 kN (# the smaller of the node strength and the strut strength), which is greater than the strut force, C_1 = 198.4 kN.

Node 2:
The width of tie T_2, $w_{T_2} = 90\,\text{mm}$ and of tie T_3, $w_{T_3} = 166\,\text{mm}$, and the angle of inclination of strut C_4, θ_4 = 25.0°; hence; the width of strut C_4, $w_{C_4}^{2} = 166 \sin\theta_4 + 90 \cos\theta_4 = 151.7\,\text{mm}$. Then, the nominal strength of the strut C_{4n} = 14.13# × 151.7 × 180 = 385.7 kN (# the smaller of the node strength and the strut strength), which is greater than the strut force, C_4 = 189.2 kN.

Nodes 3, 7, 10, and 14:
These nodes need not be checked, since all of these nodes are smeared nodes and "*the reinforcing bars passing the nodes are extended beyond the nodal zones and*

they have sufficient anchorage length." In addition, the strength of the struts connected with these nodes is governed by either the strength of the other connecting nodes or the strength of these struts.

Node 4:
The width of tie T_3, $w_{T_3} = 166\,\text{mm}$, and of tie T_5, $w_{T_5} = 143\,\text{mm}$, and the angle of inclination of strut C_2, $\theta_2 = 50.2°$; hence; the width of strut C_2,

$w^4_{C_2} = 166\sin\theta_2 + 143\cos\theta_2 = 219.0\,\text{mm}$. Then, the nominal strength of the strut,

$C_{2n} = 14.13^\# \times 219.0 \times 180 = 556.8$ kN ($^\#$the smaller of the node strength and the strut strength), which is greater than the strut force, $C_2 = 243.2$ kN.

For strut C_3, the angle of inclination is $\theta_3 = 25.0°$. Then, the width of the strut,

$w^4_{C_3} = 166\sin\theta_3 + 143\cos\theta_3 = 199.7\,\text{mm}$. Then, the nominal strength of the strut,

$C_{3n} = 14.13^\# \times 199.7 \times 180 = 508.0$ kN ($^\#$the smaller of the node strength and the strut strength), which is greater than the strut force, $C_3 = 252.6$ kN.

Node 5:
As before for nodes 1, 2, and 4, $w_{T_6} = 166\,\text{mm}$, $w_{T_7} = 143\,\text{mm}$, $\theta_4 = 25.0°$,

$w^5_{C_4} = 199.7\,\text{mm}$, $C_{4n} = 508.0$ kN, greater than $C_4 = 189.2$ kN. Also, $\theta_5 = 54.0°$,

$w^5_{C_5} = 218.4\,\text{mm}$, $C_{5n} = 555.2$ kN, greater than $C_5 = 231.2$ kN.

Node 6:
For this node, assuming that the force in the prismatic strut C_7 is equal to its nominal strength; then, $C_{7n} = 228.9\,\text{kN} = 23.55 \times w_{C_7} \times 180$, which gives $w_{C_7} = 54\,\text{mm}$. The width of tie T_6, $w_{T_6} = 166\,\text{mm}$ and the angle of inclination of strut C_3, $\theta_3 = 25.0°$; hence, the width of strut C_3, $w^6_{C_3} = 166\sin\theta_3 + 54\cos\theta_3 = 119.1\,\text{mm}$. Then, the nominal strength of the strut,

$C_{3n} = 17.66^\# \times 119.1 \times 180 = 378.6$ kN ($^\#$ the smaller of the node strength and the strut strength), which is greater than the strut force, $C_3 = 252.6$ kN.

Node 8:
Since the bearing stress has been checked before, there is no need to check it again. The width of the prismatic strut C_8 can be obtained upon assuming that the force in the strut is equal to its nominal strength; then, $C_{8n} = 273.3\,\text{kN} = 23.55 \times w_{C_8} \times 180$, which gives $w_{C_8} = 64.5\,\text{mm}$. The angle of inclination of strut C_6 is $\theta_6 = 76.6°$; hence, the width of strut C_6, $w^8_{C_6} = 100\sin\theta_6 + 64.5\cos\theta_6 = 112.2\,\text{mm}$. Then, the nominal strength of the strut,

$C_{6n} = 17.66^\# \times 112.2 \times 180 = 356.6$ kN ($^\#$ the smaller of the node strength and the strut strength), which is greater than the strut force, $C_6 = 192.0$ kN.

Node 9:
Since the bearing stress has been checked before, there is no need to check it again. The width of the prismatic strut C_8, $w_{C_8} = 64.5\,\text{mm}$, and the angle of inclination of strut C_{10} is $\theta_{10} = 76.6°$; then, the width of strut C_{10}, $w^9_{C_{10}} = 100\sin\theta_{10} + 64.5\cos\theta_{10} = 112.2\,\text{mm}$. The nominal strength of strut C_{10} is then, $C_{10n} = 17.66^\# \times 112.2 \times 180 = 356.6$ kN ($^\#$ the smaller of the node strength and the strut strength), which is less than the strut force, $C_{10} = 422.7$ kN; i.e., $C_{10n} = 0.843C_{10}$.

Node 11:
The width of tie T_8, $w_{T_8} = 166\,\text{mm}$, and of tie T_{13}, $w_{T_{13}} = 90\,\text{mm}$, and the angle of inclination of strut C_{14}, $\theta_{14} = 25.0°$; then, the width of the strut,

$w_{C_{14}}^{11} = 166\sin\theta_{14} + 90\cos\theta_{14} = 151.7\,\text{mm}$. Then, the nominal strength of the strut,

$C_{14n} = 14.13^{\#} \times 151.7 \times 180 = 385.7$ kN (#the smaller of the node strength and the strut strength), which is less than the strut force, $C_{14} = 486.5$ kN; i.e., $C_{14n} = 0.793C_{14}$.

Node 12:
As before for nodes 1, 2, and 3, $w_{T_8} = 166\,\text{mm}$, $w_{T_7} = 143\,\text{mm}$, the angle of inclination of strut C_{12}, $\theta_{12} = 54.0°$, and $w_{C_{12}}^{12} = 218.4\,\text{mm}$. The nominal strength of strut C_{12}, $C_{12n} = 555.2$ kN, which is greater than the strut force, $C_{12} = 508.9$ kN. As for strut C_{13}, the angle of inclination $\theta_{13} = 25.0°$, and $w_{C_{13}}^{12} = 199.7\,\text{mm}$, $C_{13n} = 508.0$ kN, which is greater than the strut force, $C_{13} = 485.9$ kN.

Node 13:
As before for nodes 1, 2, and 3, $w_{T_{11}} = 166\,\text{mm}$, $w_{T_{12}} = 143\,\text{mm}$, the angle of inclination of strut C_{14}, $\theta_{14} = 25.0°$ and $w_{C_{14}}^{13} = 199.7\,\text{mm}$. The nominal strength of strut C_{14}, $C_{14n} = 508.0$ kN, which is greater than the strut force, $C_{14} = 486.5$ kN. As for strut C_{15}, the angle of inclination $\theta_{15} = 54.0°$, and $w_{C_{15}}^{13} = 218.4\,\text{mm}$; then, $C_{15n} = 555.2$ kN, which is greater than the strut force, $C_{15} = 508.9$ kN.

Node 15:
As before for nodes 1, 2, and 3, $w_{T_{11}} = 166\,\text{mm}$, $w_{T_2} = 90\,\text{mm}$, the angle of inclination of strut C_{13}, $\theta_{13} = 25.0°$ and $w_{C_{13}}^{15} = 151.7\,\text{mm}$. The nominal strength of strut C_{13}, $C_{13n} = 385.7$ kN, which is less than the strut force, $C_{13} = 485.9$ kN; i.e., $C_{13n} = 0.794C_{13}$.

Node 16:
Since the bearing stress has been checked before, there is no need to check it again. The width of the prismatic strut C_{18} can be obtained by assuming that the force in the strut is equal to its nominal strength; then, $C_{18n} = 300.1\,\text{kN} = 23.55 \times w_{C_{18}} \times 180$, which gives $w_{C_{18}} = 70.8\,\text{mm}$. The angle of inclination of strut C_{16}, $\theta_{16} = 76.6°$; hence, its width,

$w_{C_{16}}^{16} = 100\sin\theta_{16} + 70.8\cos\theta_{16} = 113.7\,\text{mm}$. Then, the nominal strength of the strut,

$C_{16n} = 17.66^{\#} \times 113.7 \times 180 = 361.4$ kN, which is less than the strut force, $C_{16} = 422.7$ kN; i.e., $C_{16n} = 0.855C_{16}$.

From obtained results all model elements are safe except elements C_{10}, C_{13}, C_{14}, and C_{16}, with C_{14} the most critical, $C_{14n} = 0.793C_{14}$. Therefore, the load P should be reduced to 79.3% of the calculated value; i.e., the predicted collapse load from the *STM*, $P_{STM} = 0.793 \times 598 = 474.2$ kN. Since the measured collapse load was 555.0 kN; then, the predicted collapse load from the *STM* is 85.4% of the measured value in the test.

In the preceding calculations it has been assumed that bottle-shaped struts have adequate reinforcement to resist the transverse tension, which should be verified.

However, with the obtained beam capacity from *STM*, the forces in all elements have been reduced to 79.3% of the calculated value, which makes the check unnecessary for all struts that attained nominal strength exceeding their original forces (the ratio between the strength of struts without transverse reinforcement to that with adequate transverse reinforcement is 0.8). To this, only struts C_{10}, C_{13}, C_{14}, and C_{16} need to be checked. Upon checking these latter struts it is found that they have adequate transverse reinforcement.

REFERENCES

ACI 318-14, *Building Code Requirements for Structural Concrete and Commentary*, Detroit: American Concrete Institute, USA, 2014, 519 pp.

Barnes, R., Example 8: High wall with two openings, in: *Examples for the Design of Structural Concrete with Strut-and-Tie Models.* SP 208. ed. Reineck, K.-H. American Concrete Institute, 2002, pp. 195–211.

Chantelot, G. and Mathern, A., Strut-and-tie modeling of reinforced concrete pile caps, *Master's Thesis*, Department of Structural Engineering, Chalmers University of Technology, Publication no. 2010:51, Göteborg, Sweden, 2010.

Ghoraba, A. I., The strut-and-tie method versus nonlinear finite element analysis, *MSc Thesis*, Mansoura University, 2017.

Novak, L., and Sprenger, H., Example 4: Deep beam with opening, in: *Examples for the Design of Structural Concrete with Strut-and-Tie Models.* SP 208. ed. Reineck, K.-H. American Concrete Institute, 2002, pp. 129–143.

Schäfer, K., and El-Metwally, S. E., On the role of discontinuity regions detailing in the safety of concrete structures, *Proceedings of the Fifth International Colloquium on Concrete in Developing Countries*, Cairo, Egypt, January 2–6, 1994, pp. 43–55.

Schlaich, J. and Schäfer K., The design of structural concrete, *IABSE Workshop*, New Delhi, 1993.

Schlaich, J., and Schäfer, K., Design and detailing of structural concrete using strut-and-tie models, *Journal of the Structural Engineer*, 69(6), March, 1991.

Schlaich, J., Schäfer, K., and Jennewein, M., Toward a consistent design of structural concrete, *Journal of the Prestressed Concrete Institute*, 32(3), May-June, 1987, 74–150.

Wu, T., and Li, B., Experimental verification of continuous deep beams with openings designed using strut-and-tie modelling, *IES Journal Part A: Civil & Structural Engineering*, 2(4), November, 2009, 285–295.

5 Deep Beams

5.1 INTRODUCTION

As per ACI 318-14, each shear span, a_v, of the beam in Figure 5.1a, where $a_v < 2h$, is a D-region. If two D-regions overlap or meet as shown in Figure 5.1b, they can be considered as a single D-region for design purposes. The maximum length-to-depth ratio of such a D-region would approximately equal 2. Thus, the smallest angle between the strut and the tie in a D-region is $\tan^{-1}(1/2) = 26.56°$, rounded to 25°. If there is a B-region between the D-regions in a shear span, as shown in Figure 1c, the strength of the shear span is governed by the strength of the B-region if the B- and D-regions have similar geometry and reinforcement. This is because the shear strength of a B-region is less than that of a comparable D-region.

In most building codes, the conventional approach of shear design of deep beams is based on some empirical equations in which the nominal shear strength, V_n, includes two parts: the concrete contribution, V_c, and the steel contribution, V_s. Separate equations are introduced for both. Though this approach is easy to apply, it ignores the interaction between V_c and V_s, whereas the strut-and-tie model, *STM*, satisfies this goal.

This chapter covers the design of deep beams, either simply supported, continuous or corbels, top or bottom loaded, directly or indirectly supported. The chapter starts with the modeling of simply supported deep beams and continuous deep beams. Applications to simply supported deep beams are given to cover different types of models. Bottom loaded deep beams and deep beam with ledge are covered as well. Deep beams with indirect supports are discussed too. Applications to continuous deep beams, top or bottom loaded, are given. The final part of the chapter is devoted to brackets and corbels, where the modes of failure and modeling of different corbel problems are discussed. In addition, the detailing of critical nodes is examined, a step by step design procedure is illustrated, and the assessment of the web reinforcement of corbels is explained. An example on strength assessment of double corbel is given at the end. The failure criteria adopted by the ACI 318-14, as illustrated in Chapter 3, is applied to all the examples of this chapter.

5.2 MODELING

5.2.1 Simply Supported Deep Beams

On the basis of the shear span-to-depth ratio, three *STMs* are considered: Type I, Type II, and Type III, Figure 5.2. Types I and II cover deep and short beams, respectively, and Type III deals with slender beams. In Type I, a direct *STM* is utilized, whereas a fan- or arch-action model is used in Type II. The choice between the two types I and II in some cases is controlled by the shear span-to-depth ratio, *a*/*d*, presence of vertical web reinforcement, and the concrete strength.

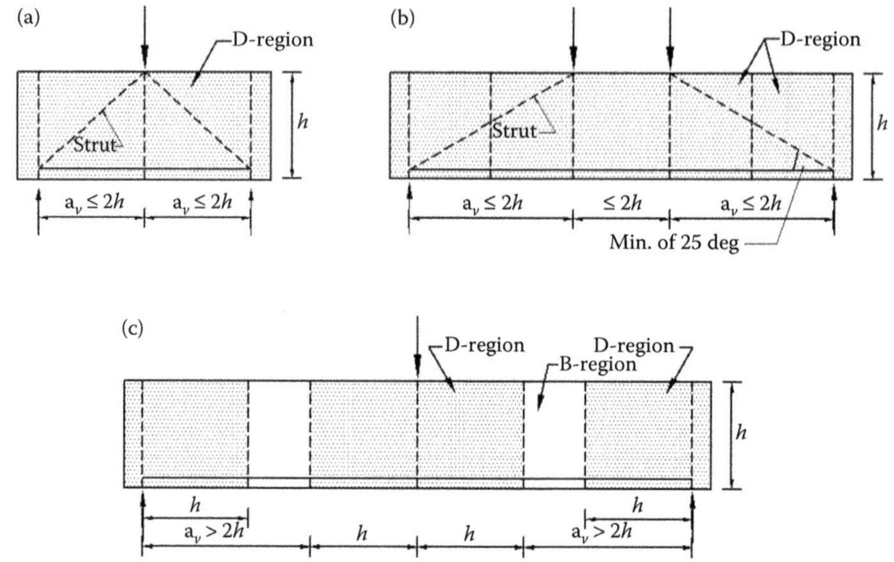

FIGURE 5.1 Description of deep and slender beams: (a, b) deep beams and (c) slender beam.

In Type I (direct model), the load is transferred from the loading point to the support directly by only one compression strut, and thus no direct account is taken for the forces that may exist in any vertical web reinforcement as shown in Figure 5.2a. The presence of such reinforcement will only contribute to the strength of the diagonal strut by participating in the resistance of transverse tension of this strut. Type II may be divided into two subtypes: arch- and fan-action, and this model is usually applied to beams where the shear span-to-depth ratio is within the range 1.5–2.0. In Type II arch-action, Figure 5.2b(i), the load is transferred by struts forming an arch mechanism with vertical web reinforcement acting as a single tie. In Type II fan-action, Figure 5.2b(ii), the load is transferred to the supports by a combination of major and minor compression struts. Therefore, hanger reinforcement is required to return the vertical components of forces developed in the minor compression struts to the top of the member. In Type III (which exists when the shear span-to-depth ratio a/d exceeds 2.0), the load is transferred across a strut-and-tie system which has an arch- or fan-action model as shown in Figure 5.2c.

5.2.2 Continuous Deep Beams

In simple deep beams, the region of high shear coincides with the region of low moment. On the other hand, in continuous deep beams, the regions of high shear and high moment coincide and failure usually occurs in these regions. Hence, the failure mechanisms of continuous deep beams are different from those of simply supported deep beams. Continuous deep beams are divided into two major groups according to their loading conditions: top and bottom loading. Top loaded continuous deep beams are commonly used in reinforced concrete buildings, while indirectly loaded or bottom loaded deep beams are widely used as cross-girders, for example, in concrete

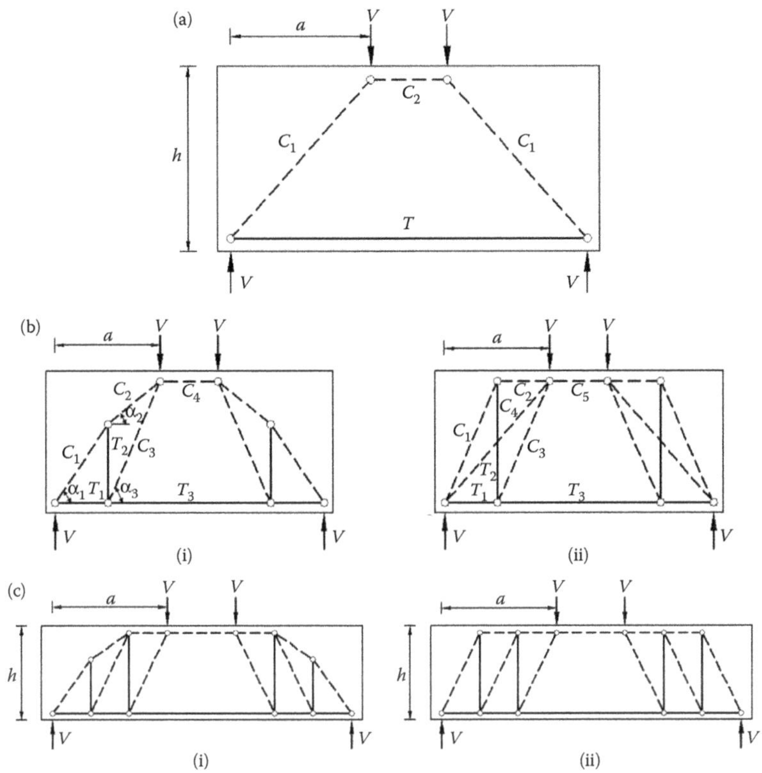

FIGURE 5.2 The strut-and-tie modeling of simple deep beams: (a) Type I direct model, (b) Type II: (i) arch-action and (ii) fan-action, and (c) Type III: (i) arch-action and (ii) fan-action.

bridges and water tanks. The two groups behave differently under the same applied loads.

Considering the effect of negative moment at interior supports, continuous deep and short beams can be simply modeled utilizing Type I, Figure 5.3, and Type II, Figure 5.4, models (El-Shora, 2005; El-Zoughiby et al., 2013). For simplicity sake and since the Type II model, Figure 5.4, requires more complex calculations, the Type I model may be the first choice in the design of continuous beams.

5.3 APPLICATIONS TO SIMPLY SUPPORTED DEEP BEAMS

5.3.1 Example 5.1: Type I Model for Strength Assessment of Beam under Two Point Loads

The normal strength concrete simple deep beam 0A0-48, Figure 5.5a, tested by Smith and Vantsiotis (1982) under two top point loads, is analyzed in this example for the collapse load. The proposed *STM* is illustrated in Figure 5.5b and c, in which the load is transferred directly from the two point loads at the top to the

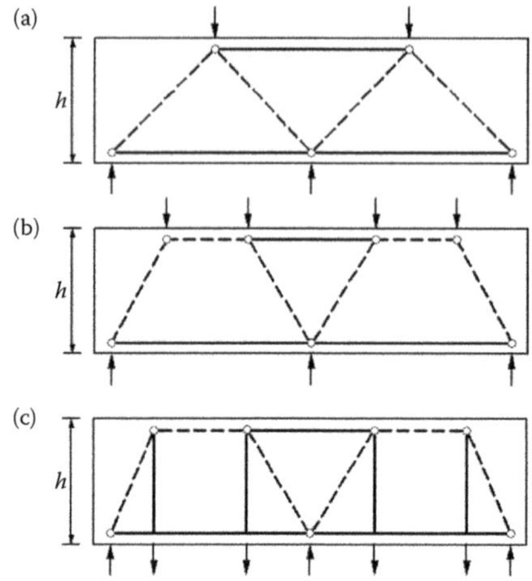

FIGURE 5.3 Simplified *STM* for continuous deep beeams, Type I, subjected to: (a) a single top point load, (b) two top point loads, and (c) bottom loads.

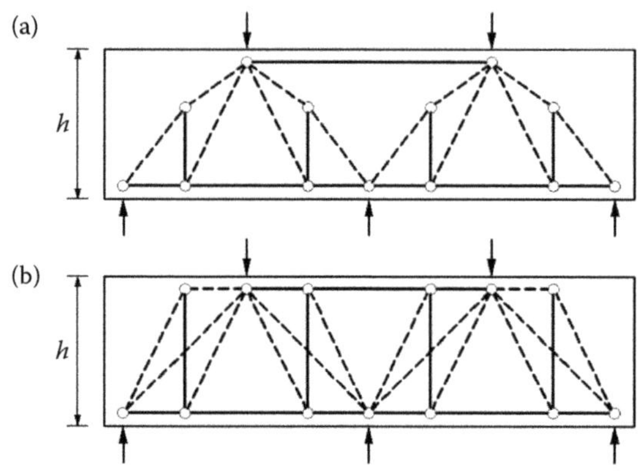

FIGURE 5.4 Type II *STM* for continuous deep beams subjected to a single top point load: (a) arch-action model and (b) fan-action model.

bottom supports through the concrete struts C_2. With reference to Figure 5.5, the beam height, $h = 356$ mm, depth, $d = 305.2$ mm, breadth, $b = 102$ mm, the width of the bearing plates, $b_1 = b_2 = 102$ mm, the shear span, $a = 304.3$ mm, and the beam effective length, $L = 811.8$ mm. The shear span-to-depth ratio, $a/d \approx 1.0$. The tension steel, $A_s = 3\phi16$ mm. The concrete cylinder strength is $f_c' = 20.9$ MPa and the steel yield stress is $f_y = 421.5$ MPa. The solution is given in the next steps, in which the beam's own weight is neglected.

Reactions:

With reference to Figure 5.5a,

$$R_1 = R_3 = V$$

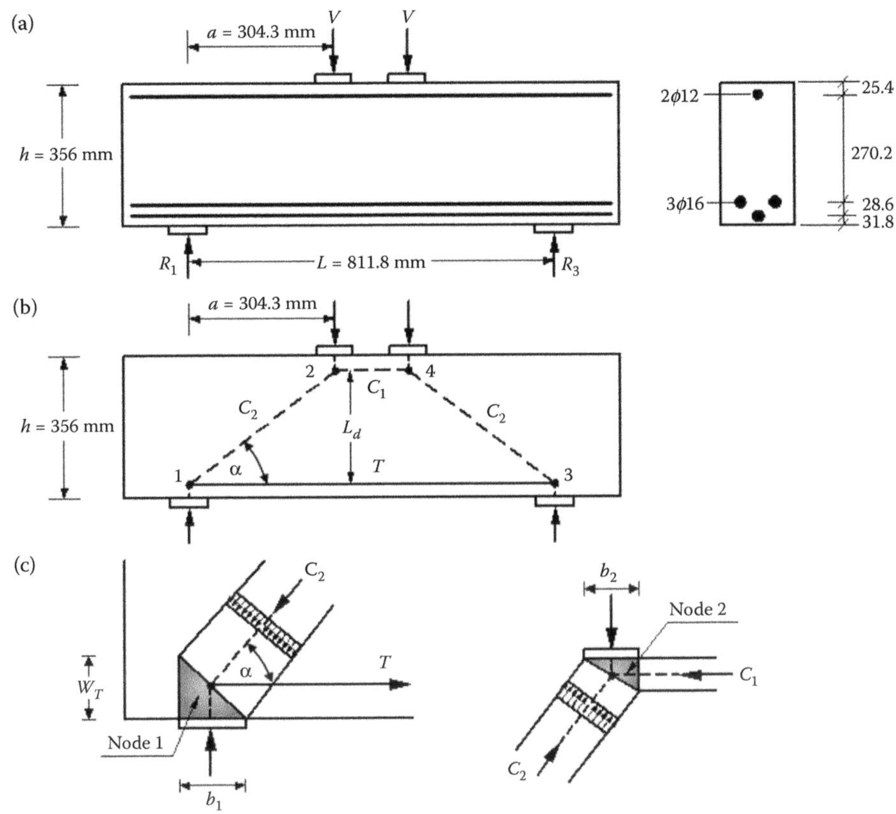

FIGURE 5.5 Details of the Type I model for a simple deep beam subjected to two top point loads: (a) beam, (b) *STM*, and (c) details of nodes 1 and 2.

Model geometry and forces:

Assuming that the tension reinforcement, 3ϕ16 mm, reached the yield stress, the tension force,

$$T = 3\times 201\times 421.5 = 254.2\,\text{kN} = C_1 = 0.85 f_c' b w_{C_1}$$

where w_{C_1} is the width of the horizontal strut C_1. The value of w_{C_1} is then 140.3 mm; thus, the lever arm, $L_d = 305.2 - 0.5 \times 140.3 = 235.1$ mm.

The angle $\alpha = \tan^{-1}(235.1/304.3) = 37.69° > \tan^{-1}(1/2)$

The force in strut $C_2 = T/\cos\alpha = 321.2$ kN and the reaction $R_1 = V = T\tan\alpha = 196.4$ kN.

Effective concrete strength of the struts:

The width of strut C_1 was calculated based the conditions that it is prismatic and is connected between two $C-C-C$ nodes; therefore, there is no need to check this strut.

For strut C_2, $f_{ce}^s = 0.85 f_c' \beta_s = 0.85\times 20.9\times 0.60^* = 10.66\,\text{MPa}$

*Since strut C_2 is a bottle-shaped stress field and there is no reinforcement to resist the transverse tension.

Effective concrete strength of the nodes:
Node 1 is a $C - C - T$ node; thus, $f_{ce}^{n} = 0.85 f_c' \beta_n = 0.85 \times 20.9 \times 0.8 = 14.21\,\text{MPa}$
Node 2 is a $C - C - C$ node; thus, $f_{ce}^{n} = 0.85 f_c' \beta_n = 0.85 \times 20.9 \times 1.0 = 17.77\,\text{MPa}$

Node 1:
The nominal value of the reaction is based on the bearing strength, which leads to $R_{1n} = 14.21 \times 102 \times 102 = 147.8$ kN

The strength of strut C_2 is based on the smaller of the node strength (14.21 MPa) and the strut strength (10.66 MPa); thus, the strut strength will govern. The height of this node is assumed $\approx$102 mm since the center of the tension T is at 50.8 mm $\approx$ 51 mm from the bottom fibers; this height is to be verified. The width of strut C_2 at this node is

$$w_{C_2}^{1} = b_1 \sin\alpha + w_T \cos\alpha = 102 \sin\alpha + 102 \cos\alpha = 143.1\,\text{mm}$$

Thus, the nominal strength of strut C_2 at this node is $C_{2n} = 155.6$ kN.

Node 2:
The nominal value of the load, $V_n = 17.77 \times 102 \times 102 = 184.9$ kN, which is less than force V. The strength of strut C_2 is based on the smaller of the node strength (17.77 kN) and the strut strength (10.66 MPa); thus, the strut strength will govern. The width of strut C_2 at this node is $w_{C_2}^{2} = b_2 \sin\alpha + w_{C_1} \cos\alpha = 102 \sin\alpha + 140.3 \cos\alpha = 173.4\,\text{mm}$. Thus, the nominal strength of C_{2n} is 188.5 kN.

Take the smaller strength of the two values of C_2 calculated at the two connecting nodes; thus, the nominal strength of strut $C_{2n} = 155.6$ kN. The results of the calculations are summarized in Table 5.1

The weakest link in the model is strut C_2 and it will govern the model strength. Therefore, in order to determine the model capacity, the model forces should be reduced to 48.4%. Hence, $V = 0.484 \times 196.4 = 95.1$ kN and the *STM* beam capacity, $P_{STM} = 2V = 190.2$ kN. The actual collapse load of the beam, $P_{EXP} = 272.2$ kN; hence,

$$P_{STM}/P_{EXP} = \frac{190.2}{272.2} = 0.70$$

TABLE 5.1
Example 5.1: Summary of Calculation Results

Member	Force (kN)	Capacity (kN)	Satisfaction
C_1	254.2	254.2	Yes
C_2	321.2	155.6	No, 48.4%
V	196.4	184.9	No, 94.2%
R_1	196.4	147.8	No, 75.3%

TABLE 5.2
Example 5.1: Summary of Calculation Results of the Iteration

Member	Force (kN)	Capacity (kN)	Satisfaction
C_1	188.6	188.6	Yes
C_2	245.3	158.0	No, 64.4%
V	156.9	184.9	Yes
R_1	156.9	147.8	No, 94.2%

According to the *STM* results, failure had occurred due to the crushing of the concrete strut C_2 and it is a diagonal shear failure.

Since there is a significant discrepancy between the ratios of the strengths to the forces in Table 5.1, it is possible to reduce such discrepancy and improve the solution accuracy by going through an iterative procedure. From the solution results, it is realized that the forces T and C_1 did not reach their nominal values; hence, the iterative procedure starts with assuming a new value for each of the two forces, which is about the average between their nominal values and their values at the end of the solution. The new assumed values are

$$T = C_1 = 0.5(1+0.484)\times 254.2 = 188.6\,\text{kN}$$

Upon redoing the calculations, $w_{C_1} = 104.1\,\text{mm}$, $L_d = 253.2$ mm, $\alpha = 39.76°$, $C_2 = 245.3$ kN, $R_1 = V = 156.9$ kN, $w_{C_2}^1 = 143.7\,\text{mm}$, $w_{C_2}^2 = 145.3\,\text{mm}$, $C_{2n} = 158.0$ kN, $V_n = 184.9$ kN, and $R_{1n} = 147.8$ kN. A summary of the calculation results is given in Table 5.2.

Finally, $V = 0.644 \times 156.9 = 101.0$ kN and the *STM* beam capacity, $P_{STM} = 2V = 202.0$ kN.

$$P_{STM}/P_{EXP} = \frac{202.0}{272.2} = 0.74$$

5.3.2 Example 5.2: Design of a Wall-Type Column

This example covers the design of a wall-type column, Figure 5.6a, which carries 13 floors, one floor at the wall bottom plus 12 floors above. The 13 floors give a total factored load as shown in the figure. The effective span between the supporting columns is 2.22 m, and therefore the height of the D-region, h, is assumed to be equal to 2.22 m.

From equilibrium, the reactions of the wall columns, R_A and R_B, can be calculated. The appropriate *STM* is shown in Figure 5.6b, but with the assumption that the full load is applied at the top of the D-region. As part of the load act at the bottom of the D-region, additional vertical reinforcement will be provided in order to carry such load to the top of the D-region. In the model, the lever arm, z, is assumed to be equal to $0.6h = 1.33$ m. The solution should proceed as usual.

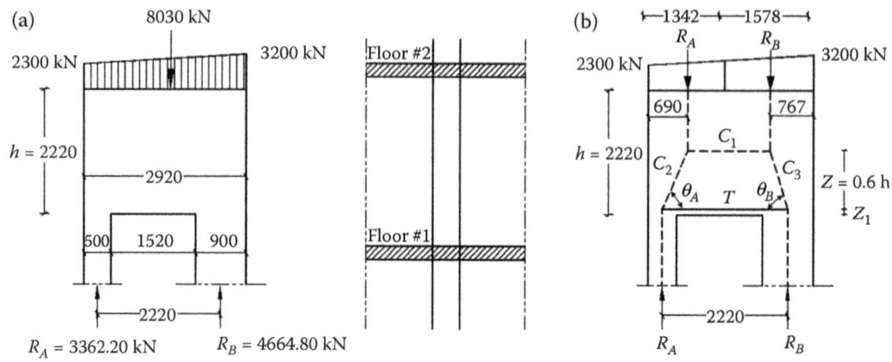

FIGURE 5.6 Example 5.2—Wall-type column: (a) elevation and (b) *STM*.

The reinforcement required to hang the load of the lowest slab can be calculated from the largest load ordinate of this floor between the two columns, 224.8 kN/m, which requires 714 mm²/m, assuming $f_y = 420$ MPa. This vertical reinforcement should be added to the vertical web reinforcement required for struts C_2 and C_3.

From the model geometry, the angles $\theta_A = 71.7°$ and $\theta_B = 76.6°$, the reactions $R_A = 3365.2$ kN and $R_B = 4664.8$ kN, the tension $T = 1113.3$ kN, and the forces in struts C_2 and C_3 are 3544.6 kN and 4795.4 kN, respectively. The tension T requires a reinforcement of 3529.7 mm² = 10ϕ22. Of course, a part of the lowest slab reinforcement in the direction of T and close to the wall can contribute to the reinforcement of this tie. The two struts C_2 and C_3 require web reinforcement to resist the transverse tension. As a result of the large values of the angles θ_A and θ_B, reliance will mainly be on the reinforcement in the horizontal direction in resisting this transverse tension. A choice of bars of diameter 22 mm every 140 mm horizontally and bars of diameter 16 mm every 200 mm vertically, on both sides, will be just sufficient to carry both the tranverse tension and to hang the load of the lowest slab.

5.3.3 Application of a Type II Arch-Action Model

In this model, Figure 5.2b(i), the influence of the horizontal web reinforcement is ignored, but the vertical web reinforcement is considered as a vertical tie at the middle of the shear span. The model is illustrated in more detail in Figure 5.7. The calculation procedure of this model starts with assuming some geometrical parameters, which is illustrated next.

With reference to Figure 5.7, the height of the tie T_3, w_{T_3}, can be assumed from

$$w_{T_3} = 2(c + \phi_{str}) + n\phi_{bars} + (n-1)s$$

where c is the concrete cover, ϕ_{str} is the diameter of stirrups, n is the number of reinforcement rows, and s is the clear spacing between rows. If the width of strut C_4, w_{C_4}, can be assumed to be equal to $0.8w_{T_3}$, the lever arm, L_d, can be assumed, $L_d = d - 0.5w_{C_4}$. The length of tie T_1, L_{T_1}, can be assumed to be equal to $0.5a$, where

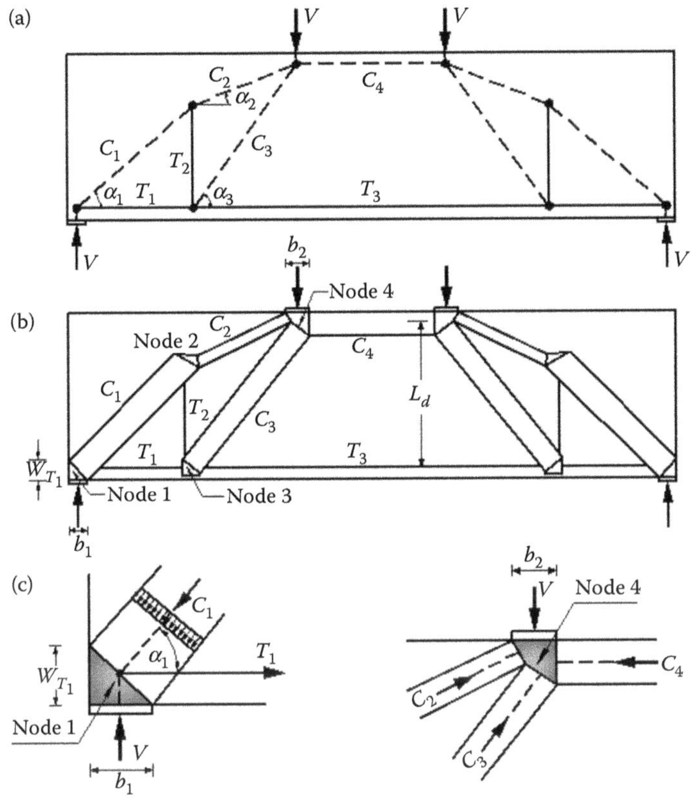

FIGURE 5.7 Details of the Type II arch-action model for a simple deep beam subjected to two top point loads: (a) *STM*, (b) details of the *STM*, and (c) details of nodes N_1 and N_2.

a is the shear span, since the location of T_2 is at the center of the vertical reinforcement within the shear span. The length of tie T_2, L_{T_2}, can be assumed as

$$L_{T_2} = kL_d$$

where k (strut factor) measures the inclination of the struts C_1 and C_2. According to Tan et al. (1997), k is given by

$$k = 0.40 + 0.20(a/d) \quad \text{for} \quad 0.50 < (a/d) < 2.5 \tag{5.1}$$

Thus, the angles α_1, α_2, and α_3 can be estimated. Alternatively, the angles α_1, α_2, C_1, and C_2 can be determined from the equilibrium of nodes 1 and 2 and from the model geometry, as will be illustrated in Example 5.3.

Figure 5.8 shows a simple reinforced concrete deep beam subjected to a single top point load along with the proposed *STM*. In this model, the tension ties T_1 and

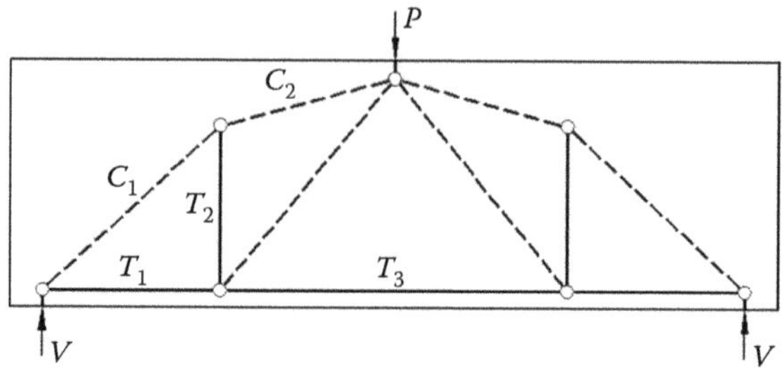

FIGURE 5.8 Type II arch-action model for a simple deep beam under a single point load.

T_3 represent the main longitudinal reinforcement, while the vertical reinforcement is represented by the tie T_2. This model is similar to the previous model after the omission of the horizontal strut. The calculation procedure is illustrated by Example 5.3 in the next section.

Figure 5.9 shows a simple deep beam subjected to two bottom point loads along with the proposed Type II arch-action model. The main longitudinal reinforcement is represented by the two ties T_3 and T_4. The tie T_1 represents the vertical reinforcement and the tie T_2 is used to transfer the bottom load to the top of the beam, to node 4. In this model, if the height of node 1 can be assumed as w_{T_3}, the width of the horizontal strut C_4 can be assumed as $w_{C_4} = 2kw_{T_3}$, where k as stated before is the strut factor which defines the inclination of struts C_1 and C_2 and is given by Equation 5.1. Nevertheless, the designer can assume any reasonable value of w_{C4} and verify it during the course of analysis. The remaining parameters, α_1, α_2, C_1, and C_2, can be determined from the equilibrium of nodes 1 and 2 and from the model geometry, as will be illustrated in Example 5.3.

5.3.4 Example 5.3: A Type II Arch-Action Model for Strength Assessment of a High Strength Concrete Deep Beam

A Type II arch-action model is applied to the HSC deep beam (B3.0-1), Figure 5.10a, tested by Foster and Gilbert (1998) under a top single point load. The beam model is illustrated in Figure 5.10b. In this model, the effect of the horizontal web reinforcement is neglected. The beam size: height, $h = 700$ mm, depth, $d = 624$ mm,

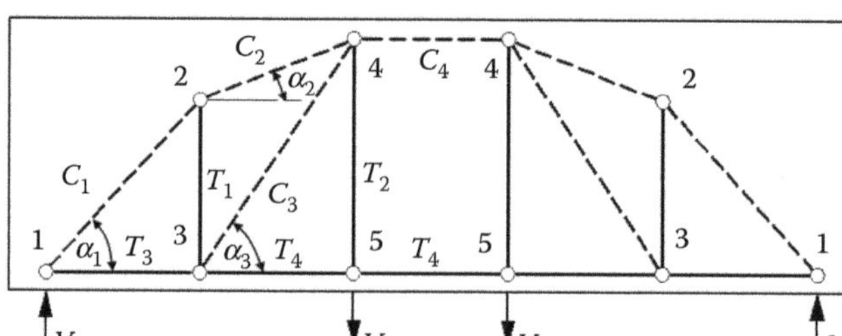

FIGURE 5.9 Details of the Type II arch-action model for a simple deep beam subjected to two bottom point loads.

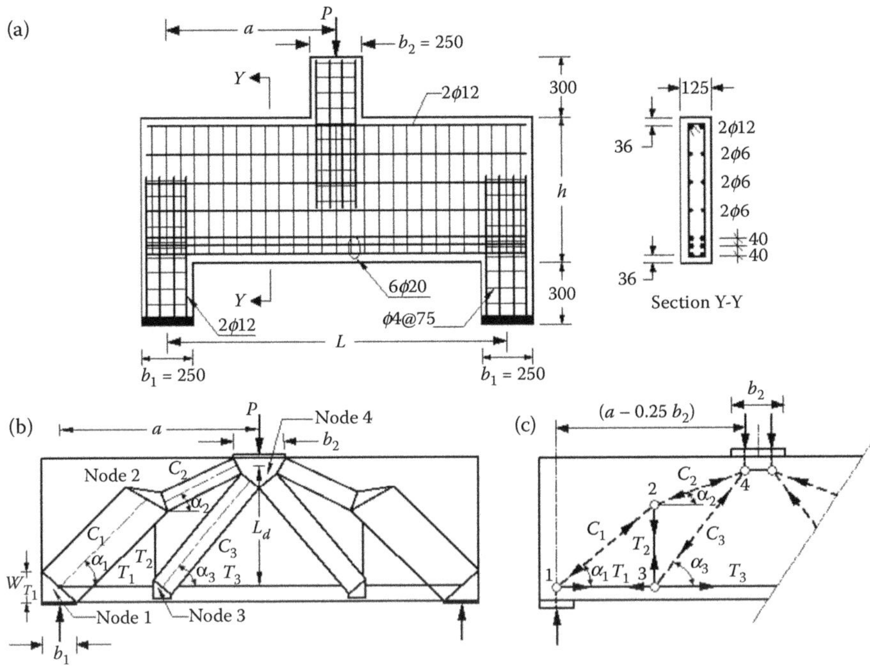

FIGURE 5.10 Example 5.3—A Type II arch-action model for a simple deep beam subjected to a top point load: (a) beam, (b) visualized *STM*, and (c) part *STM*.

breadth, $b = 125$ mm, the width of the bearing plates, $b_1 = b_2 = 250$ mm, the shear span, $a = 1175$ mm, and the beam effective length, $L = 2350$ mm. The shear span-to-depth ratio, $a/d = 1.88$. The main tension steel is $A_s = 6\phi20$ mm, the vertical web reinforcement is $A_{sv} = 13 \times 2\phi6.3$ mm (horizontal spacing = 75 mm), and the horizontal web reinforcement is $A_{sh} = 3 \times 2\phi6.3$ mm (vertical spacing = 135 mm). The concrete cylinder strength is $f_c' = 80.0$ MPa, the main steel yield stress is $f_y = 440$ MPa, and the vertical web steel yield stress is $f_{yv} = 590.0$ MPa. The solution is given in the next steps, in which the beam's own weight is neglected.

The STM Geometry and forces:

$$w_{T_1} = 2(h-d) = 2(700-624) = 152\,\text{mm}$$

or with reference to Figure 5.10, the height of node 1, or tie T_1, w_{T_1}, can be assumed as

$$w_{T_1} = w_{T_3} = 2(h-d) = 2(700-624) = 152\,\text{mm}$$

Upon assuming that the reinforcement of the tie T_3 yielded, then,

$$T_3 = f_y \times A_s = 440.0 \times 1885.0 = 829.4\,\text{kN}$$

The height of node 4, w_4, can be assumed with the aid of T_3, $w_4 = T_3/0.85 f_c' b = 97.6\,\text{mm} \approx 98\,\text{mm}$.

Then the lever arm, L_d, can be approximately assumed as, $L_d \approx d - 0.5w_4 = 624 - 0.5 \times 98 = 575$ mm.

$$\tan\alpha_3 = L_d/0.5(a - 0.25b_2) = 1.034, \text{ that is, } \alpha_3 = 45.95^\circ$$

Upon assuming that the reinforcement of the tie T_2 yielded, then,

$$T_2 = f_{yv} \times A_{sv} = 590.0 \times 374 = 220.7\,\text{kN}$$

The vertical web reinforcement is 13 stirrups, the middle stirrup is neglected and the 6 stirrups on either side contribute to T_2.

From equilibrium of node 3,

$$C_3 = T_2/\sin\alpha_3 = 307.1 \text{ kN}$$

$$T_1 = T_3 - C_3 \cos\alpha_3 = 615.9 \text{ kN}$$

The remaining parameters, α_1, α_2, C_1, and C_2, can be determined as follows.

From equilibrium of node 1,

$$C_1 = T_1/\cos\alpha_1 \tag{5.2}$$

From equilibrium of node 2,

$$C_2 = C_1\cos\alpha_1/\cos\alpha_2 \tag{5.3}$$

$$C_1\sin\alpha_1 = T_2 + C_2\sin\alpha_2 \tag{5.4}$$

Upon utilization of Equations 5.2 and 5.3 to eliminate C_1 and C_2 from Equation 5.4, the following equation is obtained

$$\tan\alpha_1 - \tan\alpha_2 = T_2/T_1 = 220.7/615.9 = 0.358 \tag{5.5}$$

From geometry,

$$\tan\alpha_1 = L_{T_2}/0.5(a - 0.25b_2) \tag{5.6}$$

where L_{T_2} is the length of tie T_2.

$$\tan\alpha_2 = \left(L_d - L_{T_2}\right)/0.5(a - 0.25b_2) \tag{5.7}$$

Upon adding Equations 5.6 and 5.7,

$$\tan\alpha_1 + \tan\alpha_2 = 2L_d/(a - 0.25b_2) = 1.034 \tag{5.8}$$

Upon solving Equations 5.5 and 5.8,

$$\alpha_1 = 34.83^\circ \quad \text{and} \quad \alpha_2 = 18.68^\circ.$$

Then the following can be obtained,

$$C_1 = 750.3\,\text{kN}, \quad C_2 = 650.1\,\text{kN} \quad \text{and} \quad V = 428.5\,\text{kN}.$$

Effective concrete design strength of the struts:
All struts are bottle-shaped stress fields; therefore, for all struts $f_{ce}^{s} = 0.85 f_c' \beta_s = 0.85 \times 80.0 \times 0.6 = 40.8\,\text{MPa}$

Effective concrete design strength of the nodes:
Nodes 1 and 2 are $C - C - T$ nodes; thus, $f_{ce}^{n} = 0.85 f_c' \beta_n = 0.85 \times 80.0 \times 0.8 = 54.4\,\text{MPa}$
Node 3 is a $C - T - T$ node; thus, $f_{ce}^{n} = 0.85 f_c' \beta_n = 0.85 \times 80.0 \times 0.6 = 40.8\,\text{MPa}$
Node 4 is a $C - C - C$ node; thus, $f_{ce}^{n} = 0.85 f_c' \beta_n = 0.85 \times 80.0 \times 1.0 = 68.0\,\text{MPa}$

Node 1:
The bearing stress $= \dfrac{428.5 \times 10^3}{250 \times 125} = 13.71\,\text{MPa} < f_{ce}^{n} = 54.4\,\text{MPa}$, which is O.K.

For the section at the interface between strut C_1 and the node, the design strength should be the smaller of the node strength (54.4 MPa) and the strut strength (40.8 MPa); thus, the strut strength will govern.

Node 1 can be considered as an extended nodal zone as a result of the heavy horizontal stirrups provided in the short column and extruded within the beam. Then the width of strut C_1 at this node $= b_1 \sin\alpha_1 + w_{T_1} \cos\alpha_1 = 267.6\,\text{mm}$. This gives a nominal strength of the strut equal to 1364.5 kN.

Nodes 2 and 3:
These two nodes need not be checked, since node 2 is a smeared node and node 3 is very wide.

Node 4:
The bearing stress $= \dfrac{2 \times 428.5 \times 10^3}{250 \times 125} = 27.42\,\text{MPa} < f_{ce}^{n} = 54.4\,\text{MPa}$, which is O.K.

The struts C_2 and C_3 connected to the node, Figure 5.10b and c, are replaced with their resultant, R_{C_2,C_3}, in order to simplify the solution.

$$\begin{aligned} R_{C_2,C_3} &= \sqrt{(C_2 \cos\alpha_2 + C_3 \cos\alpha_3)^2 + (C_2 \sin\alpha_2 + C_3 \sin\alpha_3)^2} \\ &= \sqrt{829.4^2 + 428.9^2} = 933.8\,\text{kN} \end{aligned}$$

and the angle of inclination with the horizontal, $\alpha_{C_2,C_3} = \tan^{-1} 428.9/829.4 = 27.34^\circ$. The strength of the resultant strut is the same as that of all other struts, 40.8 MPa. The required width of the strut, $w_{C_2,C_3} = (933.8 \times 10^3)/(125 \times 40.8) = 183.1\,\text{mm}$, which requires a height of node 4, $w_4 = 141.5$ mm, $w_{C_2,C_3} = 125 \sin\alpha_{C_2,C_3} + w_4 \cos\alpha_{C_2,C_3}$.

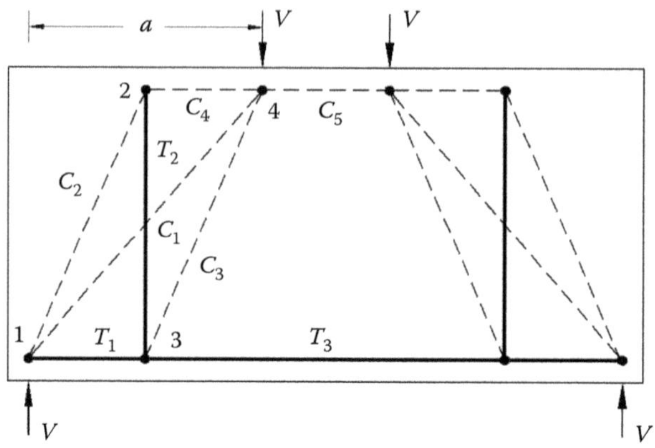

FIGURE 5.11 A Type II fan-action model for a simple deep beam subjected to two top point loads.

This required value of w_4 is larger than the assumed value, 98 mm. Therefore, an iterative process should be followed using this last obtained value of w_4 and the calculations are repeated. Upon assuming a value of $w_4 = 140.0$ mm and repeating the calculation process, the following values are obtained: $L_d = 554$ mm, $\alpha_3 = 44.88°$, $T_2 = 220.7$ kN, $C_3 = 312.8$ kN, $T_1 = 607.8$ kN, $\alpha_1 = 34.2°$, and $\alpha_2 = 17.56°$, $C_1 = 734.9$ kN, $C_2 = 637.5$ kN, and $V = 413.1$ kN. All the bearing stresses are safe. The width of strut C_1 upon considering the extended nodal zone $= 266.2$ mm. Thus, the nominal strength of strut C_1 is 1357.8 kN.

$R_{C_2,C_3} = \sqrt{829.4^2 + 413.1^2} = 926.6\,\text{kN}$ and the angle of inclination with the horizontal, $\alpha_{C_2,C_3} = 26.474°$. The required width of the strut, $w_{C_2,C_3} = 181.7\,\text{mm}$, which requires a height of node 4, $w_4 = 140.7$ mm, very close to the assumed value.

The force $V = 413.1$ kN (less than the nominal strength); hence, $P_{STM} = 2 \times 413.1 = 826.2$ kN. The measured collapse load of the beam was $P_{EXP} = 1020$ kN. Thus,

$$P_{STM}/P_{EXP} = 826.2/1020 = 0.81$$

This beam has been analyzed using the Type I model (direct model) and the obtained nominal strength P_{STM} was $0.77P_{EXP}$.

5.3.5 Application of the Type II Fan-Action Model

In this model, Figure 5.11, the influence of the horizontal web reinforcement is ignored; on the other hand, the vertical web reinforcement is accounted for by a vertical tie in the middle of the shear span. The numerical scheme for the analysis of the Type II fan-action is, in general, similar to that of the Type II arch-action. In order to start the calculation procedure, the designer has to assume the height of the tension ties T_1 and T_3, and the width of the compression strut C_5. In checking the nodes, only nodes 1 and 4 need to be checked since the size, horizontal dimension, of either node 2 or 3 is large enough to make it safe.

5.4 BOTTOM LOADED DEEP BEAMS

5.4.1 Example 5.4: Design of a Top and Bottom Loaded Deep Beam

It is required to determine the reinforcement for the simply supported beam shown in Figure 5.12a if subjected to a top and bottom load $q = 180$ kN/m. The concrete cylinder strength is $f_c' = 30$ MPa and the steel yield stress is $f_y = 420$ MPa. The beam width is 300 mm. The solution is given in the next steps, in which the beam's own weight is neglected.

Reactions:
With reference to Figure 5.12a,

$$R_A = R_B = (180 \times 2) \times (5.6/2) = 1008\,\text{kN}$$

Establish an STM:
This beam is subjected to a distributed load and the span is less than twice the depth; therefore, the entire beam is considered a D-region, that is, deep beam. The appropriate *STM* is shown in Figure 5.12b, in which the lower nodes are assumed to coincide with the centerlines of the supporting columns and are located at the expected centroid of reinforcement; in this example, the distance from the center of reinforcement to the edge of the beam is assumed to be equal to 100 mm. For this beam, the lever arm z is chosen within $0.6h - 0.7h$, where h is the beam height, but z should not be greater than $0.4l$, where l is the beam span, refer to the standard D-regions D_5 and D_6; $z = 2000$ mm. Thus, the geometry of the *STM* will be as illustrated in the figure.

With reference to Figure 5.12b,
The force in strut $C_1 = q \times (5.6/2) = 504.0$ kN
The angle $\theta_A = \tan^{-1}(2000/1400) = 55.0° > 30.0°$
The force in strut $C_3 = R_A/\sin\theta_A = 1230.4$ kN
The force in tie $T_1 = C_3 \cos\theta_A = 705.7$ kN
The force in strut $C_2 = T_1 = C_3 \cos\theta_A = 705.7$ kN
The force in tie $T_2 = q \times (5.6/2) = 504.0$ kN

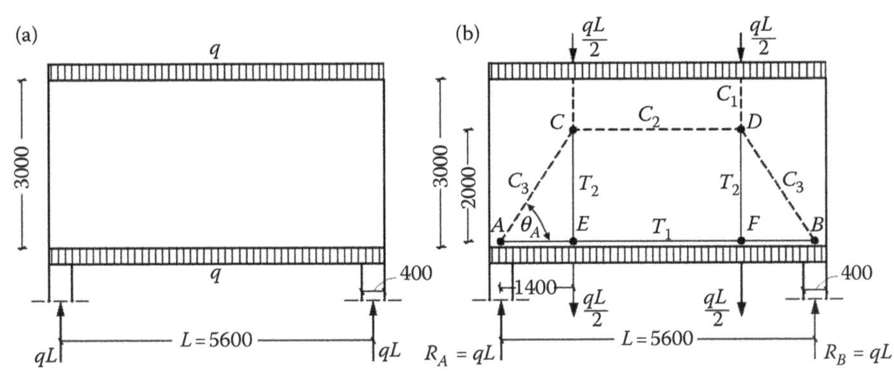

FIGURE 5.12 Example 5.4—deep beam under top and bottom distributed loads: (a) beam and loads; and (b) *STM*.

Effective concrete design strength of the struts:
For simplicity in the calculation procedure, the strength reduction factor, $\phi = 0.75$, is implemented in the design strength of the struts and the nodes, as illustrated next.

For strut C_3 $f_{ce}^s = \phi(0.85 f_c') \beta_s = 0.75 \times 0.85 \times 30 \times 0.75^* = 14.34\,\text{MPa}$

*Transverse reinforcement to resist the lateral tension will be provided since strut C_3 is a bottle-shaped stress field.

Struts C_1 and C_2 need not be checked since they are very wide.

Effective concrete design strength of the nodes:
Node A is a $C - C - T$ node; thus, $f_{ce}^n = \phi(0.85 f_c') \beta_n = 0.75 \times 0.85 \times 30 \times 0.8 = 15.30\,\text{MPa}$

Node C is very wide and therefore there is no need to check.

Node A:
The bearing stress $= (1008 \times 10^3)/(300 \times 400) = 8.40 MPa < f_{ce}^n$, which is O.K.

For the section at the interface between strut C_3 and the node, the design strength should be the smaller of the node strength (15.30 MPa) and the strut strength (14.34 MPa); thus, the strut strength will govern. The required width of strut C_3, $w_{C_3} = (1230.4 \times 10^3)/(300 \times 14.34) = 286.0\,\text{mm}$.

The bearing width of the node is 400 mm, which gives a width of the strut $= 400 \sin \theta_A = 327.7$ mm, without any consideration to the extended nodal zone which gives an even larger strut width. Thus, the node is safe.

Tie T_1:
The reinforcement required to resist the force of this tie is $(604.8 \times 10^3)/(0.75 \times 420) = 1920\,\text{mm}^2$, which can be covered with 8 bars of diameter 18 mm, 8ϕ18. This reinforcement should be extended in the node such that the anchorage length required to develop the force in the tie is satisfied.

Tie T_2:
The reinforcement required to resist the force of this tie is $(504.0\,\text{kN} \times 10^3)/(0.75 \times 420) = 1600\,\text{mm}^2$, which can be covered with vertical skin reinforcement of an area $0.57\,\text{mm}^2/\text{mm}$. Upon choosing vertical bars of diameter 12 mm every 200 mm on each side, the reinforcement of the tie and the vertical reinforcement of strut C_3 are covered.

Strut C_3:
The strut strength has been checked for stresses during the check of node A. Since the stress field of this strut is bottle-shaped, transverse reinforcement of the strut is required to resist a total force T_{C_3}, where $T_{C_3} = C_3/2 = 1230.4/2 = 615.2\,\text{kN}$. Thus, the total required reinforcement in perpendicular to the strut is $(615.2 \times 10^3)/(0.75 \times 420) = 1953\,\text{mm}^2$. Select a skin reinforcement of diameter 12 mm every 200 mm, both vertically and horizontally, on each side, equivalent to $1.13\,\text{mm}^2/\text{mm}$ vertically and horizontally. From the selected vertical reinforcement, $0.57\,\text{mm}^2/\text{mm}$ is used to hang the bottom load and the remaining $0.56\,\text{mm}^2/\text{mm}$ is utilized to

FIGURE 5.13 Bent cap.

reinforce the strut. Then the strut reinforcement will be $0.56 \times 1400 \times \sin\theta_A + 1.13 \times 2000 \times \cos\theta_A = 1938.5\ \text{mm}^2$, which is very close to the required reinforcement. Having in mind that the formula $T_{C_3} = C_3/2$ is conservative, the selected skin reinforcement is thus adequate.

5.4.2 Deep Beam with a Ledge

Inverted-T bent caps, that is, beams with ledges, are typically used to support incoming beams as shown in Figure 5.13. The use of beams with ledges allows a reduction in the overall height of the structure system, either in bridges or buildings. One of the main complications to the design of these beams is the behavior of the ledge. Traditionally, when designing the ledge of an inverted-T beam, or any other ledged member, there are five main types of failure that must be prevented, Figure 5.14:

1. Direct shear friction failure at the interface between the ledge and the web
2. Punching shear failure of the ledge at the point of loading

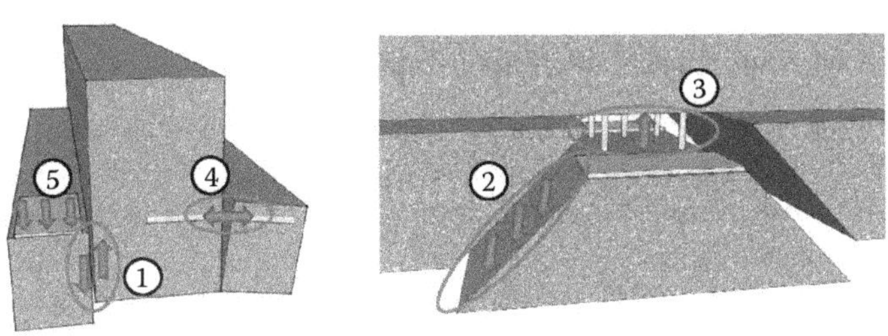

FIGURE 5.14 Cross section and longitudinal view of a beam ledge showing the different failure modes of the ledge (Garber et al., 2017): (1) shear friction, (2) punching shear, (3) yielding of hanger reinforcement, (4) ledge flexure, and (5) bearing capacity.

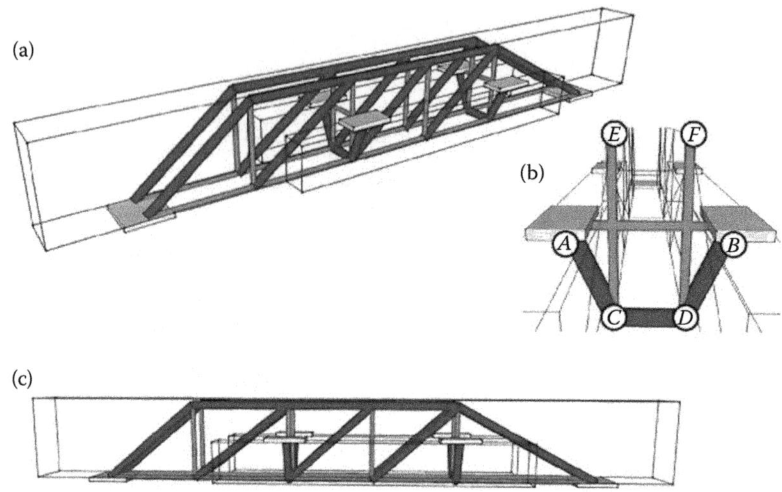

FIGURE 5.15 *STM* of a beam with a ledge (Garber et al., 2017): (a) three-dimensional strut-and-tie model for an inverted-T beam; (b) two-dimensional cross-sectional model; and (c) two-dimensional longitudinal model.

3. Failure of the hanger reinforcement (hanging the load up to the compression chord of the beam)
4. Flexural failure of the ledge reinforcement
5. Failure of concrete beneath the load point bearing

The *STM* can be used to design a ledge against these aforementioned failure mechanisms by proper modeling.

The use of a proper *STM* accounts for all of the possible failure mechanisms observed in practice. A 3D *STM* for an inverted-T deep beam is shown in Figure 5.15a. This 3D model can be broken down into two complimentary 2D models in the cross section and in the longitudinal directions, as shown in Figure 5.15b for the cross-section model and Figure 5.15c for the longitudinal model. Upon considering the safety of struts, ties, and nodes of the *STMs* including bearing strength and adequate anchorage, the design should lead to a safe solution.

5.5 DEEP BEAMS WITH INDIRECT SUPPORTS

Sometimes deep beams are supported by or support other deep beams, as in the case of the two examples in Figure 5.16. For the tee-shaped deep beam in Figure 5.16a, the beam ABC is supported at B by a rigid support and at C by the indirect support from the perpendicular beam DE. For the 3D structure in Figure 5.16b, the deep beam I is supported by the two beams II, whereas each of these latter beams is supported by a column and beam III. For illustration of the design procedure of such systems, the two examples in Figure 5.16 are discussed next.

For the Tee-shaped deep beam in Figure 5.16a, the statical system is shown in Figure 5.17a and the applied loads and reactions are illustrated in Figure 5.17b.

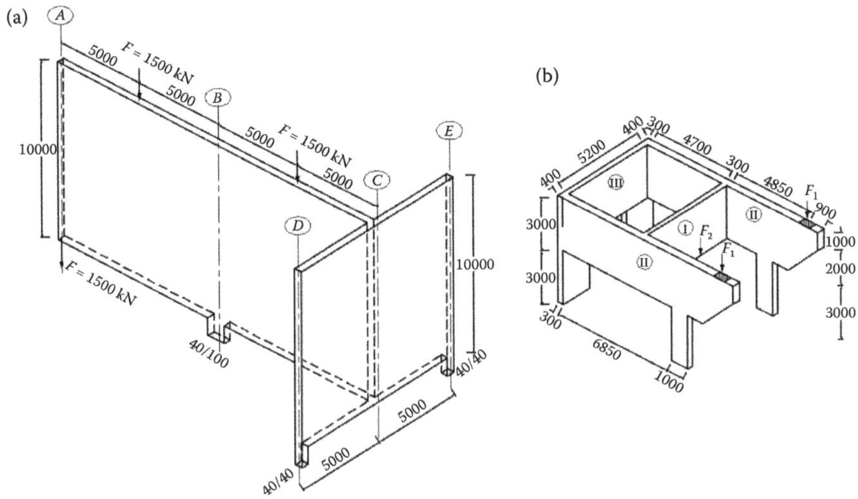

FIGURE 5.16 Examples of deep beams with indirect supports: (a) tee-shaped deep beam and (b) 3D structure. (Adapted from Schlaich, J. and Schäfer, K., The design of structural concrete, IABSE Workshop, New Delhi, 1993.)

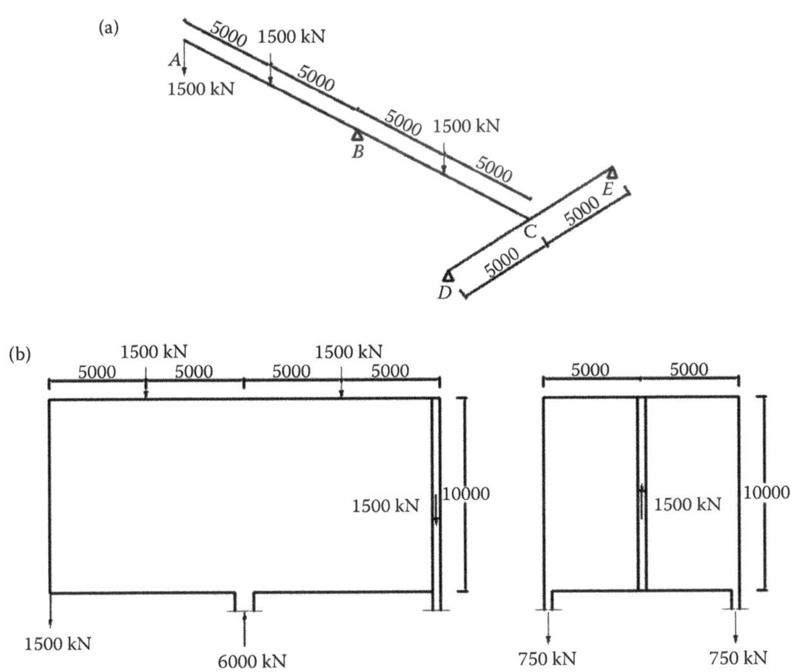

FIGURE 5.17 Statical system, loads, and reactions of the tee-shaped deep beam of Figure 5.16a: (a) statical system and loads, and (b) reactions. (Adapted from Schlaich, J. and Schäfer, K., The design of structural concrete, IABSE Workshop, New Delhi, 1993.)

In tracing the load path, Figure 5.18a, the shear forces along axis C are assumed to be uniformly distributed over the depth of the deep beam. The corresponding *STM* is given in Figure 5.18b.

For the 3D structure in Figure 5.16b, beam I is bottom loaded at the middle of the span and is indirectly supported by the two beams II. If the shear forces along the connection between beams I and II are assumed to be uniformly distributed over the depth of the beams, the load path and the corresponding *STM* will be as illustrated in Figure 5.19a. Thus, the orthogonal web reinforcement is utilized in the load transfer. Alternatively, the shear forces resultant can be assumed to act near the bottom where the bottom tension reinforcement may be extended, as shown in Figure 5.19b, leading to the *STM* in the figure. However, this model is associated with difficulty

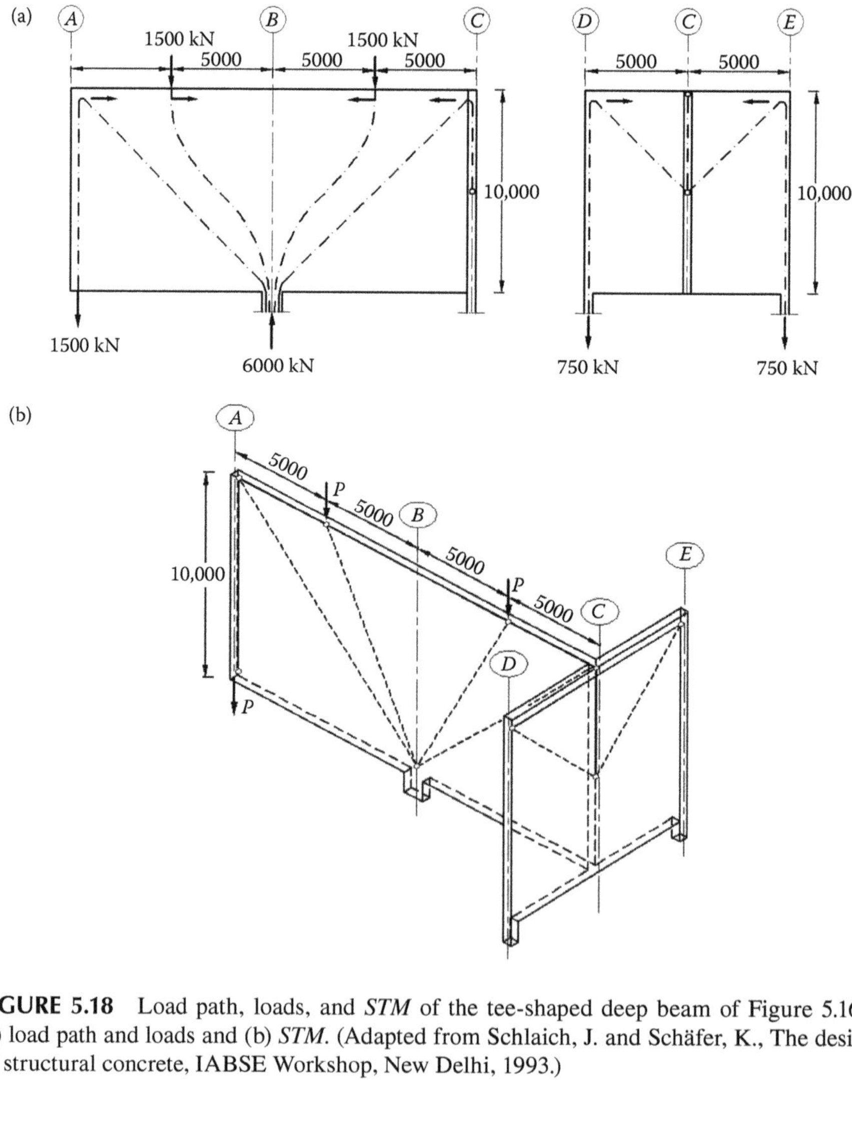

FIGURE 5.18 Load path, loads, and *STM* of the tee-shaped deep beam of Figure 5.16a: (a) load path and loads and (b) *STM*. (Adapted from Schlaich, J. and Schäfer, K., The design of structural concrete, IABSE Workshop, New Delhi, 1993.)

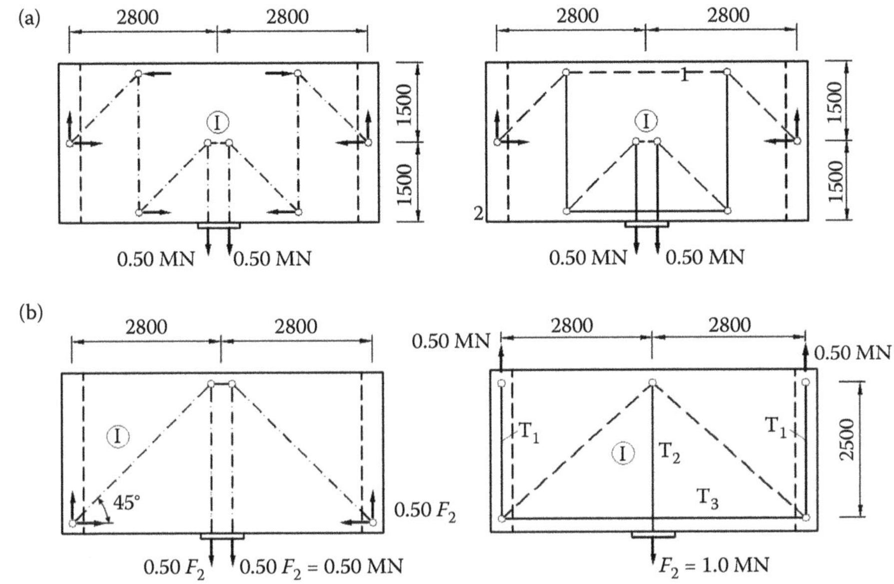

FIGURE 5.19 Deep beam I of the 3D structure of Figure 5.16b: (a) bearing forces, loads, load path and corresponding *STM*, (b) alternative load path, and corresponding *STM*. (Adapted from Schlaich, J. and Schäfer, K., The design of structural concrete, IABSE Workshop, New Delhi, 1993.)

in anchoring the reinforcement at the lower nodes, leading to the need for anchoring plates. The solutions for deep beams II and III of this three-dimensional structure are given in Figures 5.20 and 5.21.

5.6 APPLICATIONS TO CONTINUOUS DEEP BEAMS

5.6.1 Example 5.5: Strength Assessment of Top Loaded Beam Using Type I Model

The nominal strength of the NSC continuous deep beam (CDB1) subjected to a single top point load in every span, Figure 5.22a, tested by Ashour (1997) is assessed in this example. The beam height: $h = 625$ mm, depth, $d = 585$ mm, breadth, $b = 120$ mm, the width of the bearing plates $b_1 = 120$ mm, and $b_2 = b_3 = 250$ mm. The shear span, $a = 660$ mm. The shear span-to-depth ratio, $a/d = 660/585 = 1.12$. The tension steel, $A_{s1} = 4\phi12$ mm and $A_{s2} = 4\phi12 + 2\phi10$. The concrete cylinder strength is $f_c' = 30.6$ MPa and the steel yield stress is $f_y = 480$ MPa. The solution is given in the next steps.

Geometrical parameters:
The solution procedure starts with assuming some geometrical parameters. Since the reinforcement detailing allows the use of an extended nodal zone, the height of node 1, w_{T_1},

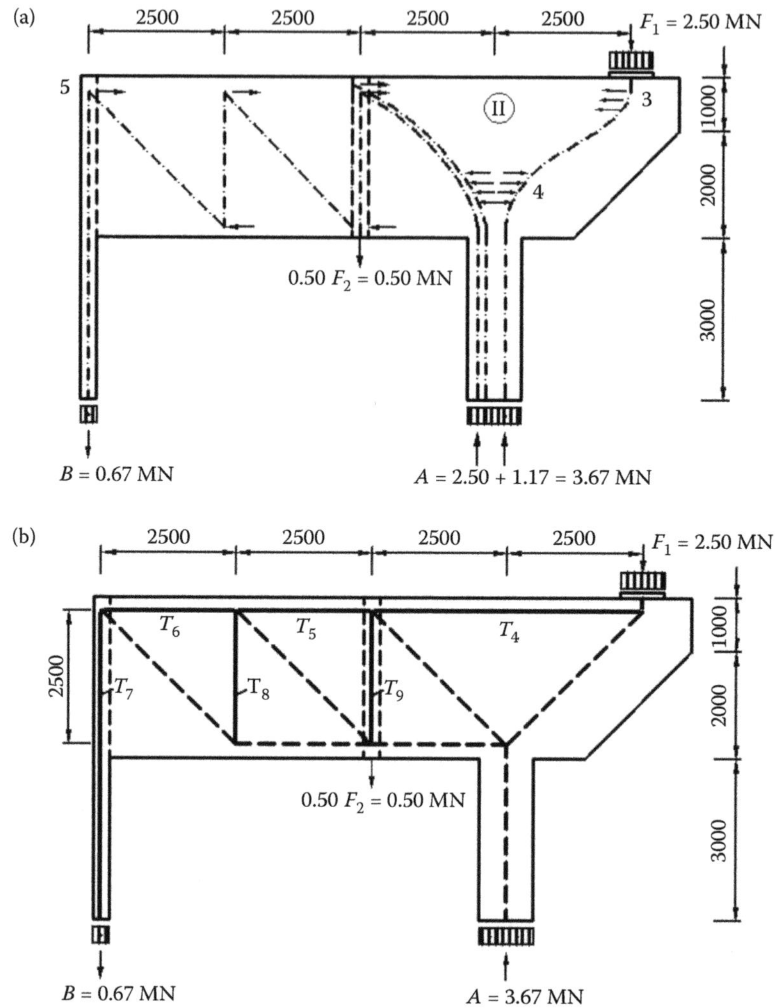

FIGURE 5.20 Bearing forces, loads, and load path and *STM* of deep beam II of the 3D structure of Figure 5.16b: (a) bearing forces, loads, and load path and (b) *STM*. (Adapted from Schlaich, J. and Schäfer, K., The design of structural concrete, IABSE Workshop, New Delhi, 1993.)

$$w_{T_1} = 2(t-d) = 2(625-585) = 80\,\text{mm}$$

or

$$w_{T_1} = 2(c+\phi_{stirrup}) + n\phi_{bars} + (n-1)s$$

The height of the tie in the upper node, node 2, w_{T_2},

$$w_{T_2} = 2(c+\phi_{stirrup}) + n\phi_{bars} + (n-1)s = 2(10+8) + (2\times12+1\times10) + 2\times10 = 90\,\text{mm}$$

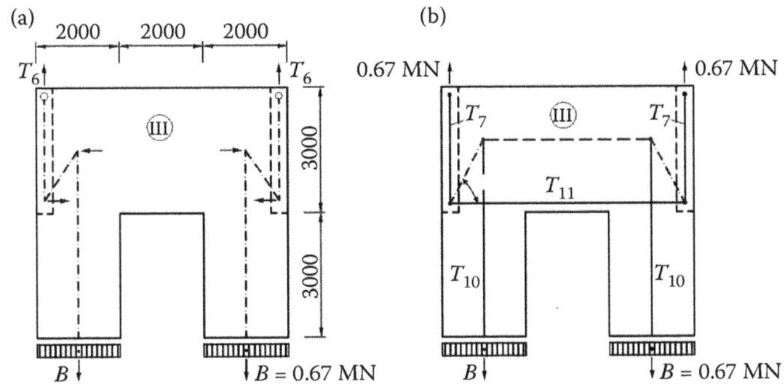

FIGURE 5.21 Load path and *STM* of deep beam III of the 3D structure of Figure 5.16b (Schlaich and Schäfer, 1993): (a) bearing forces, loads and load path; and (b) *STM*.

Then the lever arm L_d is

$$L_d \approx d - 0.5 \times w_{T_2} = 585 - 0.5 \times 90 = 540\,\text{mm}$$

The angle

$$\alpha \approx \tan^{-1} \frac{L_d}{a - 0.25 b_2} = \tan^{-1} \frac{540}{660 - 0.25 \times 250} = 42.11° > 30.0°$$

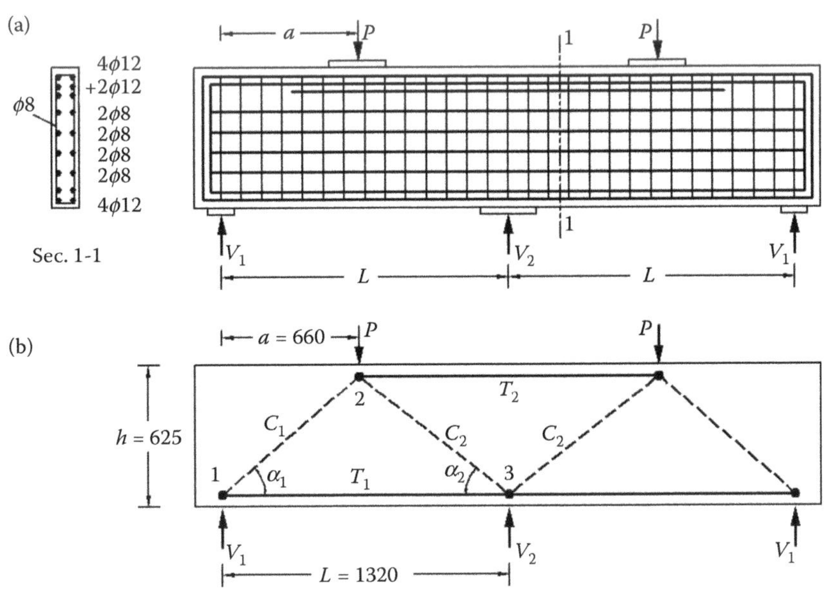

FIGURE 5.22 Details of *STM* for top loaded continuous deep beam CDB1 subjected to single point load: (a) beam, (b) *STM*. *(Continued)*

(c)

(d)

(e)

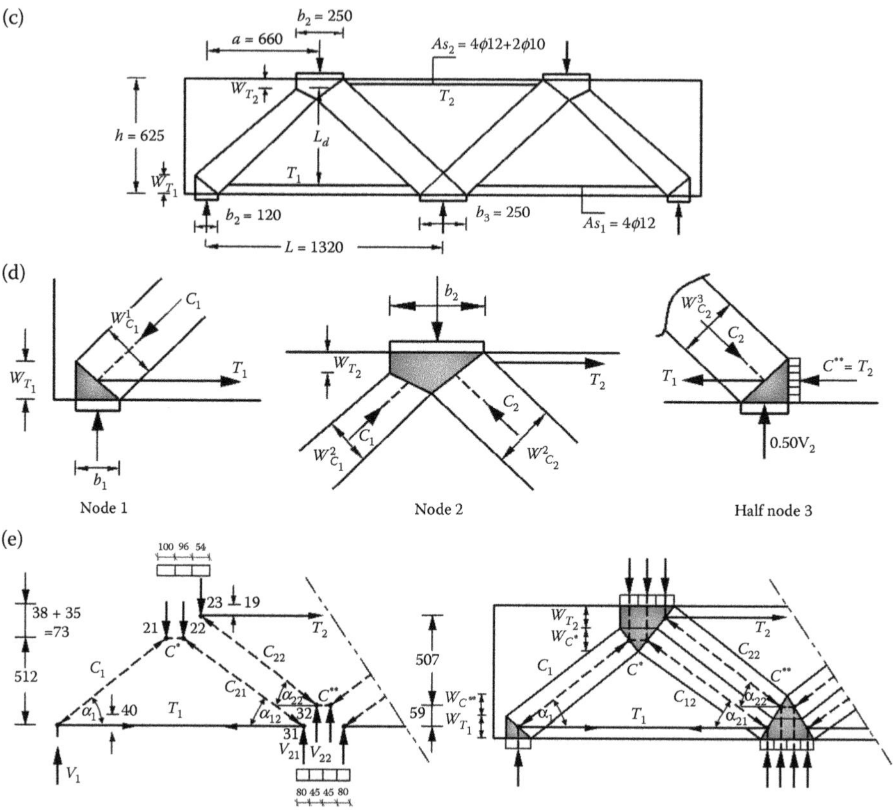

FIGURE 5.22 (Continued) Details of *STM* for top loaded continuous deep beam CDB1 subjected to a single point load: (c) details of the model, (d) details of nodes 1, 2, and half node 3, and (e) details of the detailed model.

The angle

$$\alpha_2 \approx \tan^{-1}\frac{L_d}{660-0.25\times(b_2+b_3)} = \tan^{-1}\frac{540}{660-0.25\times(250+250)} = 45.27^\circ > 30.0^\circ$$

Reactions:
From linear elastic analysis,

$$V_1 = 0.313P$$

$$V_2 = 1.375P$$

Model geometry and forces:
The model of the beam is shown in Figure 5.22b. Assuming that the tension reinforcement A_{s1} of tie T_1 and A_{s2} of tie T_2 reach their yield stress,

$$T_1 = 480 \times 4 \times 113 = 217\,\text{kN}$$

$$T_2 = 480 \times (4 \times 113 + 2 \times 78.5) = 292.4\,\text{kN}$$

The strut force $C_1 = T_1/\cos\alpha_1 = 292.5$ kN

The reaction $V_1 = C_1 \sin\alpha_1 = 196.1$ kN

Since $V_1 = 0.313P$ from linear elastic analysis, the load P is then $P = V_1/0.313 = 626.7$ kN

Then $V_2 = 1.375P = 861.7$ kN

From equilibrium of node 3, the strut force $C_2 = 0.5V_2/\sin\alpha_2 = 606.4$ kN

From equilibrium of node 2, the tie force $T_2 = C_2\cos\alpha_2 - C_1\cos\alpha_1 = 209.8$ kN, which is less than the previously calculated value assuming yielding of the tie reinforcement.

Effective concrete strength of the struts:
The effective concrete strength of a strut, $f_{ce}^{s} = 0.85 f_c' \beta_s$

For a prismatic strut, $f_{ce}^{s} = 0.85 \times 30.6 \times 1.0 = 26.01\,\text{MPa}$

For a bottle-shaped strut with sufficient reinforcement to resist the transverse tension, $f_{ce}^{s} = 0.85 \times 30.6 \times 0.75 = 19.51\,\text{MPa}$

Effective concrete strength of the nodes:
The effective concrete strength of a node, $f_{ce}^{n} = 0.85 f_c' \beta_n$

For the $C - C - C$ node, $f_{ce}^{n} = 0.85 \times 30.6 \times 1.0 = 26.01\,\text{MPa}$

For the $C - C - T$ node, $f_{ce}^{n} = 0.85 \times 30.6 \times 0.8 = 20.81\,\text{MPa}$

Check the bearing of the nodes:
For node 1, the nominal value of the reaction, $V_{1n} = 20.81 \times 120 \times 120 = 299.7$ kN, which is greater than the force V_1, $V_1 = 196.1$ kN.

For node 2, the nominal value of the load $P_n = 20.81 \times 120 \times 250 = 624.3$ kN, which is close to the force P, $P = 626.7$ kN.

For node 3, the nominal value of the reaction, $V_{2n} = 20.81 \times 120 \times 250 = 624.3$ kN, which is less than the force V_2 . $V_2 = 861.7$ kN. Therefore, the force V_2 should be reduced to the value of V_{2n}, that is, $V_2 = 624.3$ kN. Hence, the load P is then reduced to $P = V_1 + 0.5V_2 = 508.3$ kN and the strut force C_2 is subsequently reduced to $C_2 = 439.4$ kN. The force in tie T_2 is further reduced to $T_2 = 92.2$ kN.

Refinement of the model geometry and forces:
With reference to Figure 5.22c, the width of the tension tie T_2, w_{T_2}, at node 2 should not exceed a value $w_{T_2\,\max} = T_2/\left(f_{ce}^{n} b\right) = 92.2 \times 10^3/(20.81 \times 120) = 36.9\,\text{mm}$; therefore, consider $w_{T_2} \approx 38\,\text{mm}$. The width of the prismatic strut C^*, w_{C^*}, can be obtained as $C^* = T_1 = 217kN = f_{ce}^{s} \times w_{C^*} \times b = 26.01 \times w_{C^*} \times 120$, giving $w_{C^*} = 69.52$ mm ≈ 70 mm.

The strut C_2 is split into two components C_{21} and C_{22}, with C_{21} equilibrating the force C_1 and a part of the load P at node 2 and equilibrating the force T_1 and a part of the reaction V_2 at node 3. The force C_{22} equilibrates the force T_2 and a part of the load P at node 2 and equilibrates the force C^{**} and a part of the reaction V_2 at node 3. Subsequently, the vertical force V_2 is split into two vertical components V_{21} and V_{22}. This will lead to the detailed *STM* shown in Figure 5.22e, with geometrical relations and forces calculated upon applying equilibrium at the nodes. For instance, for obtaining C_{21} and α_{21}, we start by assuming $\alpha_{21} \approx \alpha_2$; then, from horizontal equilibrium at the nodes, C_{21} can be calculated. Next, from vertical equilibrium, V_{21} is calculated, followed by the assessment of the center of V_{21} at nodes 2 and 3. Next, a better estimate of α_{21} can be obtained followed by calculating C_{21}, V_{21}, and the center of V_{21} at nodes 2 and 3, until convergence takes place. Then the component V_{22} and the center of V_{22} at nodes 2 and 3 can be determined, followed by estimating α_{22}, T_2, and C_{22}.

Finally, the following forces and geometrical parameters are obtained.

$\alpha_1 = 41.19°$, $\alpha_{21} = 42.85°$, $\alpha_{22} = 43.22°$, $V_1 = 189.9$ kN, $V_{21} = 201.3$ kN, $V_{22} = 110.9$ kN, $T_1 = C^* = 217.0$ kN, $T_2 = C^{**} = 118.0$ kN, $C_1 = 288.4$ kN, $C_{21} = 296.0$ kN, and $C_{22} = 161.9$ kN.

Check of Stresses:

Node 1:

Since the bearing stress has been checked before, there is no need to check it again. For strut C_1, $w_{C_1}^1 = 120\sin\alpha_1 + 80\cos\alpha_1 = 139.2\,\text{mm}$. Then the nominal strength of the strut is $C_{1n} = 19.51^{\#} \times 120 \times 139.2 = 326.0$ kN (#the smaller of the node strength and the strut strength), which is greater than the strut force. A check of the web reinforcement will be carried out later for all struts.

Node 2:

Since the bearing stress has been checked before, there is no need to check it again.

For sub-node 21, the width of strut C_1, $w_{C_1}^{21} = 100\sin\alpha_1 + 70\cos\alpha_1 = 118.5\,\text{mm}$. Then the nominal strength of the strut is $C_{1n} = 19.51 \times 120 \times 118.5 = 277.5$ kN, which is less than the strut force, $C_{1n} = 96\% C_1$. There is no need to check the prismatic strut C^* since this step was considered during the estimate of the height of this sub-node.

For sub-node 22, the width of strut C_{21}, $w_{C_{21}}^{22} = 96\sin\alpha_{21} + 70\cos\alpha_{21} = 116.6\,\text{mm}$. Then the nominal strength of the strut is $C_{21n} = 19.51 \times 120 \times 116.6 = 273.0$ kN, which is less than the strut force, $C_{21n} = 0.92C_{21}$.

For sub-node 23, the width of strut C_{22}, $w_{C_{22}}^{23} = 54\sin\alpha_{22} + 38\cos\alpha_{22} = 64.7\,\text{mm}$. Then the nominal strength of the strut is $C_{22n} = 19.51 \times 120 \times 64.7 = 151.5$ kN, which is less than the strut force, $C_{22n} = 0.94C_{22}$.

Node 3:

For sub-node 31, the width of strut C_{21}, $w_{C_{21}}^{31} = 80\sin\alpha_{21} + 80\cos\alpha_{21} = 113.1\,\text{mm}$. Then the nominal strength of the strut is $C_{21n} = 19.51 \times 120 \times 113.1 = 264.7$ kN, which is less than the strut force, $C_{21n} = 0.89C_{21}$.

For sub-node 32, the width of strut C_{22}, $w_{C_{22}}^{32} = 45\sin\alpha_{22} + 38\cos\alpha_{22} = 58.5\,\text{mm}$. Then the nominal strength of the strut is $C_{22n} = 19.51 \times 120 \times 58.5 = 137.0$ kN, which is less than the strut force, $C_{22n} = 0.85C_{22}$.

The load $P = V_1 + V_{21} + V_{22} = 502.1$ kN, which is less than the nominal value $P_n = 624.3$ kN. Also, all bearing stresses are safe. From the obtained results the critical members are C_1, C_{21}, and C_{22}, while the other members attain nominal strength greater than their forces. Since all the bearing stresses are safe, the values of V_1, V_{21} and V_{22} should be reduced according to the nominal strength of C_1, C_{21} and C_{22}, respectively. This gives $V_1 = 0.96 \times 189.9 = 182.3$ kN, $V_{21} = 0.89 \times 201.3 = 179.2$ kN and $V_{22} = 0.85 \times 110.9 = 94.3$ kN. This leads to a value of $P = 455.8$ kN; i.e., the nominal strength from the *STM*, $2P = 911.6$ kN. Since the measured collapse load was $2P = 1100.00$ kN; then the *STM* solution is 83% of the measured value in the test.

Upon examining the previous calculations, it is noted that the critical members, which control the capacity, have different strength at their ends. Hence, in order to improve the solution results, new geometries of the nodes may be assumed in order to have the strength of each critical member at its two ends closer, then performing the calculation steps once more.

5.6.2 Type I Model for a Bottom Loaded Beam

The indirectly or bottom loaded continuous deep beams behave in a manner different from that of top loaded beams. Indirect application of loads changes the mechanism of load transfer, the failure mode, and the role of reinforcement in continuous deep beams. Bottom loaded continuous deep beams exhibit lower ultimate loads than that for top loaded beams. Introductory bars and hence compression struts C_2 in Figure 5.23 are needed to transmit the external bottom load to the top of the model.

Figure 5.23 shows a continuous deep beam (two bays) with two point bottom loads at nodes N_3 and N_4 along with the proposed *STM*. The model has three compression struts, C_1 to C_3, and five tension ties, T_1 to T_5. An additional concrete strut C is introduced for equilibrium of the model forces; in addition, it reflects the beam continuity which increases the reaction of the interior support and reduces the reactions of the exterior supports in comparison with the simple beam solution. The main bottom longitudinal reinforcement is represented by the tension ties T_1 and T_3 and the main top longitudinal reinforcement is represented by the tension tie T_2. The introductory bars

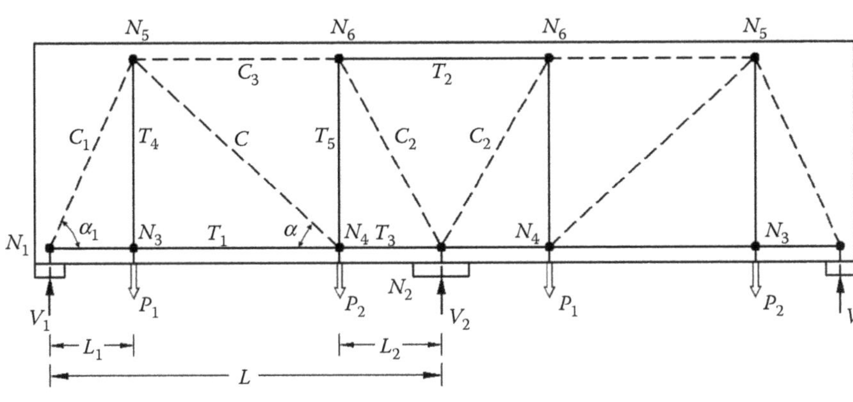

FIGURE 5.23 *STM* for a bottom loaded continuous deep beam.

are represented by the tension ties T_4 and T_5. The load is transferred indirectly from the points of load application at nodes N_3 and N_4 by these introductory bars to the external and the intermediate supports through the two struts C_1 and C_2, respectively.

5.7 BRACKETS AND CORBELS

5.7.1 Modes of Failure

Brackets, corbels, and beam ledges are widely used short cantilever members having a short shear span, and therefore they are deep structural elements. Hence, shear plays a key role in controlling the overall behavior of these elements, leaving corbels with their own particular modes of deformation and modes of failure.

Niedenhoff (Van Mier, 1987) carried out full-scale tests in order to obtain some idea about the stress distribution in reinforced concrete corbels. One example of Niedenhoff tests is illustrated in Figure 5.24. Hardly any stresses developed in the lower outside corner of the corbel, and it was found that the resultant force of the principal compressive stresses followed the diagonal from the loading point to the lower inner corner of the corbel, Figure 5.24. Tensile stresses developed in perpendicular to this diagonal.

One specimen of the tests on corbels carried out by Zeller (1983) is shown in Figure 5.25. In this test, the deterioration of the compression strength of the concrete rather than the yielding of the main tension reinforcement led to the failure of the corbel. The crack pattern depicts quite well the internal flow of forces, Figure 5.25b, which is illustrated by the *STM* in Figure 5.25d.

There are four possible failure modes of brackets and corbels, which should be controlled, Figure 5.26:

1. Direct shear friction failure at the interface between the corbel and the supporting member

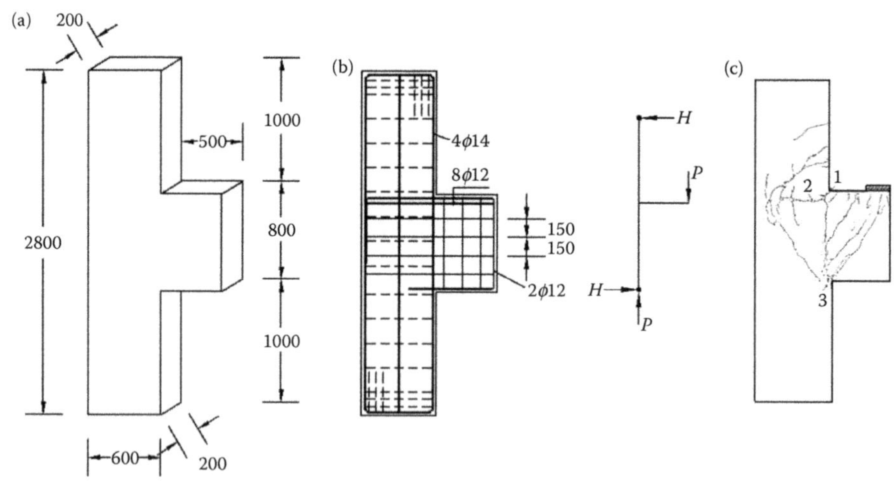

FIGURE 5.24 Niedenhoff experiment: (a) dimensions (mm), (b) reinforcement, and (c) crack pattern at failure.

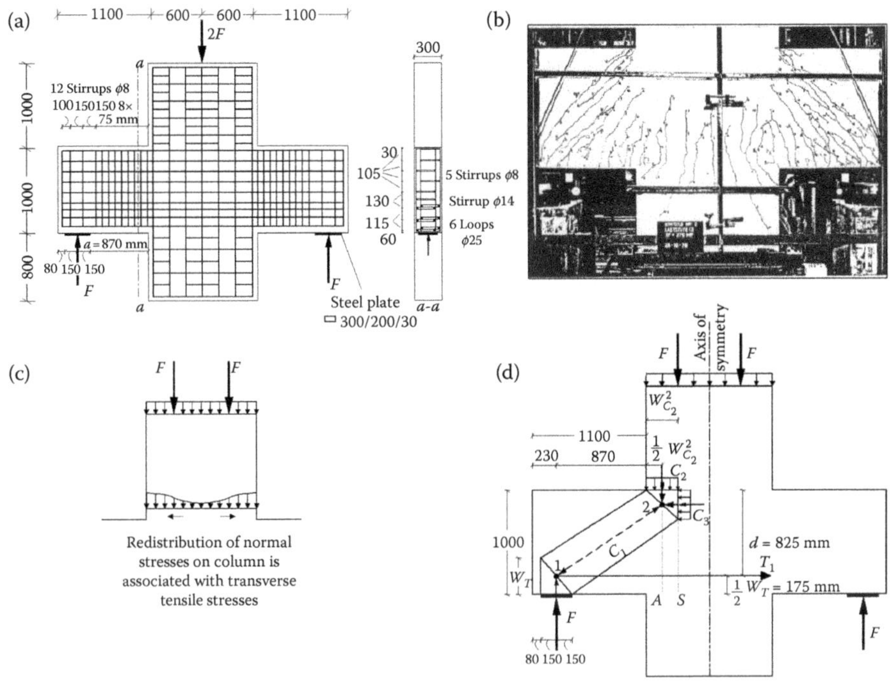

FIGURE 5.25 Zeller test: (a) test specimen, (b) crack pattern at $F = 1350$ kN (near failure), (c) stress distribution on column support, and (d) *STM* at failure ($F_u = 1425$ kN).

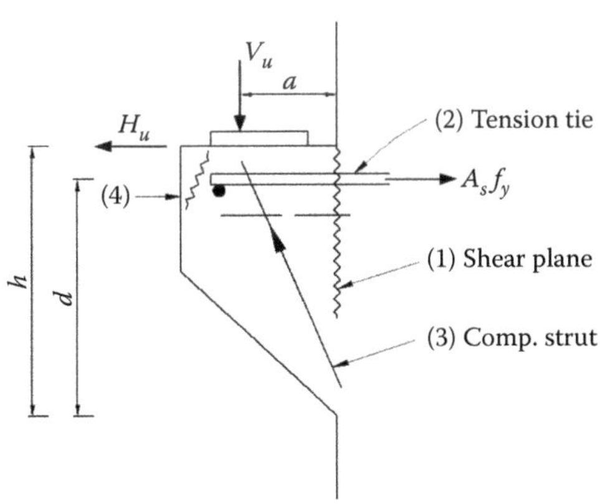

FIGURE 5.26 Failure modes of corbels.

2. Yielding of the tension reinforcement due to moment and direct tension
3. Crushing of the internal compression strut
4. Localized bearing or shear failure under the loading area

With an appropriate strut-and-tie modeling, the previous modes of failure can be controlled.

5.7.2 Strut-and-Tie Modeling

Figure 5.27 illustrates the discontinuity regions, shaded, for different corbel problems. In order to arrive at a safe design of corbels, the boundary conditions of each problem should be individually considered. Thus, each case leads to a different *STM*, Figure 5.28.

For illustration, an *STM* of a corbel subjected to the boundary conditions shown in Figure 5.29a is drawn in detail in the figure. Upon following the load path of the force C_1, it is realized that this force at node 2, Figure 5.29a, should be balanced by the vertical force C_2 and the horizontal force C_3. The force C_2 is a part of the compression stress block associated with the column moment and vertical reaction at node 2. The horizontal strut C_3 represents the horizontal stress block associated with the corbel moment at section S – S or section A – A, Figure 5.29a and f. Strut C_4, Figure 5.29a, is a part of the compression stress block required to balance the moment and the compression force applied to the column at the horizontal level of nodes 2 and 3, Figure 5.29g.

5.7.3 Nodes Detailing for Safety

In brackets and corbels, the distance between the load and the supporting face is usually too short; hence, special anchorage should be provided at the outer end of the tension reinforcement. With reference to Figure 5.29a, the anchoring of the main reinforcement (tie T_1) at node 1 could have different options: bending the reinforcement back to form a horizontal loop; a structural weld to a transverse bar of equal size (or welding to an armor angle) across the ends of the tension reinforcement beyond the edge of the loaded area; or using free anchoring, Figure 5.30.

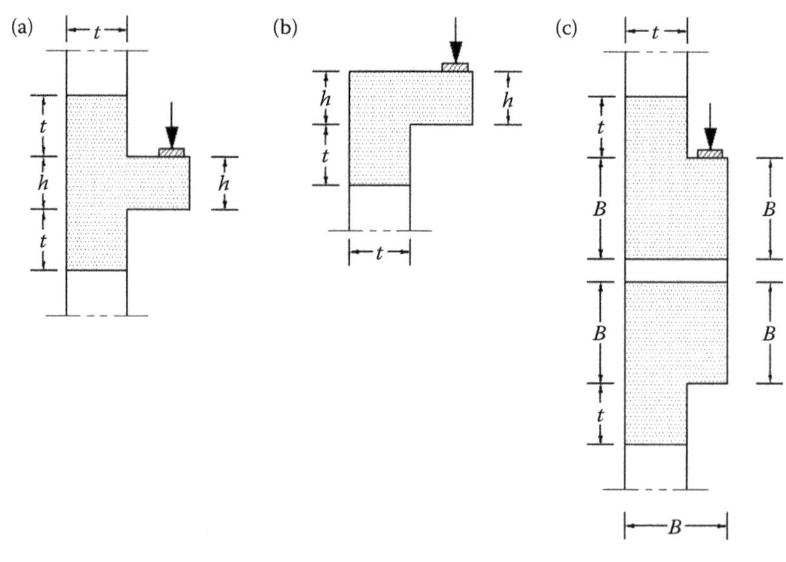

FIGURE 5.27 D-regions in examples of corbel problems.

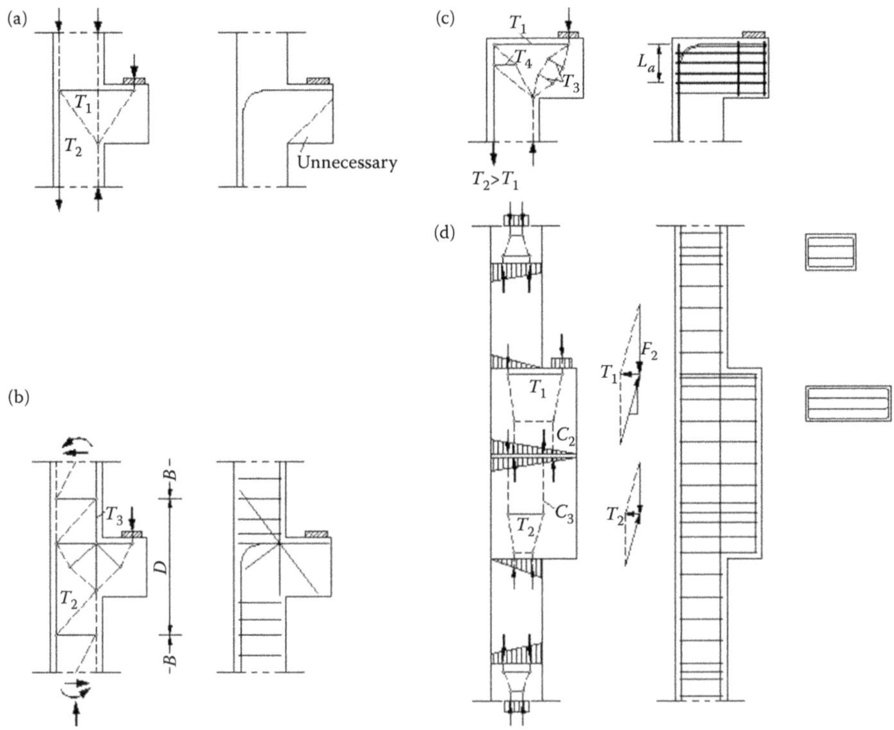

FIGURE 5.28 Different support and geometric conditions lead to different *STMs* and different reinforcement arrangements of corbels. (Adapted from Schlaich, J. and Schäfer, K., The design of structural concrete, IABSE Workshop, New Delhi, 1993.)

The use of loops results in a confining effect for the concrete within the zone of node 1, and therefore their use, rather than the use of free anchoring, is preferred. Also, the curvature of the loop results in more uniform horizontal pressure in comparison with welding the reinforcement to a transverse bar where the horizontal pressure will be essentially concentrated because of the deviation of forces associated with the right angle, and therefore the use of the loop is again preferred. If anchorage is provided by a hook or a loop, the load should not project beyond the straight portion of the hook or loop, Figure 5.30a. In case a horizontal force is expected, the bearing plate should be welded to the tension reinforcement.

The thickness of the compression stress block ($C_2 + C_4$) and the thicknesses of struts C_3 and C_5 at node 3 can be based on a value of effectiveness factor equal to $\beta_s = 1.00$ even if the tie T_4 exists, since the stirrups of T_4 are distributed over some height and enclose nodes 2, 3, and 5, Figure 5.29a, which means that the concrete between nodes 2 and 3 will generally be under a state of biaxial compressive stresses. As a result of using the same design strength for all struts at node 3, the location of this node can be determined from the thickness of strut C_4 and it will not be altered by the design of strut C_5, Figure 5.29a and g.

The detailing of node 4 should be based on a value of effectiveness factor $\beta_n = 0.80$, and the full reinforcement of tie T_1 should be curved at node 4 with the appropriate

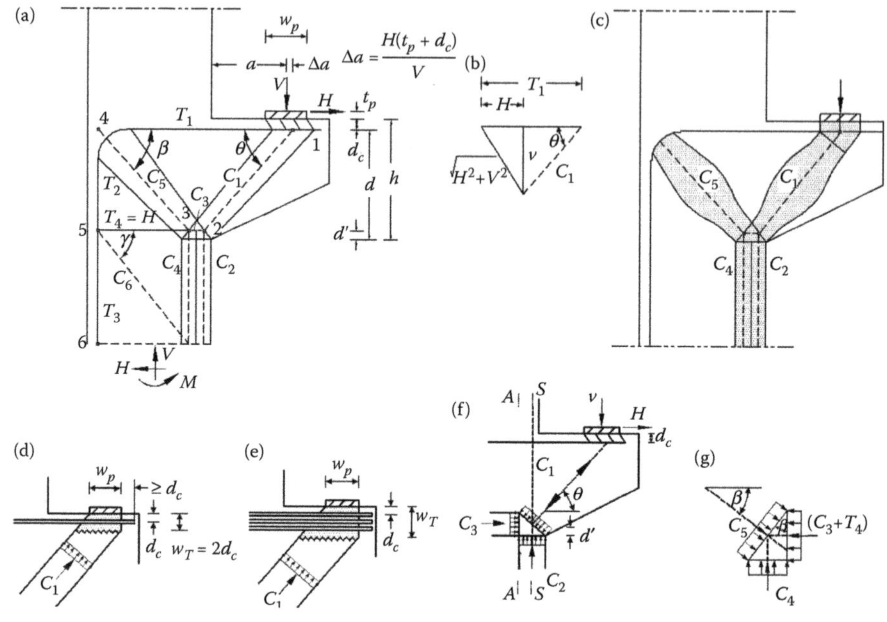

FIGURE 5.29 Details of *STM* of corbel: (a) strut-and-tie model; (b) equilibrium of node 1, (c) stress fields of struts C_1 and C_5, (d) increasing the size for a single-layer reinforcement node, (e) multi-layered reinforcement node, (f) state of equilibrium of node 2, and (g) state of equilibrium of node 3.

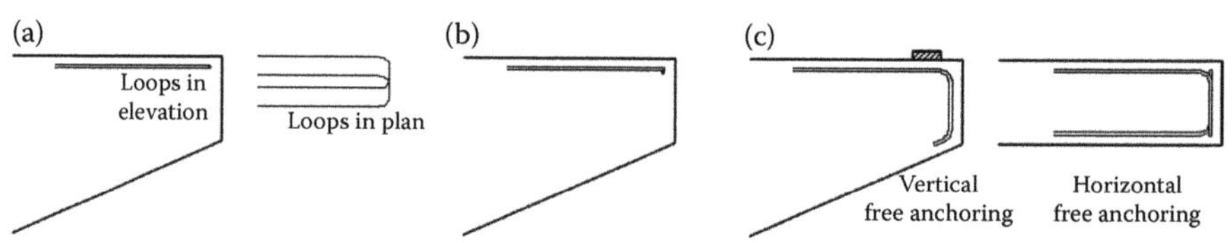

FIGURE 5.30 Possible anchorage options of reinforcement at node 1: (a) horizontal loops, (b) structural weld to a transverse bar, and (c) free anchoring.

radius of curvature, r_4. The difference between the reinforcement of T_1 and the reinforcement of T_2 can be terminated beyond node 4 provided that it is adequately anchored, with its anchorage length, as specified by the code, measured from the start of the reinforcement curvature. The radius of curvature r4 should justify the required width of strut C_5 at node 4, $w_{C_5}^4$, which can be derived from equilibrium.

$$r_4 = 0.7 w_{C_5}^4 (1 + 3\tan|45^\circ - \beta|) \text{ for } 60^\circ \geq \beta \geq 30^\circ$$

If the angle β does not lie within the range indicated in the previous equation, the corbel model should be revised. Of course, r_4 should not be less than what is allowed by the Code.

Node 1 is checked for a value of effectiveness factor $\beta_n = 0.80$ and the bearing plate is sized such that the bearing strength is justified. The size of the node depends on the node detailing, Figure 5.29d and e.

For the case of multilayered reinforcement, the geometry of the node will differ from that of a single-layer reinforcement (Schlaich and Schäfer, 1991), Figure 5.29e; therefore, the design of this node has to be revised for the respective geometry. In this case

$$w_{C_1}^{1} = w_p \sin\theta + w_T \cos\theta$$

where w_T is the height of node 1 as shown in Figure 5.29e, provided that the value of w_T when substituted for into the previous equation should be at most equal to $0.2h$, where h is the corbel height. Also, for the case of a single-layer reinforcement, Figure 5.29d, w_T could be taken as equal to $2d_c$, provided that the reinforcement is extended a distance $\geq 2d_c$ beyond the node as in Figure 5.29d, where d_c is the thickness of the concrete cover measured from the extreme tension fiber to the center of the reinforcement.

5.7.4 Step-by-step Design Procedure

With reference to Figure 5.29a and realizing that the design output may be essentially controlled by node 1, it is realistic to start with the determination of the angle θ which is prescribed by the locations of nodes 1 and 2. The location of node 1 is decided if the concrete cover, d_c, is assumed, whereas the location of node 2 can be determined by the lines of action of either struts C_1 and C_2 or struts C_2 and C_3. It is more convenient to deal with struts C_2 and C_3, as will be illustrated later, in positioning node 2.

With reference to Figure 5.29, the design procedure of corbels can be systematically achieved as follows.

1. From the equilibrium of node 2, Figure 5.29f, $C_2 = C_1 \sin\theta = V$; then the width of strut C_2 at node 2, $w_{C_2}^{2}$, can be determined independently of the value of the angle θ and based on an efficiency factor $\beta_s = 1.0$; thus, the line of action of C_2 can be determined.
2. From the geometry of node 1, the shift of the force V due to the horizontal force H at the level of node 1, Δa, for the case of a single layer of reinforcement, Figure 5.29a, is

$$\Delta a = \frac{H(t_p + d_c)}{V}$$

and for the case of multilayered reinforcement, Figure 5.29a and e,

$$\Delta a = \frac{H\left(t_p + 0.5w_T\right)}{V}$$

where t_p is the thickness of the bearing plate.

3. Assume a trial value of the corbel depth, d, unless the total height of the corbel, h, is dictated by another means; then the concrete cover d_c can be assumed.
4. Determine the moment M_{S-S} at section $S - S$ caused by the load V (in its shifted position due to the horizontal force H).

$$M_{S-S} = V\left(0.5w_{C_2}^2 + a + \Delta a\right)$$

5. Calculate the width of the compression block of M_{S-S}, $w_{C_3}^2$, which is the width of strut C_3. Then the force of strut C_3 can be obtained.

$$C_3 = \frac{M_{S-S}}{d - 0.5w_{C_3}^2}$$

   With this, the location of node 2 is determined.

6. Determine the angle θ from

$$\tan\theta = \frac{d - 0.5w_{C_3}^2}{0.5w_{C_2}^2 + a + \Delta a}$$

7. Calculate the tension force T_1 from

$$T_1 = (V / \tan\theta) + H$$

   Then choose the necessary reinforcement. Check if the assumed value of d is satisfactory and if the reinforcement is single-layer; otherwise, assume a new value of d and go to either step 2 or step 3.

8. Calculate C_1 from the equilibrium of node 1, $C_1 = V/\sin\theta$; then calculate the width of strut C_1 at node 1, $w_{C_1}^1$.
9. Calculate the width of the bearing plate, w_p, from the appropriate equation. Check if the width of the bearing plate is possible (or acceptable); otherwise, revise the design. Calculate the vertical stress at node 1, σ_v^1, and choose the other dimension of the bearing plate to satisfy the allowable bearing stresses.
10. Calculate the thickness of the compression block ($C_2 + C_4$) required to balance the moment and the compression force applied to the column cross section at the horizontal level of node 2; then, by subtracting the force C_2 from the compression block, the value and thereafter the line of action of C_4 can be determined.
11. Calculate the angle β; then calculate the force in strut C_5, $C_5 = T_1/\cos\beta$.
12. Determine the required width of strut C_5 at node 4, $w_{C_5}^4$; then detail the node such that this obtained strut width is justified.

13. Calculate the angle γ and the forces in strut C_6 and the ties T_2 to T_4; then design these ties. Check the strength of strut C_6, realizing that the thickness of the strut is from the middle distances between nodes 4 and 5 and nodes 5 and 6.
14. Determine the horizontal and vertical stirrups of the corbel as illustrated in the next section.

5.7.5 Transverse Reinforcement of Struts

In principle, the stirrup reinforcement should be placed in perpendicular to the concrete strut; nevertheless, for practicality and convenience, this reinforcement could be placed either vertically and/or horizontally. The confinement or stirrups of struts C_1 and C_5, a_{stC_1}, and a_{stC_5} are determined from,

$$a_{stC_1} = \frac{C_1/2}{\phi f_y}$$

$$a_{stC_5} = \frac{C_5/2}{\phi f_y}$$

This reinforcement can be covered by vertical and horizontal reinforcement, a_{st}^{v} and a_{st}^{h}, according to the following equations,

$$a_{stC_1} = \left(a_{st}^{v} \cos^2\theta + a_{st}^{h} \sin^2\theta\right) l_{C_1}$$

$$a_{stC_5} = \left(a_{st}^{v} \cos^2\beta + a_{st}^{h} \sin^2\beta\right) l_{C_5}$$

where a_{st}^{v} (a_{st}^{h}) is the area of selected stirrup reinforcement, to be placed vertically (horizontally), per unit length measured horizontally (vertically), and l_{C_1} and l_{C_5} are the lengths of struts C_1 and C_5, respectively.

The amount of the required stirrup reinforcement, if placed either horizontally or vertically, will be mostly governed by the reinforcement requirement from strut C_1. Therefore, if the angle $\theta \leq 45°$, the stirrups are more effective if placed vertically; otherwise, they will be more effective if placed horizontally. It should be stated, however, that if the kinematics of strut C_1 is considered in more detail, it will be realized that vertical stirrups are more beneficial than horizontal stirrups even if the angle θ is slightly greater than 45°. In practice, stirrups are placed both horizontally and vertically, and therefore for saving of reinforcement a larger percentage of the more effective reinforcement (horizontal or vertical) should be considered. The stirrups should be closed for more efficient anchorage; furthermore, closing the stirrups results in a beneficial confining effect which improves the ductility of the corbel.

As for the reinforcement of tie T_4, which exists only if the horizontal force H is present, it should be closed stirrups and centered with nodes 2 and 5.

5.7.6 Example 5.6: Strength Assessment of Double Corbel

The corbel tested by Zeller (1983), Figure 5.25, is analyzed in this section. In this example, compression failure took place at the loading node of the compression strut at a failure load, $F = 1425$ kN (the strain in the tension reinforcement was significantly below the yield strain). The concrete cylinder strength was $f_c' = 26.3\,\text{MPa}$ and the steel yield stress was $f_y = 452.0$ MPa. The breadth of the corbel was $b = 300$ mm, while the dimensions of the bearing plate were 300 mm width and 200 mm breadth. Thus, the limiting value of bearing stress given by ACI 318-14, in this case, is; $f_b = 0.85 \times 26.3 \times 0.8 \times \sqrt{400 \times 300/300 \times 200} = 25.30\,\text{MPa}$. The bearing stress at failure was $\sigma_b = (1425 \times 10^3)/(300 \times 200) = 23.75$ MPa, which is less than the limiting value of bearing stress given by the ACI 318-14.

In this example, the compressive stresses on the column due to the force $2F$, if assumed uniform, will be equal to 7.9 MPa, which is much less than the concrete strength. This allows for shifting the node location as close as possible to the knee of the corbel in order to develop the maximum possible carrying capacity. Shifting the node for maximum strength position would require nonuniform normal stress distribution on the column section as the section location becomes closer to node 2, Figure 5.25c. The redistribution of normal stresses is possible as long as the transverse tension associated with this redistribution is within the transverse tensile strength of the column. Upon examining the tensile strength of the column using an *STM*, it is found that an equivalent compression block of an intensity equal to $0.85 f_c' = 22.36\,\text{MPa}$ is possible for idealizing the normal stress distribution in the column near node 2, strut C_2, as shown in the appropriate *STM* of the corbel in Figure 5.25d.

In this example, node 1 is treated as an extended nodal zone since the reinforcement at this node is horizontal loops with sufficient projection beyond the bearing plate. From the reinforcement layout in Figure 5.25, the height $w_T = 350$ mm; hence, $d = h - 0.5w_T = 825$ mm. The width of strut C_2 at node 2, $w_{C_2}^2 = F_u/0.85 f_c' b = 212.4\,\text{mm}$. The moment at section $S - S$, $M_{S-S} = F_u\left(870 + 0.5 w_{C_2}^2\right) = 1391\,\text{MN.mm}$, which requires a compression stress block of depth $w_{C_3}^2 = 309.3\,\text{mm}$; this is the width of strut C_3. Then the angle of inclination of strut C_1, $\theta = \tan^{-1}\left[\left(d - 0.5 w_{C_3}^2\right)/\left(870 + 0.5 w_{C_2}^2\right)\right] = 34.48°$. Upon imposing this angle on Figure 5.25b of the corbel under consideration, it is noticeable that the value of θ fits reasonably with the recorded cracking pattern at failure.

The next step in the verification process is to check the strength of the compression strut C_1 at nodes 1 and 2. From the equilibrium of node 1, the force in strut $C_1 = V/\sin\theta = 2517.14$ kN. From the geometry of nodal zone 1, the width of the strut $w_{C_1}^1 = w_p \sin\theta + w_T \cos\theta = 458.63\,\text{mm}$. Then the stress in strut C_1 at node 1 is $\sigma_{C_1}^1 = 18.3\,\text{MPa}$. The strength limit in this case is the smaller of the node strength and strut strength; the node strength is $0.85 f_c' \times 0.8 = 17.89\,\text{MPa}$ and the strut strength is $0.85 f_c' \times 0.75 = 16.77\,\text{MPa}$; assuming the adequacy of transverse reinforcement of the strut, this latter value of strength governs. Thus, the nominal strut strength is 2307.4 kN, which is 92% of the strut force, the same as the ratio (16.77/18.30).

From the geometry of nodal zone 2, the width of strut C_1, $w_{C_1}^2 = w_{C_2}^2 \sin\theta + w_{C_3}^2 \cos\theta = 375.21\,\text{mm}$. The strength of the strut at this node is based on the smaller of the node strength, 22.36 MPa, and the strut strength, 16.77 MPa; this latter value governs. Then the nominal strut strength is 1887.7 kN, which is 75% of the strut force.

From the equilibrium of node 1, the tension force in the tie $T = (V/\tan\theta) = 2074.94$ kN; hence, the steel stress = 353.3 MPa, corresponding to a steel strain = 1.77×10^{-3} (less than the yield strain as detected in the experiment).

In conclusion, the predicted strength of the corbel using the *STM* is 75% of the recorded value. However, the calculations can be refined in order to obtain a better estimate of the corbel capacity.

The improvement of the strength prediction can be approached upon noting that the strength of strut C_1, controlling the capacity, is weaker at node 2 than at node 1, due to the small size of the strut at node 2. The strength of node 2 is 22.36 MPa while the strut strength is 16.77 MPa; hence, the strength of node 2 should be revised. This can be achieved by using an average value of the node and strut strengths (19.6 MPa) in the assessment of the node height ($w_{C_3}^2$), in other words increasing the node size. Upon using this assumed value and redoing the calculations, the following results are obtained.

The width of strut C_2, $w_{C_2}^2 = 242.4\,\text{mm}$. The moment $M_{S-S} = 1412.42$ MN.mm. The width of strut C_3, $w_{C_3}^2 = 377.6\,\text{mm}$. The angle $\theta = 32.7°$. For node 1, the force in strut $C_1 = 2638.0$ kN, and the width of the strut, $w_{C_1}^1 = 456.6\,\text{mm}$. The nominal strength of the strut at this node, $C_{1n} = 2297.2$ kN. For node 2, the width of strut C_1, $w_{C_1}^2 = 448.7\,\text{mm}$; thus, the nominal strength of the strut at this node, $C_{1n} = 2257.3$ kN. Then the nominal strength of C_1 is the smaller of the obtained two values; i.e., $C_{1n} = 2257.3$ kN, which is 85.6% of the force in the strut. As for the tie *T*, the tension force $T = 2219.63$ kN; hence, the steel stress = 377.9 MPa (less than the yield stress as detected in the experiment). In a final conclusion, the accuracy of the prediction is 85.6%.

The transverse reinforcement required to resist the transverse tension of strut C_1 can be assessed according to ACI 318-14 since the concrete cylinder strength $f_c' = 26.3\,\text{MPa} < 44.0\,\text{MPa}$. The actual stirrups consist of a horizontal reinforcement of an area = 1438 mm^2 and a vertical reinforcement of an area = 1207 mm^2. Upon calculation, $\Sigma \dfrac{A_{si}}{bs_i} \sin\gamma_i = 0.0072 > 0.003$, which means the adequacy of the transverse reinforcement, as assumed in the solution.

REFERENCES

ACI 318-14, *Building Code Requirements for Structural Concrete and Commentary*, Detroit: American Concrete Institute, USA, 2014, 519 pp.

Ashour, A. F., Tests of reinforced concrete continuous deep beams, *ACI Structural Journal*, 94(1), 1997, 3–12.

El-Shora, A. T., Design and detailing of deep beams, *PhD Thesis*, Mansoura University, Egypt, 2005.

El-Zoughiby, M. E., El-Metwally, S. E., El-Shora, A. T., and Agieb, E. E., Strength prediction of simply supported R/C deep beams using the strut-and-Tie method, *The Arabian Journal for Science and Engineering*, 38-8, 2013, 1973–1991.

El-Zoughiby, M. E., El-Metwally, S. E., El-Shora, A. T., and Agieb, E. E., Strength prediction of continuous R/C deep beams using the strut-and-Tie method, *The Arabian Journal for Science and Engineering*, 39-3, 2014, 1685–1699.

Foster, S. J. and Gilbert, R. I., Experimental studies on high-strength concrete deep beams, *ACI Structural Journal*, 95(4), 1998, 382–390.

Garber, D. B., Varney, N. L., Gómez, E. F., and Bayrak, O., Performance of ledges in inverted-T beams, *ACI Structural Journal*, 114(2), March-April, 2017, 487–498.

Schlaich, J. and Schäfer, K., Design and detailing of structural concrete using strut-and-Tie models, *The Structural Engineer*, 96(6), 1991, 113–125.

Schlaich, J. and Schäfer, K., The design of structural concrete, IABSE Workshop, New Delhi, 1993.

Schlaich, J., Schäfer, K., and Jennewein, M., Toward a consistent design of structural concrete, *Journal of the Prestressed Concrete Institute*, 32(3), 1987, 74–150.

Smith, K. N. and Vantsiotis, A. S., Shear strength of deep beams, *ACI Structural Journal*, 79(3), 1982, 201–213.

Tan, K. H., Weng, L. W., and Teng, S., A strut-and-Tie model for deep beams subjected to combined top-and-bottom loading, *The Structural Engineer*, 75(13), 1997, 215–222.

Van Mier, J., Examples of nonlinear analysis of reinforced structures with DIANA, *HERON*, 32(3), *Delft University of Technology, Delft, Netherlands*, 1987, 147.

Zeller, W., Bruchversuche an Stahlbetonkonsolen bei Veraenderung des Bewehrungsgrades, (Failure Tests on Reinforced Concrete Corbels with Different Reinforcement Ratios), Abschlussbericht zum Forschungsvorhaben, Institut f. Massivbau und Baustofftechnologie, Universitaet Karlsruhe, 1983.

6 Openings in Shallow and Deep Beams

6.1 INTRODUCTION

Inserting openings in the web of a reinforced concrete beam is not only associated with a sudden change in the dimensions of its cross-section, but also with a concentration of stresses at the corners of the openings. In addition, openings may induce transverse cracks and reduce the stiffness of the beam leading to excessive deformations and considerable redistribution of forces. The effect of an opening on the flow of forces and hence the capacity of a beam depends largely on the size and location of the opening.

In shallow beams, an opening (circular, square, or nearly square in shape) can be considered large when its depth (or diameter) is greater than 40% the beam thickness (Mansur and Tan, 1999). The introduction of a large opening in a reinforced concrete beam would normally reduce its load-carrying capacity considerably. In practice, openings are located near supports where shear is predominant. In such a case, tests have shown that a beam with insufficient reinforcement and improper detailing around the opening region fails prematurely in a brittle manner.

This chapter presents a brief but comprehensive treatment of the design of reinforced concrete simple and continuous shallow and deep beams that contain transverse openings (small or large) through the web. The chapter starts with the treatment of shallow beams with small openings, where the term small opening is defined followed by the strut-and-tie modeling of such a beam. Next, shallow beams with large openings are treated, where the term large opening is defined and the appropriate location, size and reinforcement detailing of the opening zone are discussed. Strut-and-tie modeling of shallow beams with large circular or rectangular openings at different locations of the beam, within the span or depth, is discussed thoroughly. The appropriate locations of openings in deep beams, simply supported or continuous, are examined based on the flow of forces. The different schemes of modeling deep beams, simply supported or continuous, with web openings, small or large, with accounting for practical reinforcement detailing, are discussed.

6.2 SHALLOW BEAMS WITH SMALL OPENINGS

6.2.1 Small Openings

When the size of an opening is concerned, many researchers have used the terms small and large; nevertheless, the essence of such classification lies in the structural response of the beam. When the opening is small enough to maintain the beam-type behavior or, in other words, if the usual beam theory applies, then the opening may be termed as a small opening. In such a case, the beam action may be assumed to prevail. Therefore, the analysis and design of a shallow beam with small openings may follow the similar course of action as that of a solid beam. Small openings are thus defined as openings which are small enough and located in such a way that an *STM* is able to jump over the openings without causing additional vertical or horizontal struts in the chords above and below the openings (Schlaich and Schäfer, 1993). When beam-type behavior ceases to exist due to the presence of openings, the opening may be classified as a large opening. Thus, beams with small and large openings need separate treatments in design.

The assumption that an *STM* is able to pass by openings without any additional struts or ties in the bottom- or top-chord might be interpreted in this case as an indication of almost linear strain distribution in the cross-section of the opening, Figure 6.1a (Schlaich et al., 1987). In this case, the opening is considered small; otherwise, it is considered large, Figure 6.1b. In either case, the presence of openings produces discontinuities or disturbances in the normal flow of stresses, thus leading to stress concentration and early cracking around the opening region. Similar to any discontinuity, special reinforcement, enclosing the opening close to its periphery, should therefore be provided in sufficient quantity for crack control and to prevent a possible premature failure of the beam.

In beams, shear is always associated with a bending moment, except for the section at inflection point. When a small opening is introduced in a region subjected to predominant shear and the opening is enclosed by reinforcement, as shown by solid lines in Figure 6.2, the beam may fail in two distinctly different modes, beam-type

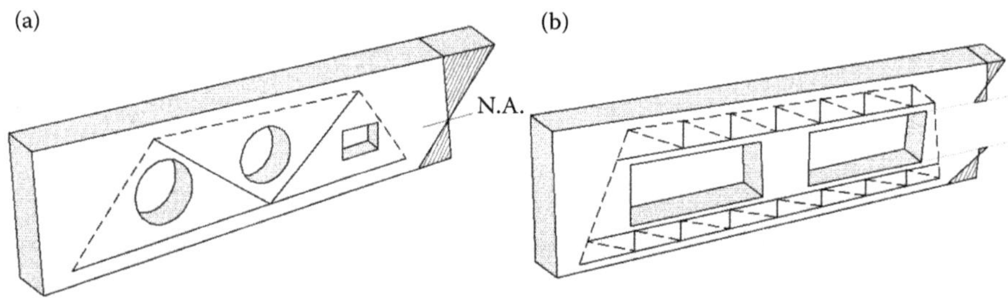

FIGURE 6.1 Definition of small and large openings. (Schlaich and Schäfer, 1993): (a) beam with small openings and (b) beam with large openings.

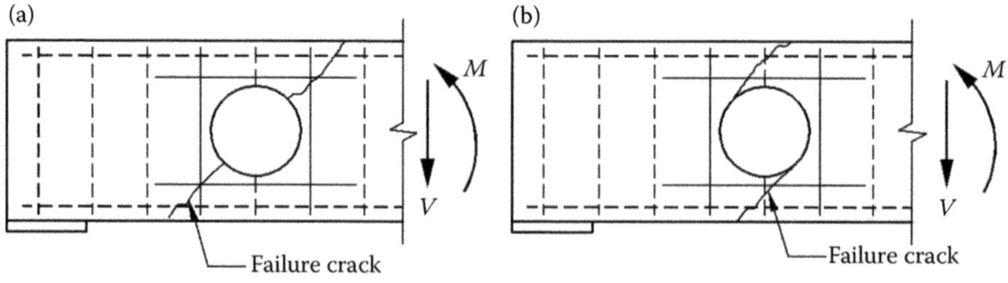

FIGURE 6.2 Failure modes of a beam with a small opening in a region of predominant shear: (a) beam-type shear failure and (b) frame-type shear failure.

failure and frame-type failure (Hanson, 1969; Somes and Corley, 1974; Salam, 1977; Tan et al., 2001). The first type is typical of the failure commonly observed in solid beams, except that the failure plane passes through the center of the opening, Figure 6.2a. In the second type, the formation of two independent diagonal cracks, one in each member bridging the two solid beam segments, leads to the failure, Figure 6.2b.

Tan et al. (2001) tested seven T-beams with circular web small openings which were designed for moderate to high shear force. They were tested in an inverted position to simulate the conditions that exist in the negative moment region of a continuous beam, Figure 6.3. The results of the test indicated that crack control and preservation of ultimate strength are achievable through providing reinforcement around the opening. It was found that diagonal bars reduce the high stresses in the compression chord and avoid premature crushing of the concrete. However, the test

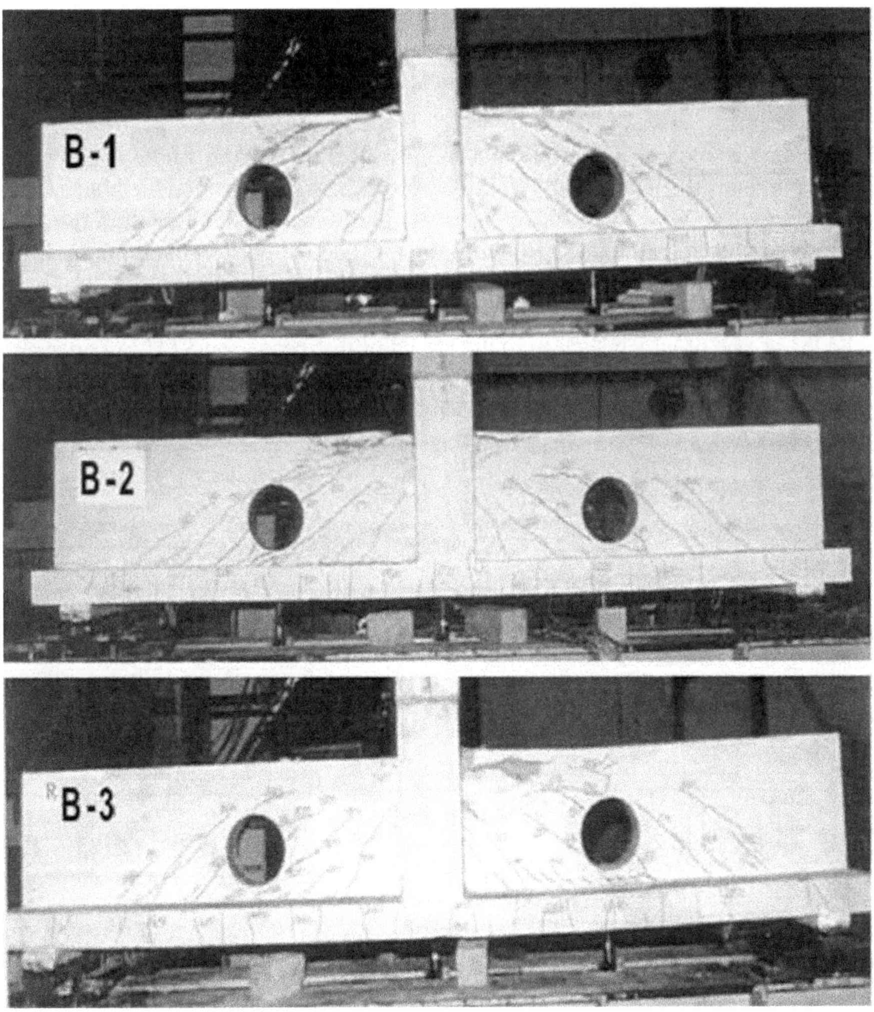

FIGURE 6.3 Cracking patterns of beams tested by Tan et al. (2001).

results illustrated that the provision of transverse openings alters the simple beam behavior into a more complex behavior.

6.2.2 Strut-and-Tie Modeling of Beams with Small Openings

Two parts of beams with two different patterns of small openings in the web are shown in Figure 6.4a and c (Schlaich and Schäfer, 1991). The truss model of the beam with the first pattern of openings, Figure 6.4a, shows how much shear reinforcement is necessary; the corresponding reinforcement layout is shown in Figure 6.4b. A check of the concrete stresses within the struts should be carried out based on a strength $\leq \phi(0.85 f_c' \beta_s)$, with $\beta_s = 0.6$. Thus, the maximum possible size of openings can be determined. The preceding statement also applies to the beam with the second pattern of openings, Figure 6.4c; the reinforcement layout of this beam is shown in Figure 6.4d.

It should be noted that the truss model clearly shows where the pattern of openings should be placed. Of course, in these examples, the standard shear design can never be applied.

For a single small circular hole in a shallow beam, $D < h/3$, where D is the hole diameter and h is the beam height, it is easy to trace an *STM* around the hole as illustrated in Figure 6.5 for three different cases. These include a hole within the beam span, a hole within the beam span but in the presence of concentrated load near the hole, and a hole close to the beam support.

The beam in Figure 6.6 with a small circular hole, $D < h/3$, carries two concentrated loads, one of which is close to the hole. For this beam, the *STM* can go around the hole as illustrated in the figure. In this model, the inclined ties around the hole can be carried by either inclined or equivalent vertical and horizontal reinforcement.

For shallow beams with small openings, a suitable reinforcement detail for the opening zone is illustrated in Figure 6.7.

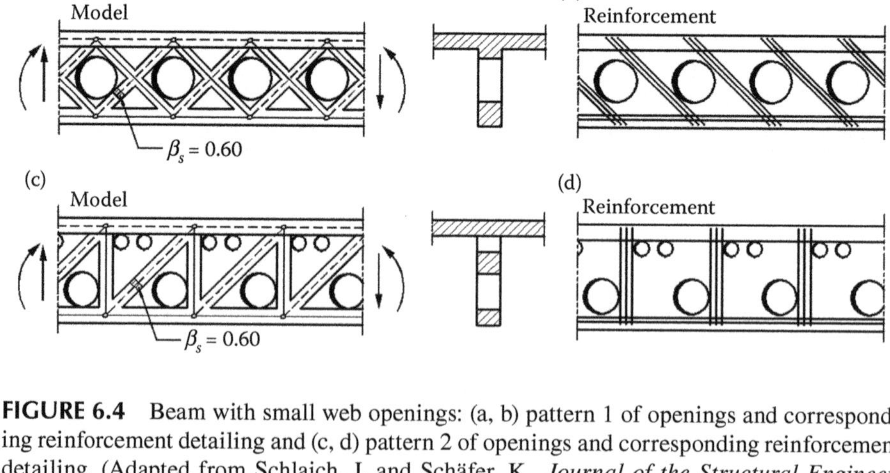

FIGURE 6.4 Beam with small web openings: (a, b) pattern 1 of openings and corresponding reinforcement detailing and (c, d) pattern 2 of openings and corresponding reinforcement detailing. (Adapted from Schlaich, J. and Schäfer, K., *Journal of the Structural Engineer*, 69(6), 1991, 113–125.)

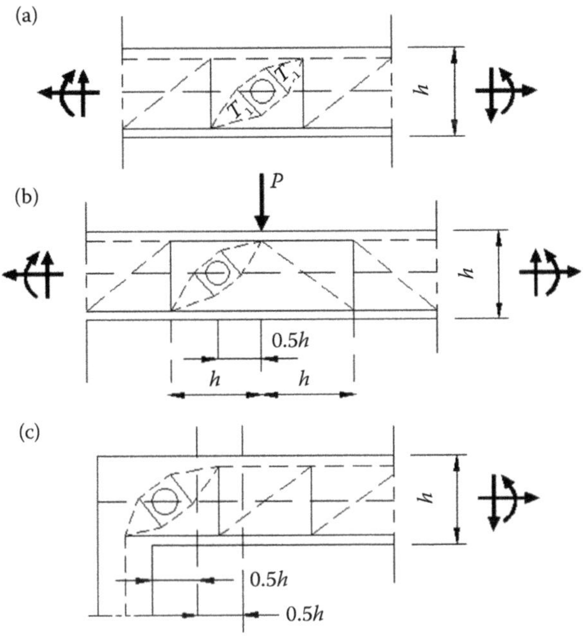

FIGURE 6.5 *STM* of a shallow beam with a small circular hole, $D < h/3$, for three different cases: (a) hole within the beam span, (b) hole within the beam span and in the presence of concentrated load near the hole, and (c) hole close to the support.

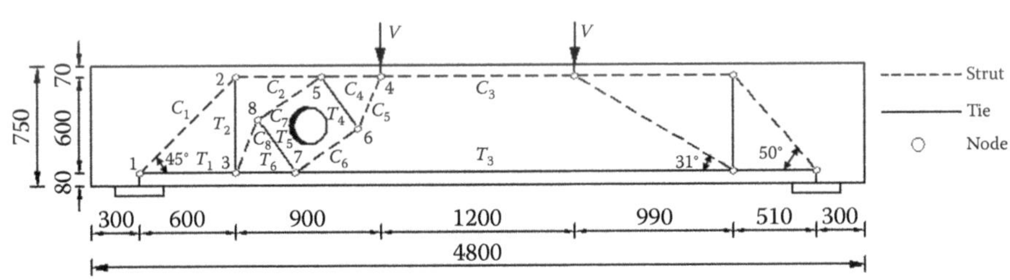

FIGURE 6.6 *STM* of a beam with a small circular hole, $D < h/3$.

6.3 SHALLOW BEAMS WITH LARGE OPENINGS

6.3.1 Large Openings

In analogy to a beam with small openings, large openings can be defined as an opening that requires additional vertical and horizontal struts in the chords above and below the opening (Schlaich and Schäfer, 1993). Bernoulli's hypothesis of plane strain distribution is invalid concerning the whole cross-section through a large opening as in Figure 6.1b.

Tan et al. (1996) tested 15 specimens (each simulating either the negative or positive moment regions of a reinforced concrete continuous T-beam) with large

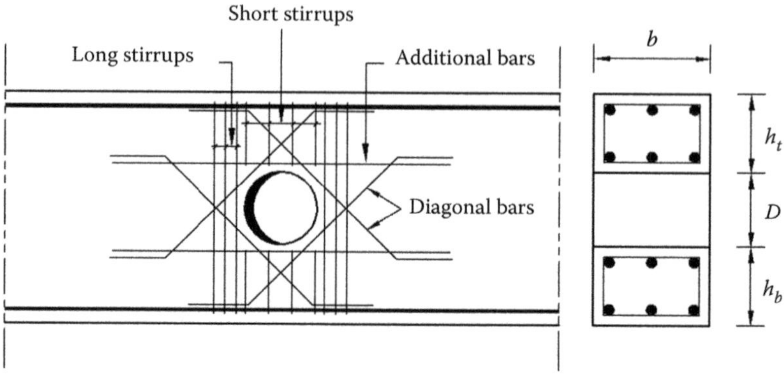

FIGURE 6.7 Reinforcement details around a small web opening.

FIGURE 6.8 Failure of a beam with multiple rectangular large openings. (Adapted from Tan, K. H. et al., *ACI Structural Journal*, 93(3), 1996, 404–411.)

openings through the web, to failure, Figure 6.8. The test results indicated that the presence of web openings leads to a reduction in both the cracking and ultimate strength as well as the post cracking stiffness. For the same passageway, beams with multiple openings were found to perform better in terms of strength and serviceability than those with a single opening. Test results confirmed the Vierendeel panel behavior at the opening segment of the beam. From their study, Tan et al. (1996) concluded as follows:

1. The width of the post between adjacent openings should not be less than one-half the overall beam depth, and the post should be adequately reinforced to avoid premature failure.
2. A continuous T-beam containing a large rectangular opening behaves similarly to a Vierendeel panel at the opening segment. Under combined bending and shear, the chord members bend in double curvature with contra-flexure points located approximately at their mid-span.
3. The total applied shear may be apportioned between the top and bottom chords in accordance with their flexure stiffnesses, based on either gross or

cracked transformed sections. This distribution applies at both the service load and ultimate conditions, irrespective of whether the opening is located within the positive or negative moment.

Considering continuous beams, reduction in stiffness due to the presence of openings causes a redistribution of internal forces and moments, the amount of which needs to be properly evaluated to achieve a satisfactory design.

From tests on unstrengthened large openings made in existing concrete beams, Siao and Yap (1990) indicated that the beams failed prematurely by the sudden formation of a diagonal crack in the compression chord when no additional reinforcement was provided to the members above and below the opening (chord members). The experiments showed that an increase in the opening size decreases the strength as well as the stiffness of the beam. However, the eccentricity of the opening has a minor influence on both the strength and the stiffness of the beam.

For shallow beams with large openings, a suitable reinforcement detail (consisting of additional longitudinal bars near the top and bottom faces of the bottom and top chords, respectively) to resist the combined effect of moment and axial force in each chord member should be designed (Mansur and Tan, 1999). Such reinforcement can be symmetrically detailed, with the addition of short stirrups in both chords, Figure 6.9a, to resist the forces carried by the chords. At each vertical edge of the opening (the critical section for cracking), a combination of vertical stirrups and diagonal bars should be used. At least 50% of the shear resistance is provided by the diagonal bars; thus, the chords behave in a manner similar to a Vierendeel panel and failure occurs in a ductile manner (Mansur and Tan, 1999), Figure 6.9b. Clearly, the failure mechanism consists of four hinges, one at each end of the top and bottom chords.

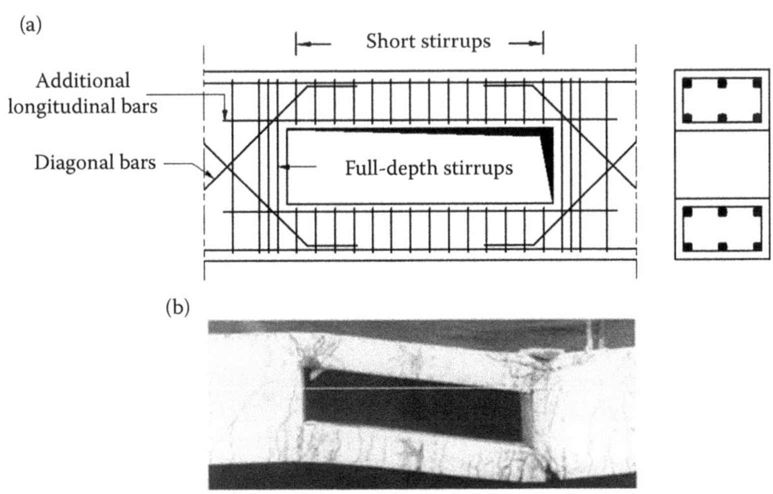

FIGURE 6.9 Reinforcement details around large openings in a shallow beam and expected failure mode: (a) suitable reinforcement detail and (b) corresponding failure mode.

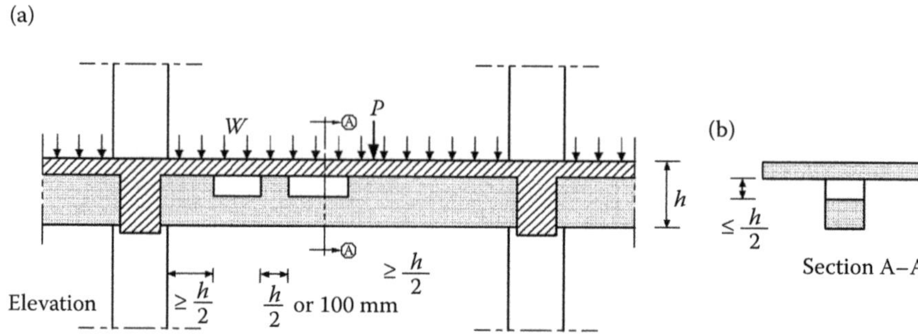

FIGURE 6.10 Size and location of web openings.

The following requirements (Mansur and Tan, 1999) can help to facilitate the selection of size and location of web openings in shallow beams, Figure 6.10:

1. For tee beams, openings should preferably be positioned flushed with the flange.
2. For rectangular beams, openings are commonly placed at mid-depth of the beam section, and they may be placed eccentrically if situation dictates.
3. For both tee- and rectangular beams, care must be exercised to provide sufficient concrete cover to the reinforcement for the chords above and below the opening. The compression chord should also have sufficient concrete area to develop the ultimate compression stress block in flexure as well as sufficient depth to provide effective shear strength.
4. Openings should not be located closer than $0.5h$ from the supports to avoid the critical region of shear and reinforcement congestion. Similarly, positioning of an opening closer than $0.5h$ to any concentrated load should be avoided.
5. The depth of an opening should be limited to $0.5h$.
6. The factors that limit the length of an opening are the stability of the compression chord member and the deflection requirement. It is preferable to use multiple openings providing the same passageway instead of using a single long opening. For multiple openings to behave independently, the width of the post separating two adjacent openings should not be less than $0.5h$ or 100 mm, whichever is larger.

6.3.2 Modeling

Modeling the shallow beam with an eccentric large opening in Figure 6.11a is illustrated by Schlaich et al. (1987) as explained in the following. Only the regions affected by the eccentric opening of the beam in Figure 6.11a, D_2, D_3, B_2, and B_3, are examined in this example with regard to the moment and shear force diagrams shown in the figure. The boundary forces of the B_1- and B_4-regions have to be applied as loads to the D_2- and D_3-regions as shown in Figure 6.11b. The B_1-region is treated by the standard truss model, $C_2 = (V_1/\sin\theta)$, $C_1 = (M_1/z) - (0.5V_1\cot\theta)$, and $T_1 = (M_1/z) + (0.5V_1\cot\theta)$, which yields $\theta = 31°$ and a lever arm $z = 1.15$ m, $T_1 = 3.92F_u$. The B_3-region obviously has a very small bending stiffness, and therefore

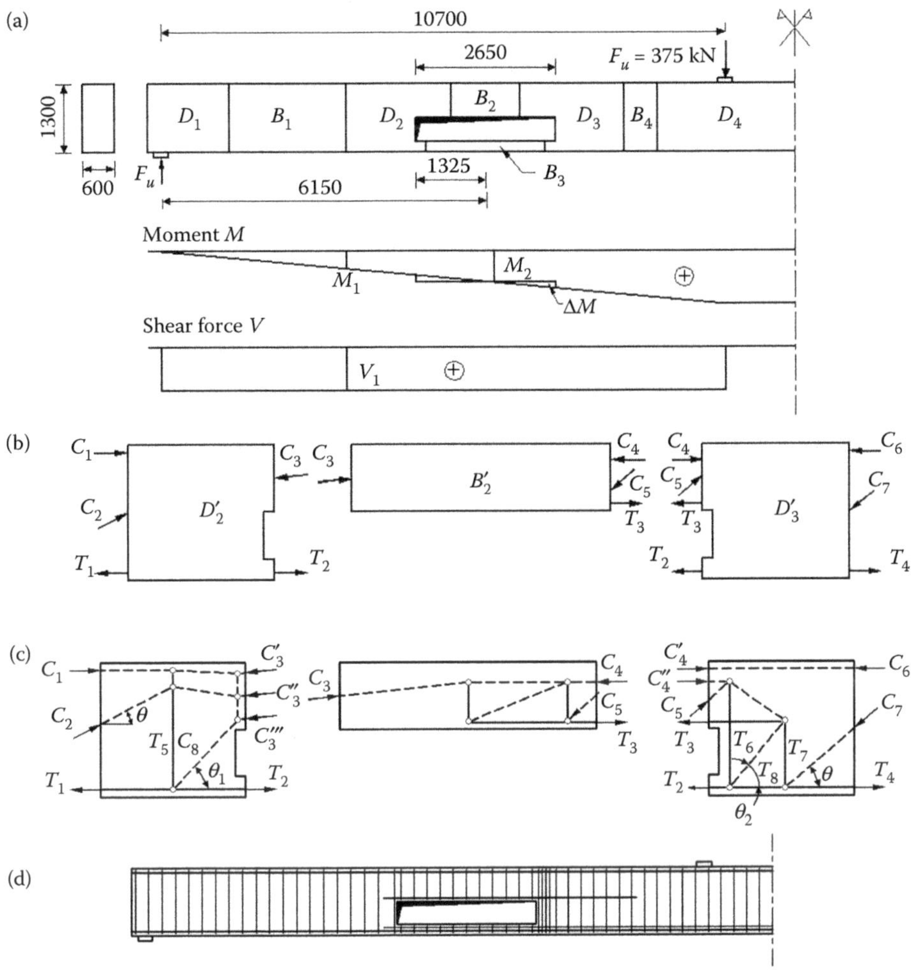

FIGURE 6.11 Shallow beam with an eccentric large opening: (a) B- and D-regions and moment and shear diagrams, (b) reduced D-regions at the opening ends with boundary forces from the B-regions, (c) *STMs* of D_2- and D_3-regions, and (d) reinforcement layout. (Adapted from Schlaich, J. et al., *Journal of the Prestressed Concrete Institute*, 32(3), 1987, 74–150.)

it should be designed for a constant axial tension T_2, where $T_2 = M_2/z = 5.35F_u$. The lever arm $z = 1.15$ m is determined by standard methods based on the assumption of linear strain distribution of the cross-section at the opening. Therefore, the B_2-region has to carry the axial compressive force $C = T_2$ (eccentrically with respect to the axis of the B_2-region) plus the differential moment $\Delta M = M - M_2$ plus the total shear force V. Under the combined action of these forces, the B_2-region shows the transition from the column type B-region on its left end (resultant C_3) to the truss type B-region at the other end (C_4, C_5, and T_3). For simplicity, the model of the B_2-region is extended somewhat into the D_2- and D_3-regions leaving over for modeling only the D_2- and D_3-regions in Figure 6.11b (Schlaich et al., 1987).

Upon knowing the boundary conditions from B_1, B_2, and B_3, the model for the D_2-region can be developed, Figure 6.11c. Looking for the counterparts of C_1, C_2, and T_1 at the opposite side of the D_2'-region, it helps in establishing the load paths to split up C_3 into three forces: $C_3', C_3'',$ and C_3''', where the horizontal components of these forces balance those of C_1, C_2, and T_1 ($C_{3h}' = C_1$, $C_{3h}'' = C_{2h}$ and $C_{3h}''' = T_2 - T_1$). Vertical equilibrium in the D_2' -region is established by two forces, a vertical tension tie T_5 and by the vertical component of a compression strut C_8, the magnitudes of which depend on the choice of their position. Knowing that T_5 and C_8 represent transverse stresses which are inside the D_2-region and that these stresses tend to fill the available space, the resultant tension T_5 is chosen in the middle of the D_2'-region and the resultant compression C_8 is chosen at the right end of the D_2' (which is inside D_2). Then $T_5 = (T_2 - T_1) \tan \theta_1 = 1.48\ F_u$, where $\theta_1 = 46°$. The tie force T_5 may be interpreted as the transverse tension necessary to anchor the differential force of the beam's tension chord, $T_2 - T_1$.

In a similar way, the D_3'-region at the other end of the opening may be treated. The transverse tension forces are: $T_7 = V = F_u$ and $T_6 = (T_8 - T_2) \tan \theta_2 = (T_4 - T_2 - V \cot \theta) \tan \theta_2 = 2.12F_u$, where $\theta_2 = 50°$.

The results of this example indicate that the stirrup forces in some places considerably exceed the "normal shear" reinforcement of a beam without an opening.

Modeling a shallow beam with a large opening depends on many factors such as the shape, size, and location of the opening within the beam span and within the beam depth. Also, the presence of a concentrated load near the opening is very influential in the modeling. The *STMs* of a shallow beam with a large circular opening of diameter, $0.3h < D \leq 0.5h$, where h is the beam height, are illustrated in Figure 6.12 for different cases. These include the opening location, either within the span or near the support, either in the middle height or shifted to the bottom of the beam.

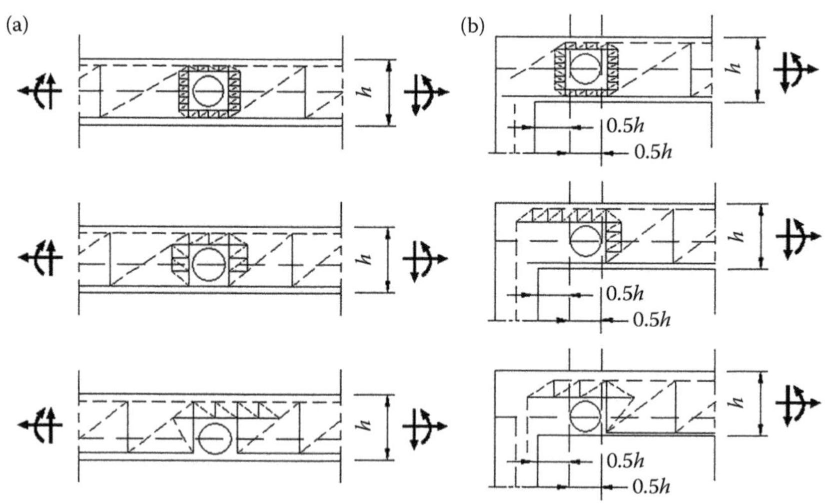

FIGURE 6.12 *STMs* of a shallow beam with a large circular opening, $0.3h < D \leq 0.5h$, at different positions within the beam depth for: (a) an opening within the span and (b) an opening near the support.

Other models similar to those in Figure 6.12a are given in Figure 6.13, but with the presence of a concentrated load near the opening.

For the case of a rectangular opening, the opening geometry and location, in addition to the presence of concentrated loads near the opening, are the main factors affecting the beam modeling. In Figure 6.14, two *STMs* are illustrated for a shallow beam with an opening of height $a < h/3$, and width $b \leq 3a$, where h is the beam height, one model for the case of opening centered with beam axis and the other if the opening is shifted toward the beam tension reinforcement. If the opening height is increased up to $a = 0.5h$, the models will be as illustrated in Figure 6.15. In tracing the truss model above and around a rectangular opening, as given in Figures 6.14 and 6.15, it is appropriate to assume that the angle of inclination of the truss struts is around 45°; then the number of truss panels can be easily determined.

Figure 6.16 illustrates the *STMs* of shallow beams with rectangular openings centered with the beam axis, with opening height, $a \leq 0.5h$, where h is the beam height, and the rectangularity of the opening, $b/a \leq 3$, where b is the opening width. The influence of both the opening location within the beam span and the presence of a concentrated load near or far from the opening are very obvious.

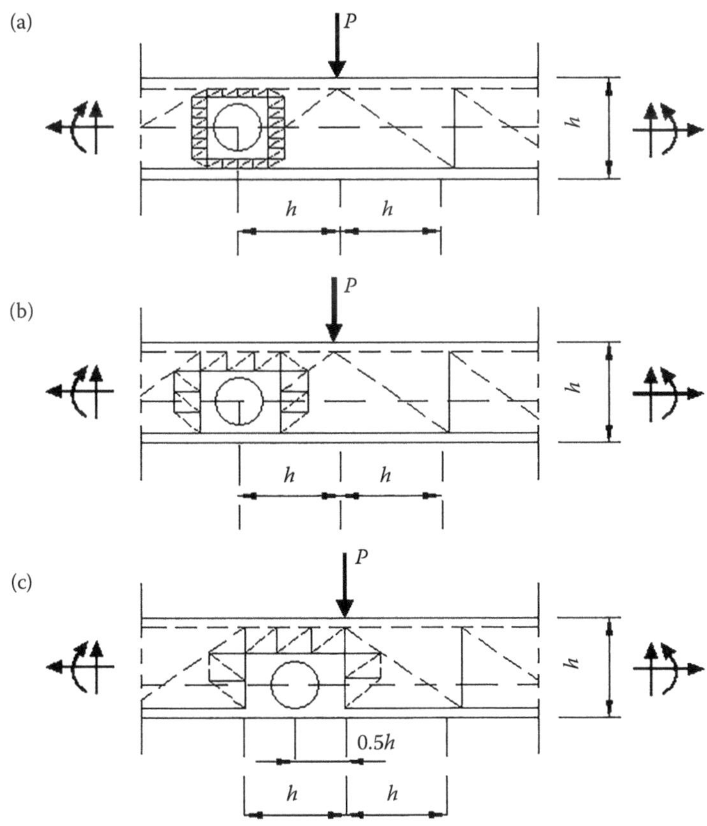

FIGURE 6.13 *STMs* of a shallow beam with a large circular opening, $0.3h < D \leq 0.5h$, in the presence of a concentrated load near the opening at different positions within the beam depth.

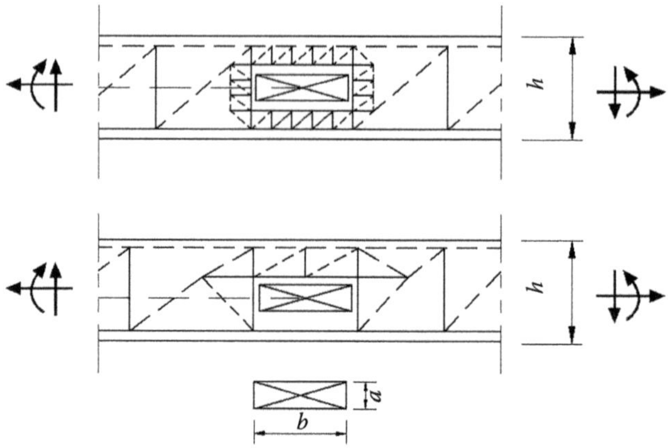

FIGURE 6.14 *STM* of a shallow beam with a rectangular opening, $a < h/3$ and $b/a \leq 3$, at different positions within the beam depth.

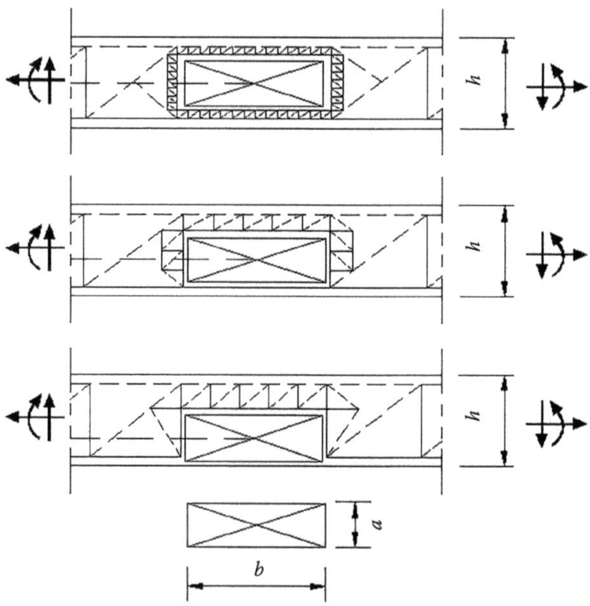

FIGURE 6.15 Modeling of a beam with a large rectangular opening, $a \leq 0.5h$ and $b/a < 3$.

6.4 SIMPLY SUPPORTED DEEP BEAMS WITH WEB OPENINGS

6.4.1 Modeling

In the modeling of a deep beam, either the openings are the choice of the designer or are dictated by other design specialty. If it is up to the designer to decide the location of the openings, the *STM* can help in this respect, Figure 6.17. Otherwise, the *STM* should go around the openings in order to ensure a path for load transfer, Figure 6.18.

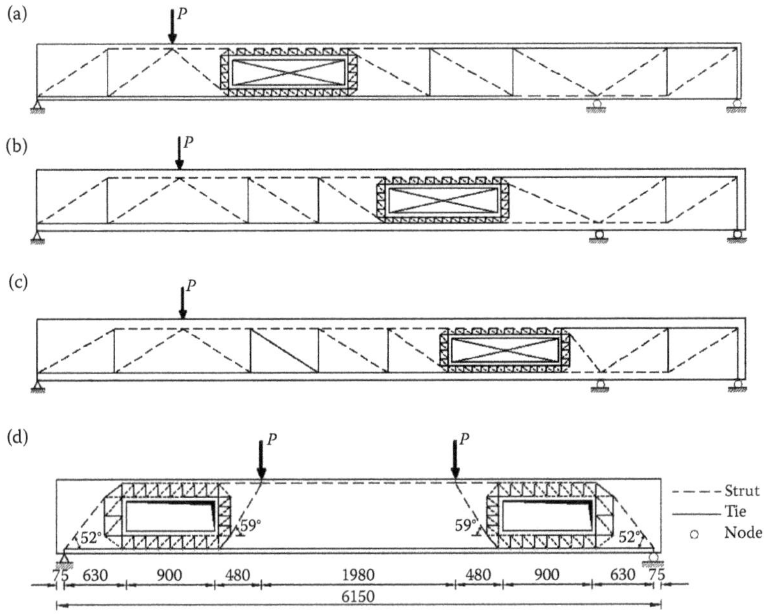

FIGURE 6.16 *STMs* of different beams with rectangular openings centered with the beam axis and subjected to concentrated loads near or far from the opening; the opening height, $a \leq 0.5h$, and the rectangularity of the opening, $b/a \leq 3$.

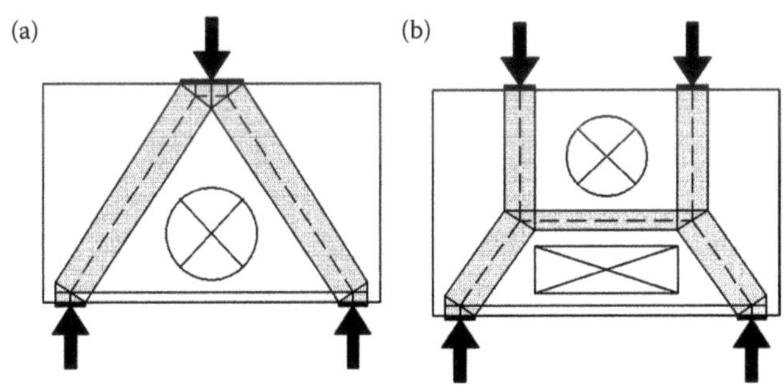

FIGURE 6.17 Examples where the *STM* guides in the choice of the opening location in deep beams.

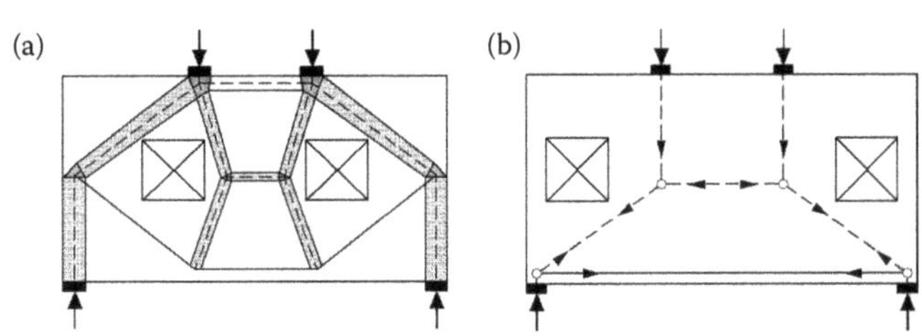

FIGURE 6.18 Examples where the *STM* should go around the openings in deep beams.

In the development of the *STM*, it is up to the designer to either use the load path method or to rely on a linear elastic analysis for tracing the stress trajectories and hence develop the model, Figure 6.19. Nevertheless, in some cases, different *STMs* can be developed from the same stress trajectories, as for the deep beam with two symmetric openings in Figure 6.20a with the stress trajectories in Figure 6.20b, where the two *STMs* in Figure 6.20c and d are valid. In addition, owing to the practicality of reinforcement detailing, other models are valid as well, Figure 6.20e and f.

6.4.2 Example on Strength Assessment of a Deep Beam with a Large Opening

The beam in Figure 6.20a, with the reinforcement details in Figure 6.21, had been tested by El-Azab (2007), beam DSON3. The concrete cylinder strength of the beam, $f_c' = 30.45$ MPa, the yield stress of the 16 mm and 10 mm bars longitudinal steel, $f_y = 410.0$ MPa, of the 6 mm vertical web bars, $f_{yv} = 244.5$ MPa, and of the 8 mm horizontal web bars, $f_{yh} = 260.2$ MPa. The beam failed under a load $P_{EXP} = 2V = 140.0$ kN.

Four different *STMs* of beam DSON3 are proposed as illustrated in Figure 6.20c–f. As stated in the previous section, all models were initiated from the stress trajectories but the two models in Figure 6.20c and d utilize inclined web reinforcement, whereas the two alternative models in Figure 6.20e and f utilize vertical and horizontal web reinforcement near the edges of the openings.

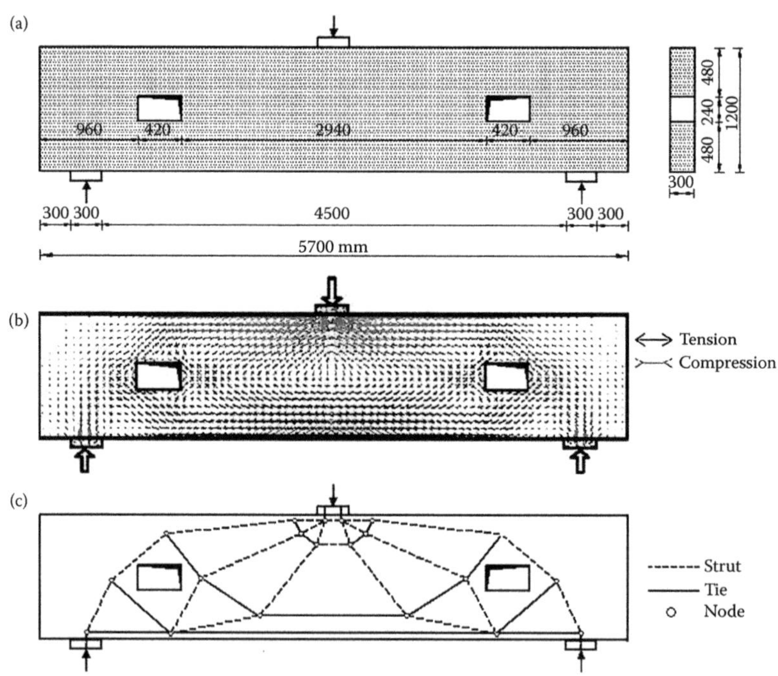

FIGURE 6.19 Example where the stress trajectories provide convenience in developing an *STM*: (a) a deep beam with two symmetric openings, (b) the stress trajectories, and (c) the *STM*.

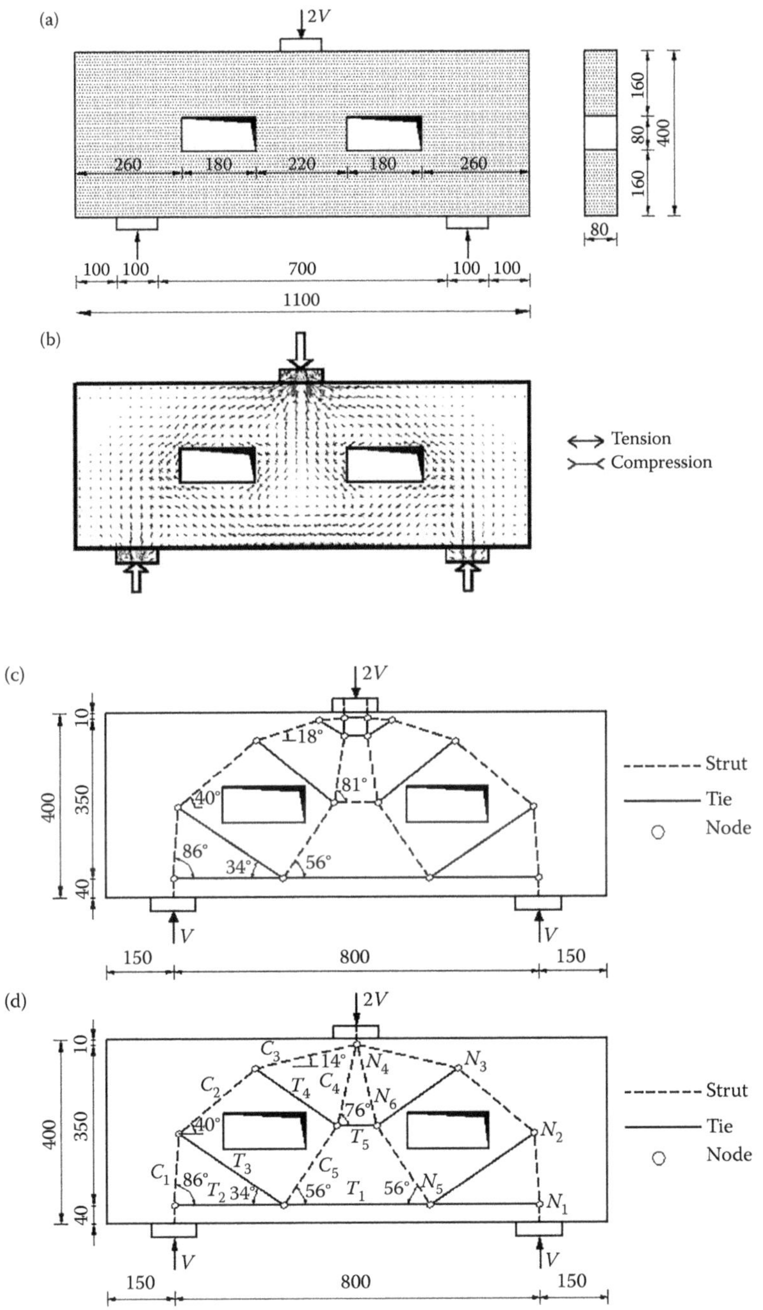

FIGURE 6.20 Different *STMs* for the same problem—a deep beam with two symmetric openings: (a) beam geometry, (b) stress trajectories, and (c) *STM*, (d) alternative simplified *STM* *(Continued)*

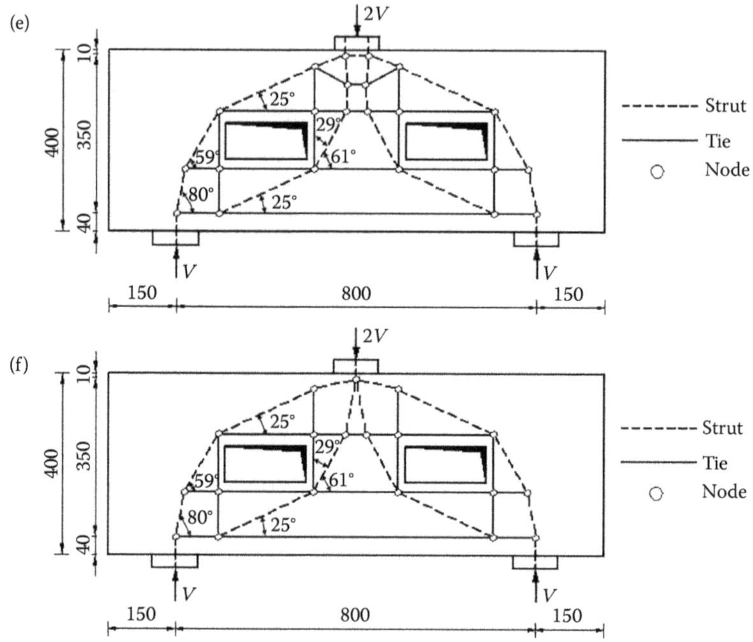

FIGURE 6.20 (Continued) Different *STMs* for the same problem—a deep beam with two symmetric openings: (e, f) alternative refined and simplified *STMs* considering convenience in the reinforcement layout.

Both the simplified model in Figure 6.20d and the alternative simplified model in Figure 6.20f were utilized by El-Sawi et al. (2015) to assess the strength of beam DSON3 by employing the ACI 318-14 failure criteria. The simplified model in Figure 6.20d gave a nominal strength of the beam, $P_{STM} = 121.0$ kN $= 0.86P_{EXP}$, whereas the alternative simplified model in Figure 6.20f gave a nominal strength, $P_{STM} = 130.0$ kN $= 0.93P_{EXP}$. The alternative simplified model with vertical and horizontal ties near the opening edges, Figure 6.20f, is better than that with inclined ties, Figure 6.20d, because it better reflects the reinforcement detailing of the beam, and therefore it has resulted in a larger capacity of the beam. Using the same analogy, the alternative refined model in Figure 6.20e is better than the refined model in Figure 6.20c.

6.4.3 Example on Design of a Deep Beam with Eccentric Large Openings

This example had been presented by Schlaich and Schäfer (1993) and Schlaich et al. (1987) and it has been discussed in Chapter 2. The geometry and loading of the deep beam are shown in Figure 6.22a. The beam data, stress trajectories, modeling, and reinforcement details have been presented in Chapter 2. Nevertheless, this example is presented here again because of its illustrative function. The *STM* in Figure 6.22b demonstrates the significance of combining both inclined web reinforcement and horizontal and vertical reinforcement near the edges of the opening in obtaining

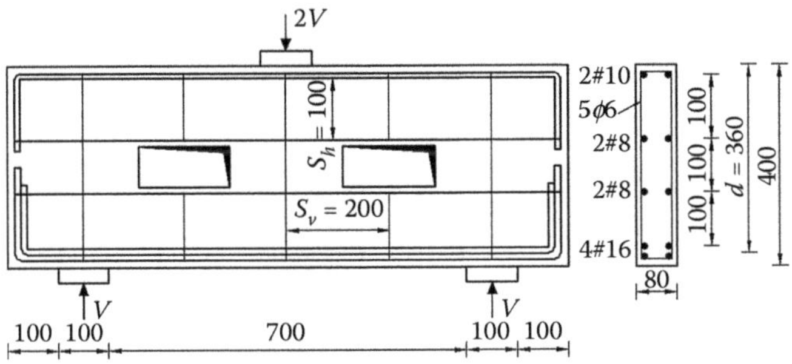

FIGURE 6.21 Example on strength assessment of a deep beam with large openings—reinforcement details.

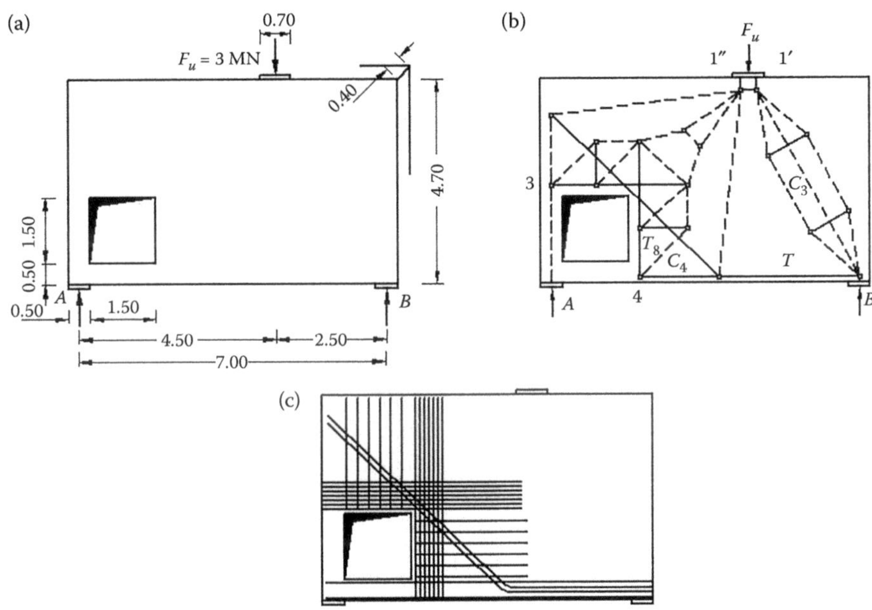

FIGURE 6.22 Example of a deep beam with an eccentric large opening (Schlaich et al., 1987; Schlaich and Schäfer, 1993): (a) geometry and loads, (b) *STM*, and (c) reinforcement layout.

better performance of the beam. Hence, the reinforcement layout in Figure 6.22c is tailored according to this combined model.

6.5 CONTINUOUS DEEP BEAMS WITH WEB OPENINGS

6.5.1 Example on a Continuous Deep Beam with a Small Opening

The analysis of the reinforced concrete continuous deep beam with small openings in Figure 6.23a resulted in the stress trajectories shown in Figure 6.23b. Upon utilization of the load path method or the stress trajectories, it is a straightforward matter

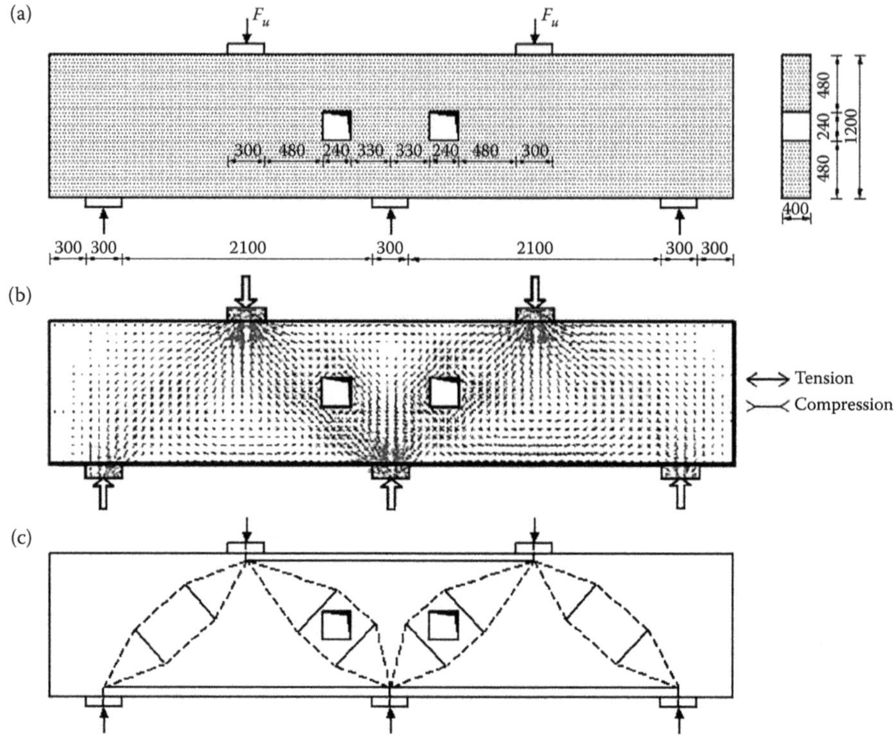

FIGURE 6.23 Example of a continuous deep beam with small openings: (a) geometry and loading, (b) stress trajectories, and (c) *STM*.

to obtain the appropriate *STM* in Figure 6.23c. In the model, the strut framing from the loading point to the outer support has been detailed as shown in order to account for web reinforcement.

6.5.2 Modeling of Continuous Deep Beams with Large openings

Continuous deep beams, may function as transfer girders in multi-story frames, pile caps, foundation wall structures, etc. The strut-and-tie method offers different options

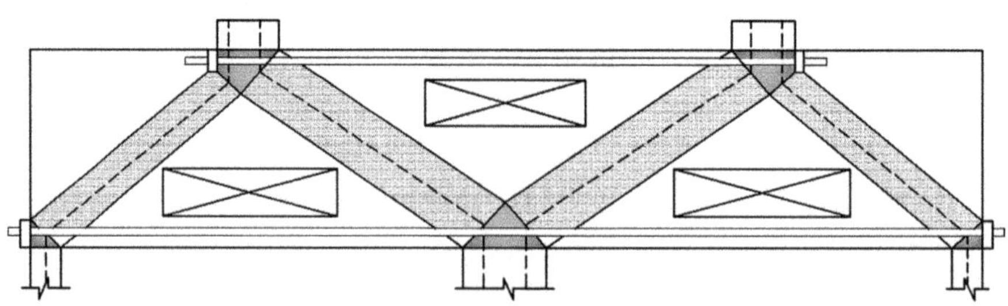

FIGURE 6.24 Modeling of continuous deep beam showing the possible locations of openings without disturbing the flow of forces.

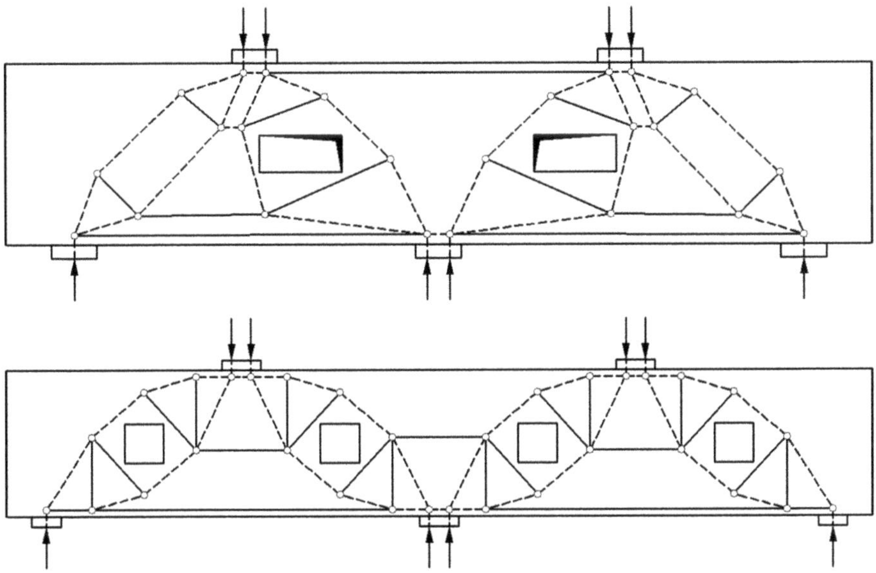

FIGURE 6.25 Examples on modeling of continuous deep beam with large openings utilizing inclined web reinforcement.

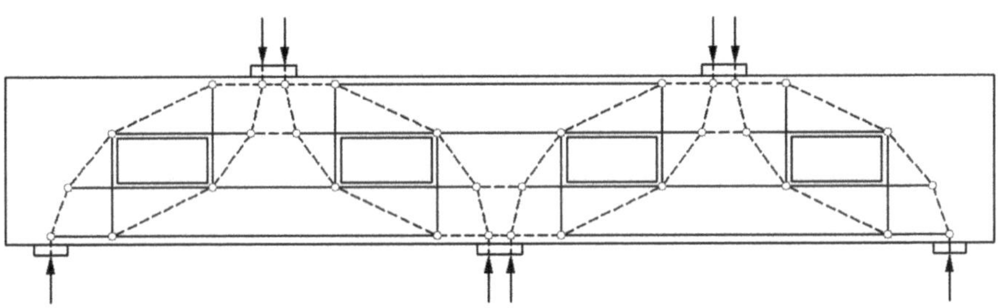

FIGURE 6.26 Example on modeling of continuous deep beam with large openings utilizing orthogonal web reinforcement around the openings.

for opening locations without jeopardizing the beam strength. The example in Figure 6.24 demonstrates this vision, where the appropriate *STM* of a continuous beam provides such convenient locations of openings. Nevertheless, the locations of openings are often dictated by engineering disciplines other than the structural design. In such a case the model has to go around the openings in order to find means for load transfer, Figures 6.25 and 6.26. The models of the beams in Figure 6.25 mainly utilize inclined web reinforcement, whereas the model of the beam in Figure 6.26 adhere with orthogonal web reinforcement details around the openings, which is frequently used in practice.

REFERENCES

ACI 318-14, *Building Code Requirements for Structural Concrete and Commentary*, Detroit: American Concrete Institute, USA, 2014, 519 pp.

El-Azab, M. F., Behavior of reinforced high-strength concrete deep beams with web openings, *M.Sc. Thesis in Structural Engineering*, Faculty of Engineering, Mansoura University, El-Mansoura, Egypt, 2007.

El-Sawi, W. E. E., El-Metwally, S. E., El-Zoughiby, M. E., and Ghaleb, A. A., Behavior of RC shallow and deep beams with openings via the strut-and-tie model method and non-linear finite element, *The Arabian Journal for Science and Engineering*, 41(2), 2015, 401–424.

Hanson, J. M., *Square Openings in webs of Continuous Joists*, Portlant Cement Association, 1969, pp. 1–14.

Kiag-Hwee, T., Mansur, M. A., and Huang, L-M., Reinforced concrete T-beams with large web openings in positive and negative moment regions, *ACI Structural Journal*, 93(3), 1996, 277–289.

Kiang-Hwee, T., Mansur, M. A., and Wei, W., Design of reinforced concrete beams with circular openings, *ACI Structural Journal*, 98(3), 2001, 407–415.

Mansur, M. A., and Tan, K. H., *Concrete Beams with Openings: Analysis and Design*, National University of Singapore, Singapore, 1999, pp. 1–70

Mansur, M.A., Tan, K. H. and Weng, W., Effects of creating an opening in existing beams. *ACI Structural Journal*, 98(3), 2001, 407–415.

Salam, S. A., Beams with openings under different stress conditions, *Conference on Our World in Concrete and Structures*, Singapore, August 25–26, 1977, pp. 259–267.

Schlaich, J. and Schäfer, K., Design and detailing of structural concrete using strut-and-tie models, *Journal of the Structural Engineer*, 69(6), 1991, 113–125.

Schlaich, J. and Schäfer, K. The design of structural concrete, *IABSE Workshop*, New Delhi, February, 1993.

Schlaich, J., Schäfer, K., and Jennewein, M., Toward a consistent design of structural concrete, *Journal of the Prestressed Concrete Institute*, 32(3), 1987, 74–150.

Siao, W. B., and Yap, S. F., Ultimate behaviour of unstrengthened large openings made in existing concrete beams, *Journal of the Institution of Engineers*, 30(3), May-June, 1990, 51–57.

Somes, N.F. and Corley, W.G. (1974). Circular openings in webs of continuous beams. *Shear in Reinforced Concrete, Special Publication SP-42*, American Concrete Institute, Detroit, 359–398.

Tan, K. H., Mansur, M. A., and Huang, L.-M., Reinforced concrete T-beams with large web openings in positive and negative moment regions, *ACI Structural Journal*, 93(3), May-June, 1996, 277–289.

Tan, K. H., Mansur, M. A., and Wei, W., Design of reinforced concrete beams with circular openings, *ACI Structural Journal*, 98(3), 2001, 407–415.

7 Beam–Column Connections

7.1 INTRODUCTION

Beam–column connections in framed structures represent geometric and stiffness discontinuity and hence regions of force concentration. The regions at and near the joints have been observed to be common points of distress in frame buildings. When the framing beams are stressed near or beyond the yield, particularly in case of cyclic loads, the joints of the structure will be subjected to stresses which, if not properly provided for, can result in undermining the load-carrying capacity of the columns and beams forming the joints.

It is usually assumed in some codes that beam–column joints will perform satisfactorily using any detailing of the reinforcement where the anchorage requirements for the reinforcement are satisfied. This conception is completely wrong since beam–column connection is a discontinuity region which experiences disturbance in the flow of forces associated with nonlinear strain distribution. Hence, conventional design is not applicable to this region and a different methodology, based on realistic physical models which can be easily understood, should be adopted. This leads to the *STMs* method as the appropriate approach for the design of such structural concrete element. In addition, the desired quality of performance regarding either strength or ductility requirements can be attained in only a well-detailed connection.

In this chapter all types of frame joints are treated. These include knee corner joints under opening and closing moments, obtuse corner joints, wide beam supported on narrow column and vice versa, and exterior, tee and interior beam-column connections. The treatment includes the examination of the joint behavior and a discussion on the role of detailing in controlling the joint performance. Different strut-and-tie models for every type of connection are proposed with the corresponding reinforcement detailing based on the boundary forces of the joint region. In order illustrate the calculation procedure the strength of selected examples of beam-column connections that have been tested experimentally is assessed using the method of *STM*.

7.2 KNEE CORNER JOINTS UNDER OPENING MOMENTS

7.2.1 Joint Behavior

Corners subjected to opening loads have been found in experimental tests to be far more critical than those subjected to closing loads (Yuan et al., 1982). The distribution of elastic stresses before cracking is illustrated in Figure 7.1. Large tensile stresses occur at the reentrant corner and in the middle of the joint, Figure 7.1b. As a

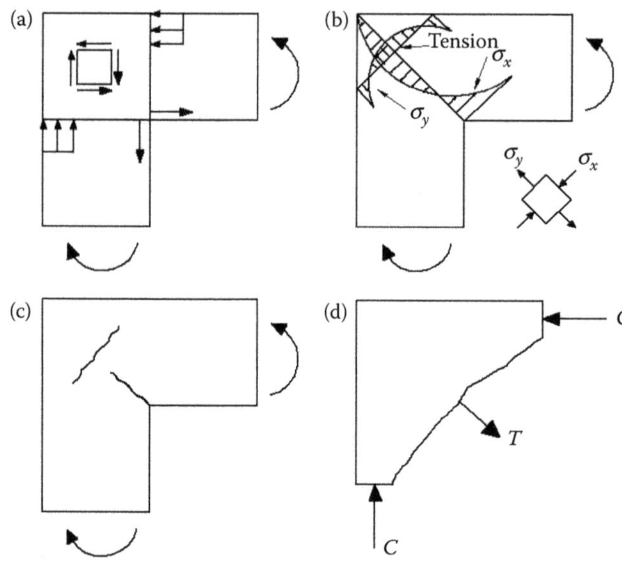

FIGURE 7.1 Behavior of corner joint due opening moment: (a) internal forces of the joint, (b) elastic stresses, (c) cracking pattern, and (d) free body diagram of the portion outside the diagonal crack.

result, cracking develops as shown in Figure 7.1c. A free body diagram of the portion outside the diagonal crack, as shown in Figure 7.1d, indicates that the compression forces near the outer corner give rise to a resultant that tends to push off the triangular portion of the joint. The internal tension force, *T*, is necessary for equilibrium. If reinforcement is not provided to develop this force, the joint will fail almost immediately after the development of the diagonal crack (MacGregor and Wight, 2005).

In tests at the University of Nottingham (Park and Paulay, 1975), the full moment capacity was attained with steel content 0.75%. On the other hand, joints studied by Swann (Park and Paulay, 1975) with steel content 3.0% failed at a load less than 80% of the theoretical ultimate values, derived from the flexural capacity of the adjoining section. Thus, it can be concluded that the full flexural capacity of the adjoining members at the face of the joint was attained when the steel content was small. For higher values of steel content, a brittle splitting failure occurred at less than the full strength of adjoining members (Park and Paulay, 1975).

7.2.2 Role of Detailing

When the knee corner joint is subjected to an opening moment, the effect of reinforcement detailing on the strength is significantly pronounced. Figure 7.2a compares the measured efficiency of a series of corner joints reported by Balint and Taylor (1972) and Nilsson and Losberg (1976). The efficiency is defined as the ratio of the failure moment of the joint to the moment capacity of the members entering the joint. The reinforcement was detailed as in Figure 7.2b–e. The solid curved line corresponds to the computed moment at which diagonal cracking is expected

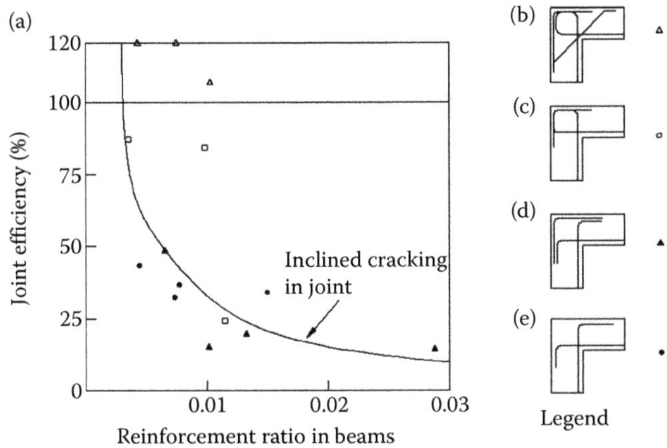

FIGURE 7.2 Influence of detailing on the behavior of corner subjected to opening moment: (a) joint efficiency, (b)–(e) details.

to occur in such a joint. Typical beams have reinforcement ratios about 1%. At this reinforcement ratio, the very common joint details shown in Figure 7.2d and e can transmit only 25%–35% of the moment capacity of the beams.

Nilsson and Losberg (1976) have shown experimentally that a joint reinforced as shown in Figure 7.2b will develop the needed moment capacity without excessive deformations. The joint consists of two hooked bars enclosing the corner and diagonal bars having a total cross-sectional area of half that of the beam reinforcement. The tension in the hooked bars has a component across the diagonal crack, helping to provide the force in Figure 7.1d. The inclined bar limits the growth of the crack at the reentrant corner, slowing the propagation of cracking in the joint. The symbols in Figure 7.2a show the efficiency of the joints shown in Figure 7.2b–e, both with and without the diagonal corner bar. It can be seen that the corner bar is needed to develop the full efficiency in the joint. The inclined reinforcement should have a total cross-sectional area approximately 50% of the largest main reinforcement.

7.2.3 Strut-and-Tie Modeling

In the design of a reinforced concrete moment-resisting frame structure, the corner geometry can be defined from the dimensions of the structural elements meeting at the joint, beam, and column. The first step in the design of connection using the method of *STM* is to identify all forces acting on the D-region (connection). Based on the observed joint behavior and the proposed reinforcement detailing, the appropriate *STM* can be derived.

The different *STMs* shown in Figure 7.3a were suggested by Schlaich et al. (1987) for frame joints under an opening moment. In order to circumvent the tensile chord reinforcement and prevent cracking of the compression chord due to radial tensile stresses, either the chord reinforcement must be extended as a loop around the corner or inclined stirrups must be adequately arranged. For further explanation of the joint behavior, other *STMs* are suggested as illustrated in Figure 7.3b.

(a)

(b)

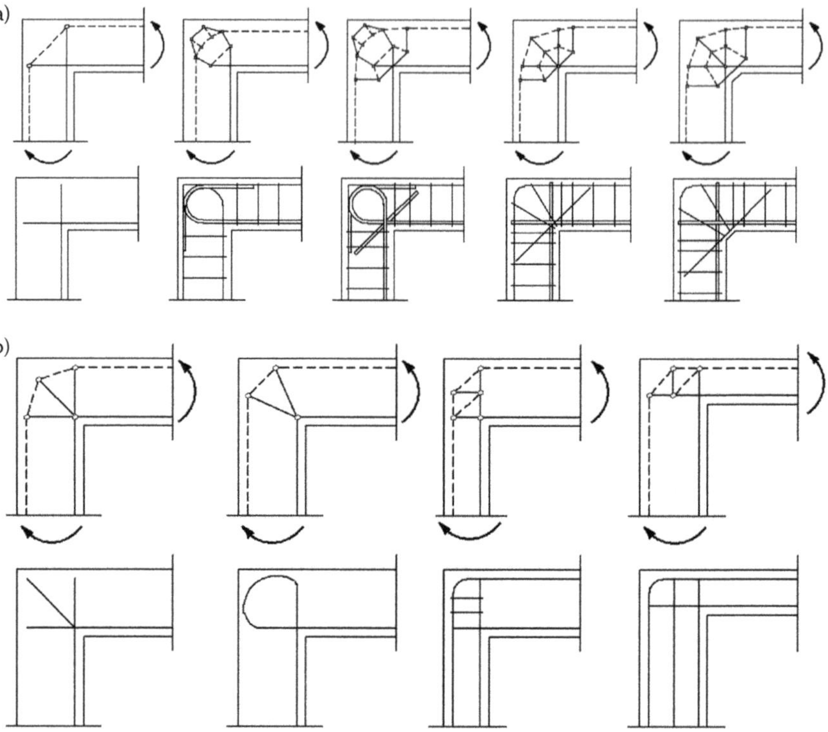

FIGURE 7.3 Strut-and-tie modeling of an opening corner joint: (a) models proposed by Schlaich et al. (1987) and (b) additional suggested models.

7.2.4 Example 7.1: Strength Assessment of an Opening Corner

The strength of the reinforced concrete opening corner in the frame specimen tested by Nilsson and Losberg (1971), shown in Figure 7.4a, with the reinforcement detailing of Figure 7.4b, is assessed via the strut-and-tie method.

In this specimen, the concrete cylinder strength was $f_c' = 24\,\text{MPa}$, whereas all reinforcement bars were 13 mm diameter deformed bars with a yield stress $f_y = 400$ MPa. The specimen failed at a moment equal to 22.27 kN.m, when a diagonal crack inside the joint ran along the reinforcement into the compression zone and the outside of the corner was pushed off.

The nominal moment capacity of the smallest member entering the joint, $M_u = A_s f_y(d - 0.5a) = 32.95$ kN.m, where $d = 170$ mm, $a = A_s f_y/(0.85 f_c' b) = 29.7\,\text{mm}$. The nominal strength is much larger than the failure moment of the joint, which means that failure occurred within the joint itself and that the stresses in the beam and the column were within or near the elastic limits.

The appropriate *STM* of the joint is shown in Figure 7.4c. In this model, the lever arm of the column is smaller than that of the beam, and therefore the calculations start with the column.

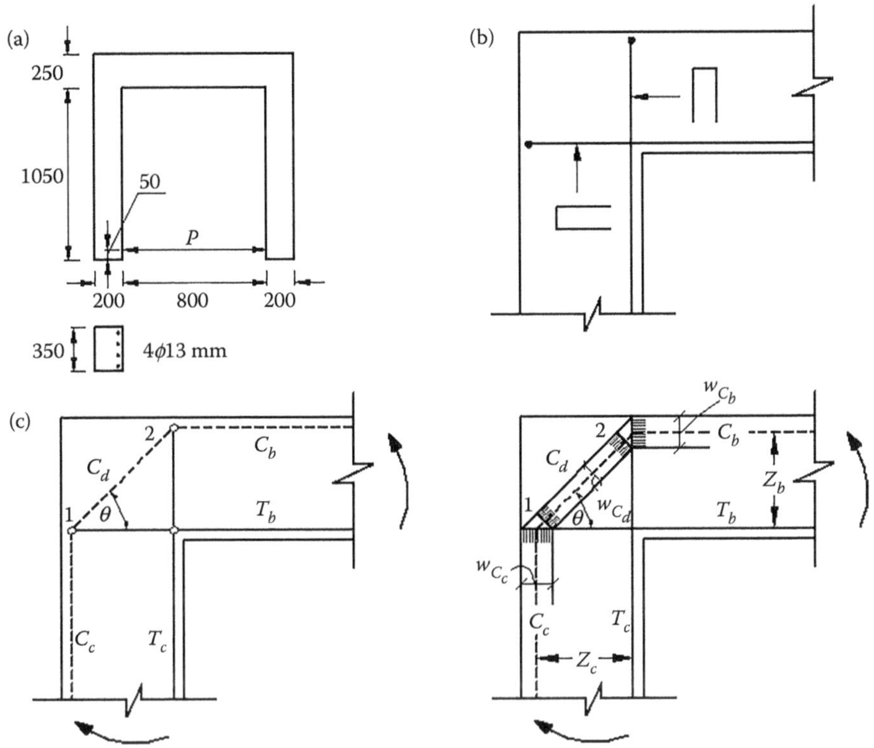

FIGURE 7.4 Example 7.1—corner subjected to an opening moment (a) test specimen, (b) reinforcement details, and (c) the *STM* details.

Effective strength of nodes and struts

Both nodes 1 and 2 are $C-C-T$ nodes; hence, the effective strength of either node of the model, $f_{ce}^{n}=0.85f_c'\beta_n=0.85f_c'\times0.8=16.32\,\text{MPa}$

Both struts C_c and C_b are prismatic struts; hence, the effective strength of either strut, $f_{ce}^{s}=0.85f_c'\beta_s=0.85f_c'\times1.0=20.4\,\text{MPa}$

Strut C_d is a bottle-shaped stress field with no transverse reinforcement; hence, $f_{ce}^{s}=0.85f_c'\beta_s=0.85f_c'\times0.6=12.24\,\text{MPa}$

STM forces:

With reference to Figure 7.4c,

$$T_c=A_sf_y=212.4\,\text{kN}$$

$$C_c=T_c=212.4\,\text{kN}$$

The strength of strut C_c is the smaller of the strut strength and the strength of node 1; the latter governs. $C_c=212.4\,\text{kN}=16.32\times w_{C_c}\times350$, giving the width of the strut $w_{C_c} = 37.2\,\text{mm}$. Then the lever arm $z_c = 170 - 0.5 \times 37.2 = 151.4$ mm. The moment capacity of the column is therefore $M_{cn} = 212.4 \times 151.4 = 32.16$ kN.m.

From equilibrium, the moment of the beam is $M_b = 32.16$ kN.m.

$M_b = C_b(d_b - 0.5w_{C_b}) = 16.32 \times 350 \times w_{C_b}(220 - 0.5w_{C_b})$ giving $w_{C_b} = 27.28\,\text{mm}$, $C_b = T_b = 155.82$ kN and lever arm $z_b = 206.36$ mm.

The angle $\theta = \tan^{-1}(z_b/z_c) = 53.74°$.

Then the force in the diagonal strut $C_d = C_c/\sin\theta = 263.43$ kN.

Node 1:

There is no need to check strut C_c since the strut size was calculated based on the appropriate strength.

For strut C_d, the width of the strut is $w_{C_d} = w_{C_c}\sin\theta = 30.0\,\text{mm}$. The effective strength is the smaller strength of the node and the strut strength; the latter governs. Then the nominal strength of the strut is $C_{dn} = 12.24 \times w_{C_d} \times 350 = 128.52\,\text{kN}$, which is 49% of the strut force.

Node 2:

There is no need to check strut C_b since the strut size was calculated based on the appropriate strength.

For strut C_d, the width of the strut is $w_{C_d} = w_{C_b}\cos\theta = 16.1\,\text{mm}$. This width will give a much smaller strength than that at node 1. Therefore, the width of strut C_b can be increased in order to increase the capacity of the diagonal strut at node 2 to become close to that at node 1. Of course, the lever arm z_b will be reduced and the force C_b will increase.

Assuming $z_b = 195.0$ mm, the force of strut $C_b = 164.9$ kN and the width $w_{C_b} = 50\,\text{mm}$. Then the angle $\theta = \tan^{-1}(z_b/z_c) = 52.17°$. The force in the diagonal strut $C_d = 268.9$ kN and the width of the strut is 29.4 mm at node 1 and 30.7 mm at node 2. Then the smaller nominal strength of the diagonal strut is that at node 1, $C_{dn} = 126.0$ kN, which is 47% of the strut force. Therefore, the joint capacity should be reduced to $M_n = 0.47 \times 32.16 = 15.11$ kN.m.

The obtained strength from *STM* is 68% of the measured value. This obtained value can be optimized upon increasing the size of nodes 1 and 2 to become $w_{C_c} = 52\,\text{mm}$, $w_{C_b} = 70\,\text{mm}$, $z_c = 144$ mm, $z_b = 185$ mm, $\theta = 52.1°$, and $C_b = 165.3$ kN. Then the force in the diagonal strut $C_d = 269.2$ kN and the width of the strut are 41 mm at node 1 and 43 mm at node 2. The nominal strength of C_d is then 175.0 kN, which is 65.3% of the strut force. Then the joint capacity should be $M_n = 0.653 \times 212.4 \times 144 = 20.0$ kN.m. The obtained strength from *STM* is 90% of the measured value.

7.3 KNEE CORNER JOINTS UNDER CLOSING MOMENTS

7.3.1 Joint Behavior

The elastic stresses in a closing corner joint subjected to a negative bending moment are exactly opposite to those in an opening corner joint, Figure 7.1b. The forces at the ends of the members forming the joint are illustrated in Figure 7.5. It is assumed that these forces are introduced into the joint core in the form of uniform shear stresses resulting from an anchorage bond. As a result, cracking of such a joint occurs with a

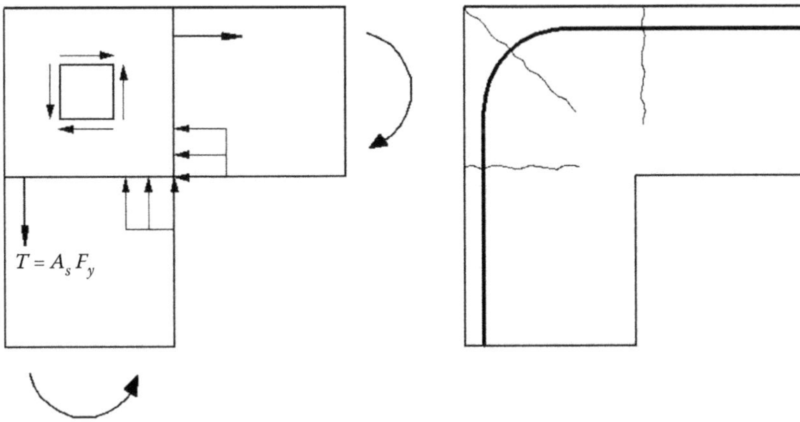

FIGURE 7.5 Stresses and cracking in a closing corner joint.

major crack on the diagonal. The diagonal crack can be avoided by limiting the diagonal tensile stress within the concrete tensile strength, f_t, (MacGregor and Wight, 2005).

$$\sigma_t = T/bd = A_s f_y/bd = \rho f_y \leq f_t$$

where σ_t is the diagonal tension stress, T is the tensile force, b and d are the breadth and depth of the cross-section, respectively, A_s is the cross-sectional area of the main steel reinforcement, f_y is the yield stress of steel reinforcement, and ρ is the flexural steel content. This condition would limit the flexural steel content to

$$\rho = f_t/f_y$$

If the diagonal tension stress, σ_t, in the closing corner joint is not carried by the concrete tensile strength, f_t, secondary reinforcement must be used (Park and Paulay, 1975), Figure 7.6.

7.3.2 Role of Detailing

Problems arise due to the bearing inside the bent bars in the corner since these bars must transfer a resultant force to the concrete on the diagonal of the joint. The resultant force is relatively large and acts on an arc of a length equal to $\frac{1}{2}\pi r$, where r is the radius of the bend. A smaller value of the radius r will be associated with a higher intensity of bearing stresses from the diagonal compression and hence larger splitting tensile stresses. For this reason, the radius of this bend should satisfy the following equation (El-Metwally, 1992),

$$r = \frac{C}{\sqrt{2} f_{ce}^{n} b}$$

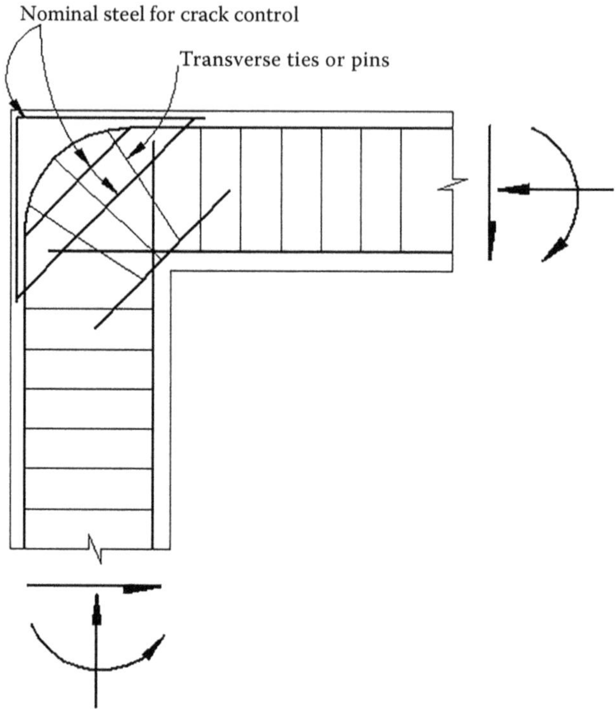

FIGURE 7.6 Secondary reinforcement at a closing corner joint.

where C is the compressive force of the diagonal strut, and f_{ce}^{n} is the node design strength. For this case, according to ACI 318-14, $f_{ce}^{n} = \phi(0.85 f_c')\beta_n = 0.75 \times 0.85 \times f_c' \times 0.6 = 0.383 f_c'$, where ϕ is the strength reduction factor.

Secondary reinforcement, as shown in Figure 7.6, is required for a corner joint of large structural members having substantial reinforcement content, as stated before.

7.3.3 Strut-and-Tie Modeling

Different *STMs* suggested by Schlaich and Schäfer (1991) are shown in Figure 7.7a. Other *STMs*, illustrated in Figure 5.7b, are suggested to explain the joint behavior.

7.3.4 Example 7.2: Strength Assessment of a Closing Corner

The channel-shaped frame specimen tested by Yuan et al. (1982) for closing moments, shown in Figure 7.8a, is examined in this example for strength assessment. In this specimen, the concrete cylinder strength was $f_c' = 31.6$ MPa and the steel bars were 9.53 mm diameter deformed bars with a yield stress $f_y = 420$ MPa. The radius of bars bend was 38.1 mm.

The test set up is shown in Figure 7.8b. The specimen failed due to yielding of reinforcement at a load $P = 16.8$ kN, which corresponds to a moment equal to

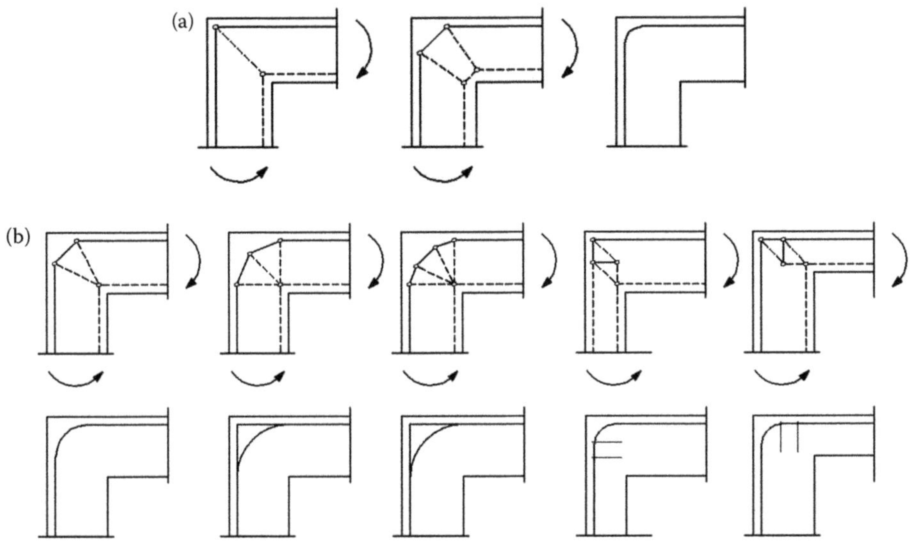

FIGURE 7.7 Strut-and-tie modeling of a closing corner joint: (a) models proposed by Schlaich and Schäfer (1991) and (b) additional suggested models.

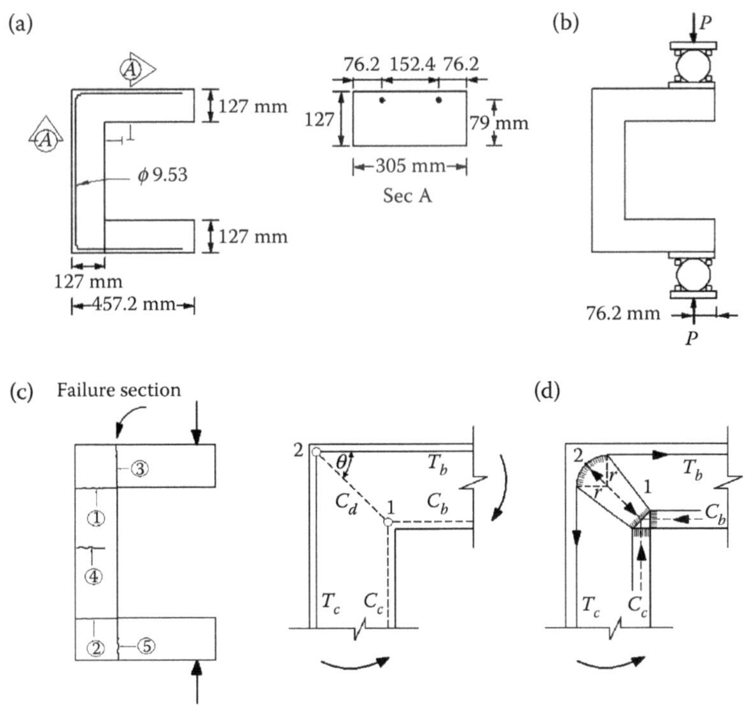

FIGURE 7.8 Example 7.2 - corner subjected to a closing moment: (a) channel specimen, (b) test setup, and (c) rack pattern, (d) *STM*.

5.33 kN.m. The specimen developed cracks at the joint-member interfaces of the channel-shaped frame, and the representative crack pattern is shown in Figure 7.8c, but no cracks developed across the corner.

The nominal moment capacity of the smallest member entering the joint, $M_n = A_s f_y (d - 0.5a) = 45.14$ kN.m, where $d = 79$ mm, $a = A_s f_y/(0.85 f_c' b) = 7.3\,\text{mm}$.

The nominal strength is less than the measured strength, which means that failure may not have occurred within the joint. Therefore, the nodes and the diagonal strut of the joint are checked first. The appropriate *STM* is shown in Figure 7.8d.

Effective strength of nodes and struts

Node 1 is a $C - C - C$ node; hence, the effective strength of the node is $f_{ce}^n = 0.85 f_c' \beta_n = 0.85 f_c' \times 1.0 = 26.86\,\text{MPa}$

Node 2 is a $C - T - T$ node; hence, the effective strength of the node is $f_{ce}^n = 0.85 f_c' \beta_n = 0.85 f_c' \times 0.6 = 16.12\,\text{MPa}$

Both struts C_c and C_b are prismatic struts; hence, the effective strength of either strut is $f_{ce}^s = 0.85 f_c' \beta_s = 0.85 f_c' \times 1.0 = 26.86\text{MPa}$

Strut C_d is a bottle-shaped stress field with no transverse reinforcement; hence, $f_{ce}^s = 0.85 f_c' \beta_s = 0.85 f_c' \times 0.6 = 16.12\,\text{MPa}$

STM forces:

With reference to Figure 7.8d, the solution will start with investigating the diagonal strut. Upon assuming yielding of the tension reinforcement, $T_c = T_b = A_s f_y = 59.9$ kN. The force in the diagonal strut $C_d = \sqrt{2} \times 59.9 = 84.7\,\text{kN}$. The width of this strut at node 2 is $w_{C_d}^2 = \sqrt{2} r = 53.9\,\text{mm}$, where r is the radius of bend. Then the nominal strength of this strut is $C_{dn} = 16.12 \times w_{C_d}^2 \times 305 = 265.0\,\text{kN}$ at node 2, which is larger than the strut force. For the diagonal strut to carry its force at node 1, it requires a width at this node, $w_{C_d}^1 = (84.7 \times 10^3)/(16.12 \times 305) = 17.2\,\text{mm}$, which corresponds to a width of either strut C_c or C_b equal to 12.2 mm. The width of C_c or C_b corresponds to a nominal strength 26.86 × 12.2 × 305 = 99.94 kN, which is larger than the force of either strut, 59.9 kN. Then the lever arm $z_c = z_b = 79 - 0.5 \times 12.2 = 72.9$ mm, giving a nominal moment, $M_{cn} = M_{bn} = 59.9 \times 72.9 = 4.4$ kN.m.

The obtained nominal moment is 83% of the measured moment. This result can be slightly improved by reducing the width of the diagonal strut at node 1 to about 0.5(1 + 0.83) = 0.92 times the previously used width, 17.2 mm, and redoing the calculations.

Thus, failure could not take place due to the failure of either strut C_d or the nodes, and the only possible failure mode is flexural failure by yielding of tie T_c or T_b. Crushing of struts C_c and C_b is not possible since on one side node 1 is stressed to less than its strength, and on the other side the beam and column forming the frame are under-reinforced.

7.4 OBTUSE CORNER JOINTS

At the obtuse corner joint, the internal forces generated in the joint must be considered when the members entering into the joint are being detailed, Figure 7.9a.

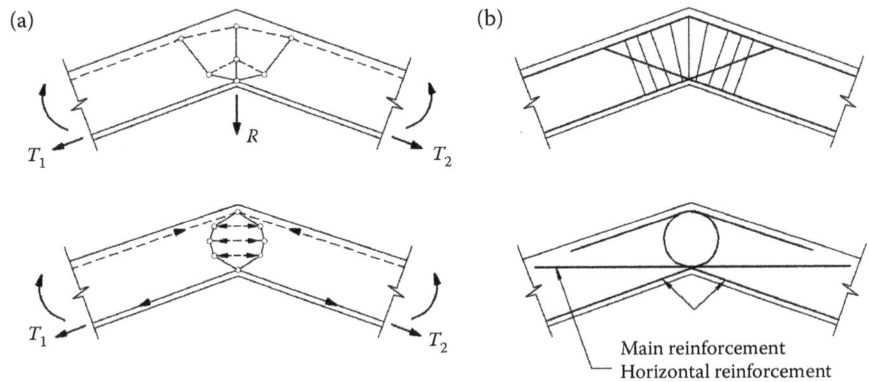

FIGURE 7.9 *STM* and the corresponding reinforcement details for an obtuse corner joint under an opening moment: (a) *STMs* and (b) corresponding reinforcement details.

The two tensile forces T_1 and T_2 generated at the kink of the reinforcing bars are not unidirectional, which results in a third force R that should be resisted by stirrups.

From the results of the experimental investigation carried out by Abdul-Wahab and Ali (1989) on obtuse corners formed by joining two walls or slabs of equal or varying stiffness, and subjected to a positive (opening) moment, the following remarks have been pointed out:

1. The efficiency of the joint detail is improved when inclined bars are added to take the tensile force in the inner corner. Loop bars with inclined bars resulted in the highest efficiency of 139%, and appear to be the most suitable detail for a continuous corner between lightly reinforced slabs.
2. The efficiency of a corner is greatly improved when the thicknesses of the adjoining members were different. The efficiency increased to 197% when the thickness of one leg was increased from 100 to 300 mm. The mode of failure was also changed from diagonal tension failure to flexural failure as a result of the difference in the stiffness between the two legs.
3. The increase in the length of one leg resulted in a gradual change in the cracking and failure pattern from a typical diagonal tension to an extensive flexural cracking with secondary diagonal tension failure. The efficiency of the corner also increased by 32% when the length ratio was doubled.

The appropriate *STM* for an obtuse joint is illustrated in Figure 7.9a, with the corresponding reinforcement. In this model, the longitudinal bars carry the tension forces T_1 and T_2, while the stirrups carry all other tension forces. An alternative model is illustrated in Figure 7.9b with the corresponding reinforcement, where the main steel takes the U-shape and carries all the tension forces; nevertheless, horizontal reinforcement, as illustrated, is added for crack control.

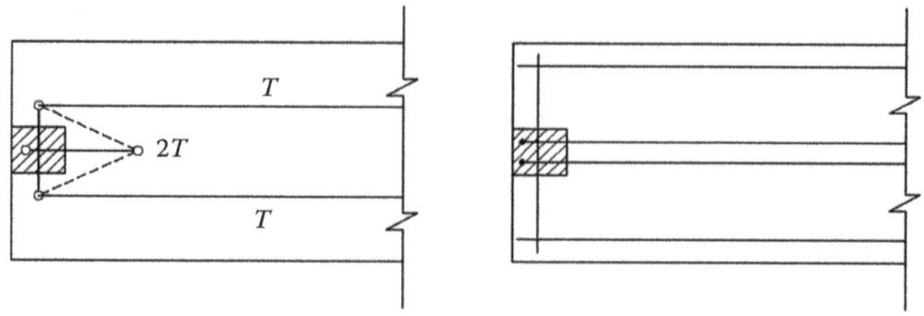

FIGURE 7.10 *STM* and the corresponding reinforcement of a wide beam-narrow column joint.

7.5 WIDE BEAM SUPPORTED ON A NARROW COLUMN AND VICE VERSA

An experimental study on frames of wide beams supported on narrow columns had been conducted by Metawei (1990). From the observed experimental behavior, the following conclusions were obtained:

1. Upon increasing the percentage of reinforcement concentration in strip width equal to column width, the initial cracking load decreased and failure load increased.
2. Central deflection of a beam with steel concentration showed a smaller deflection than a beam with uniform steel distribution. On the other hand, edge deflection of both beams was the same up to the cracking load of the section near the support, but after the cracking load, the steel concentration has increased the deflection till failure.
3. Crack patterns at the bottom surface of the frame with uniform distribution of steel indicated a bigger number and longer cracks than those for the frame with steel concentration and vice versa for crack patterns at top surfaces. Crack widths measured for the frame with steel concentration were bigger than the width for the frame with uniform steel distribution at the same loading level.

The appropriate *STM* for this joint is shown in Figure 7.10, with the corresponding reinforcement.

7.6 EXTERIOR BEAM–COLUMN CONNECTIONS

7.6.1 Joint Behavior

The system of internal forces generated around the joint zone of the exterior beam–column connection in the cast-in-place frame construction is indicated in Figure 7.11a. In the connection zone, axial compression stresses from the column axial load act in combination with high shear stresses caused by the rapid change of the bending moment across the beam depth. This system leads to bending moments being transmitted into the connection zone by a couple formed from the tensile force

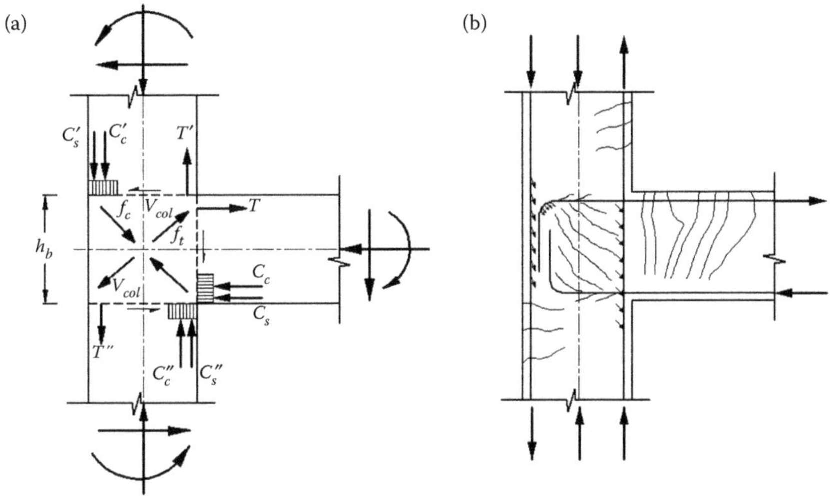

FIGURE 7.11 Internal forces and cracking of exterior beam column connection: (a) internal forces and (b) cracking pattern.

in the main beam reinforcement and an equal and opposite force on the compression side of the beam.

From the position of the stress resultants, apparently diagonal tension stress f_t and compression stress f_c are induced in the panel zone of the joint. The diagonal tension may be high when the ultimate capacity of the adjoining members is developed. This can lead to extensive diagonal cracking. The severity of diagonal tension is influenced by flexural steel content and the magnitude of the axial compression load on the column (Park and Paulay, 1975).

As a consequence, bond stresses are developed along the beam tension steel as the load is transferred into the concrete. For the load pattern in Figure 7.11a, assuming that the axial compression on the column is small, the surrounding concrete is subjected to sedimentation and it is exposed to transverse tension. Usually a splitting crack forms along these bars at a relatively early stage of the loading. Repeated loading will aggravate the situation and a complete loss of bond up to the beginning of the bent portion of the bar may occur (Park and Paulay, 1975; Scott et al., 1994). Consequently, high bearing stresses generated in the bend arise due to the bending of the beam steel either down or up into the column, or into a U-bar. Thus, there is a combination of stresses acting in the joint zone of the beam–column connection—axial compression, shear, bond, and bearing stresses—which makes the behavior of this zone complex and difficult to resolve (Scott et al., 1994).

The straight vertical portion following the bend must be sufficiently long if the full strength of the top bar is to be developed. It may be noted that bending the top steel into the joint induces concrete hoop forces along the "right" direction. The forces transmitted from the bars to the concrete by the bearing or bond are indicated by the small arrows in Figure 7.11b (Park and Paulay, 1975).

The bottom beam bars, in compression, enter the joint in a region of ideal bond conditions, since the surrounding concrete is also in compression transversely to the bars. Owing to reversed loading and subsequent possible tensile yielding of these bars, however, serious bond deterioration can occur here too. The straight portion of the bars beyond the bend remains largely ineffective for compression loads. Therefore, after a few cycles of reversed seismic loading, serious anchorage losses can occur, particularly when the beam frames into a shallow column (Park and Paulay, 1975).

The outer column bars are subjected to perhaps the most severe bond conditions. Over the depth h_b of the beam, a total bond force needs to be transferred to the concrete in the joint, if the internal forces at the critical sections across the column are to be sustained. If the code recommendations are to be adhered too, the available anchorage length h_b is grossly inadequate. Moreover, the entire bond force is to be transferred into the panel zone of the joint and not, as one might be tempted to assume, partly into the cover and partly into the core of the joint. The extremely high bond stresses along the outer column bars can be the cause of vertical splitting cracks. These might interconnect and in turn cause the cover to separate itself. Unfortunately, the failure planes along the splitting cracks around these bars coincide during reversed loading (Park and Paulay, 1975).

Under severe lateral loads, ductile moment-resisting reinforced concrete frames will be subjected to large loads and displacements. These deformations result in large shear and bond stresses that must be resisted by the joint. Cyclic loading in cross-cracked concrete causes a repeated opening and closure of cracks. Owing to the dominance of shearing action across the joint, movements parallel to the open cracks will also occur. When the cracks become large, because the transverse reinforcement has yielded, the process of grinding and progressive splitting due to uneven concrete bearing begins. A complete disintegration of the concrete within the body of the joint may result. This is associated with a drastic volumetric increase in the core unless containment is provided (Park and Paulay, 1975).

A study of exterior beam–column connections was undertaken by Hanson and Conner (1967). They demonstrated that without transverse reinforcement, exterior beam–column connections could not sustain much load after the third moderate cycle of reversed loading, where the concrete burst and the column bars buckled. In joints that contained hoop reinforcement equal to that required for confinement in the column above and below, the hoop steel stresses increased during cyclic alternating loading till yielding occurred. This clearly showed the important role of transverse joint reinforcement for seismic-type loading.

Compression from the column axial load, when it acts in combination with high shear, bond, and bearing stresses, can be expected to improve joint behavior and reduce the demand for joint shear reinforcement. Such effect has been studied experimentally by Scott (1992). It was found that high column axial load specimens had developed, or had closely approached, their full theoretical moment of resistance when initial joint cracking occurred, but low column axial load specimens had reached only about 50% of their theoretical moment of resistance when the joint first cracked. High column axial load specimens with beam tension steel ratio of 1%, had failed by the development of a plastic hinge in the beam at the face of the column; therefore, high reinforcement strains were recorded due to gross yield of the beam bars. A low column axial load with similar specimens had resulted in failure by extensive joint cracking and low strains were recorded.

7.6.2 Role of Detailing

Tests on 15 reinforced concrete exterior beam–column connections were studied by Scott (1992). Three reinforcement details were used (bending beam tension steel up or down into the column and the U-bar detail, Figure 7.12) in the conjunction between the column and the beam. Also, the beam depth, beam tension steel percentage, and column axial load were varied. Specimens with 1% beam tension reinforcement bent down into the column, or bent into a U-bar, failed by development of a plastic hinge in the beam at the face of the column when a high column load was used. High reinforcement strains were recorded due to the gross yield of the beam bars. All other specimens with the beam tension reinforcement bent up detail, failed by extensive joint cracking and reinforcement strains, were lower and sometimes in the elastic range. Up to joint cracking, load transfer in all three beam details (up, down, and U-bar detail) was predominantly by bond stresses developed at the bend. After joint cracking, bars bent down and the U-bars compensated for loss of bond at the bend by developing bond stresses over an increasing part of their length, giving substantial load increments between joint cracking and failure. The bars bent up detail were unable to do this, resulting in brittle failure. Overall, the beam tension bars bent down into the column and U-bars details performed significantly better than the beam tension bars bent up detail. However, if ductility is a prime consideration, use of the beam tension bars bent down detail would appear preferable.

Park and Paulay (1975) have suggested some reinforcement detailing of exterior beam–column connections which are made to satisfy the requirement of anchorage. As shown in Figure 7.13a, the development length of the beam reinforcement should be computed from the beginning of the 90° bend, rather than from the face of the column, because of the inevitable loss of bond at an exterior joint. In wide columns, any portion of the beam bars within the outer third of the column could be considered for computing the development length as shown in Figure 7.13b. In case of narrow columns, the use of stub beams, as shown in Figure 7.13b, will be imperative. A large diameter bearing bar fitted along the 90° bend of the beam bars should be beneficial in distributing bearing stresses. In deep columns and whenever straight beam bars are preferred, mechanical anchorages, as shown in Figure 7.13c,

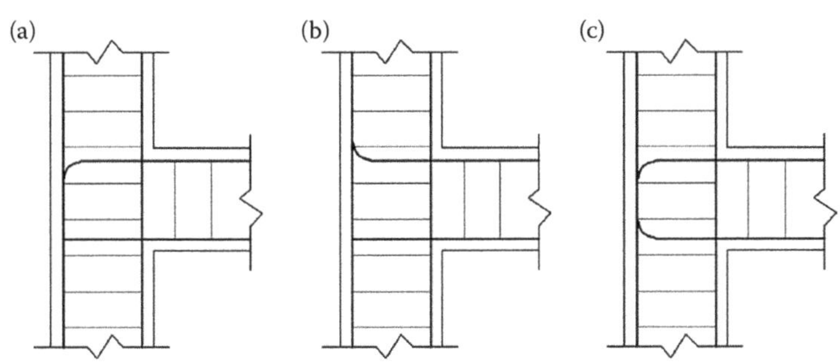

FIGURE 7.12 Reinforcement details for exterior beam column connection: (a) beam bars bent down into the column, (b) beam bars bent up into the column, and (c) "U" bar detail.

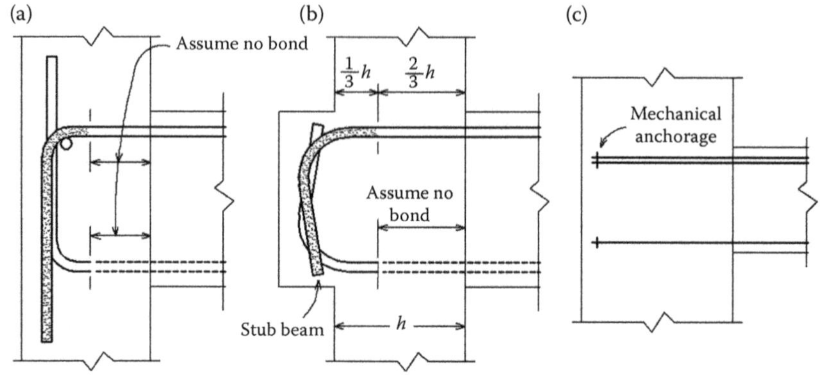

FIGURE 7.13 Suggested anchorage of bars in columns for exterior beam column connection using: (a) bent-up bars, (b) bent-up bars in stub beam, and (c) mechanical anchorage.

could be advantageous. The top bars in a beam passing through holes in a bearing plate may be welded to a steel plate. Joint ties should be so arranged that the critical outer column bars and the bent-down portions of the beam bars are held against the core of the joint.

7.6.3 Strut-and-Tie Modeling

In Figure 7.14, different *STMs* of the exterior beam–column connection are shown for different stiffness of the adjacent members connected with the joint. The simple *STM* as shown in Figure 7.14a suggests that the shearing and compression forces resulting from the particular load pattern are largely transmitted by a diagonal strut across the joint. In fact, there are several struts separated by diagonal cracks, which can be represented by the truss model in Figure 7.15. It would be extremely optimistic to assume that the full compression strength could be approached in these struts. Not only are they subjected to indeterminate eccentricities, but also they are exposed to transverse tensile strains. In this biaxial state of stress, a considerable reduction of compressive strength ensues (Park and Paulay, 1975).

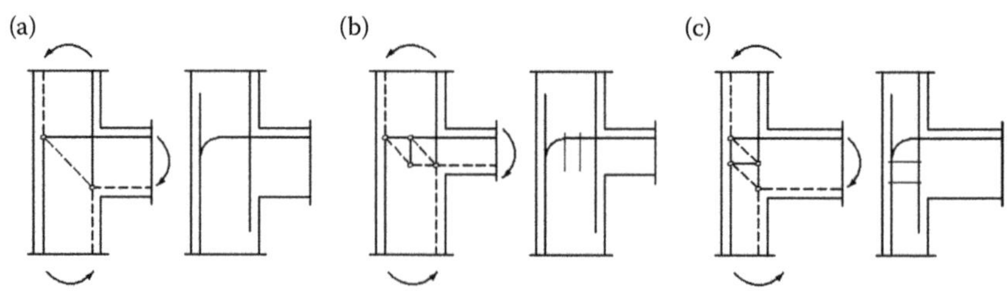

FIGURE 7.14 Suggested *STMs* and corresponding details for exterior beam–column connection with beam to column thickness: (a) between 0.67 and 1.5, (b) less than 0.67, and (c) greater than 1.5.

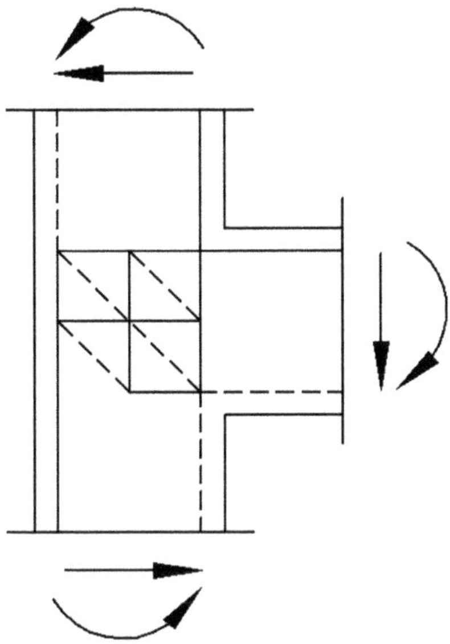

FIGURE 7.15 Truss model for exterior beam–column connection.

7.6.4 Example 7.3: Strength Assessment of an Exterior Joint

In this example, the strength capacity of a reinforced concrete exterior beam–column connection tested by Abrams (1987) is assessed via the method of *STM*. The test specimen is a planner exterior beam–column assembly, Figure 7.16a. Nonlinear flexural behavior was expected to occur at the ends of the beam while the columns were expected to crack but not to yield. The connection upper and lower columns are 343 mm wide and 457 mm overall depth with 8ϕ16 reinforcing bars as shown in Figure 7.16a. The connection beam is 343 mm wide and 343 mm overall depth with

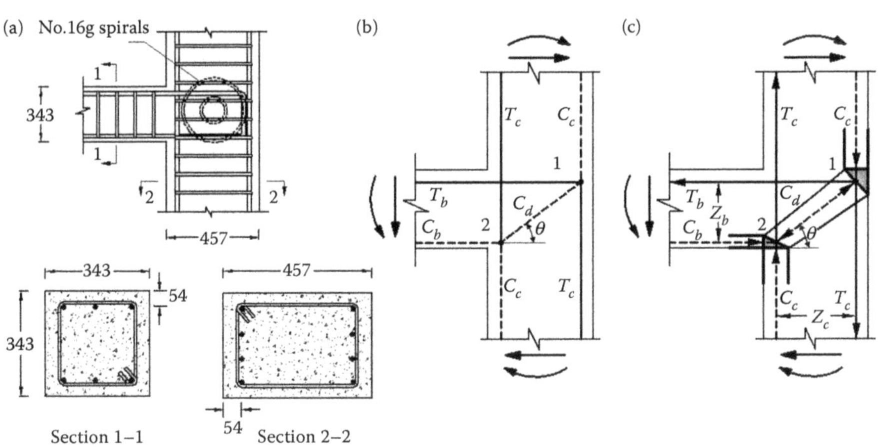

FIGURE 7.16 Example 7.3—an exterior beam–column connection: (a) connection, (b) *STM*, and (c) *STM* details.

$3\phi19$ top and bottom reinforcing bars. The beam depth $d_b = 289$ mm and the column depth $d_c = 403$ mm. The steel yield stress, $f_y = 470$ MPa and the concrete cylinder strength, $f_c' = 32.6$ MPa.

It is obvious that the beam is the weakest member, though it carries a moment equal to the summation of the moments of the columns. Therefore, the solution starts with investigating the beam. The appropriate *STM* is illustrated in Figure 7.16b.

Geometry of the STM and Forces:
The beam tension steel is $3\phi19 = 850.6$ mm². Upon assuming the beam tension reinforcement yields, the tension force of the beam, $T_b = 470 \times 850.6 = 399.8$ kN. The compression force of the beam, $C_b = 399.8\,\text{kN} = (0.85 f_c' \beta_s) w_{C_b} b$. Though strut C_b is a prismatic strut, it is acting at node 2, which is a $C - C - T$ node, and therefore it is better to take β_s equal to β_n of node 2, that is, 0.8. Then $w_{C_b} = 52.6$ mm.

The beam lever arm, $z_b = 289 - 0.5 \times 52.6 = 262.7$ mm. Then the nominal moment of the beam, $M_{bn} = 399.8 \times 262.7 = 105.0$ kN.m. Thus, the column moment is $M_c = 0.5 M_{bn} = 52.5$ kN.m.

The column reinforcement on either side is located at 54 mm from the outer edge of the column. Therefore, the width of either C_c or T_c can be taken as $2 \times 54 = 108$ mm, giving a lever arm, $z_c = 403 - 54 = 349.0$ mm. The width of T_b can be taken as $2 \times (343 - 289) = 108$ mm. As for the width of strut C_b, it has been calculated before as 52.6 mm. These geometrical relations are illustrated in Figure 7.16c. The force $C_c = T_c = M_c/z_c = 150.4$ kN. The angle $\theta = \tan^{-1} z_b/z_c = 36.97°$. The force in the diagonal strut is $C_d = T_b/\cos\theta = 500.4$ kN.

Effective concrete strength of the nodes and the struts:
Node 1 is a $C - T - T$ node; thus, $f_{ce}^n = 0.85 f_c' \beta_n = 0.85 \times 32.6 \times 0.6 = 16.63$ MPa

Node 2 is a $C - C - T$ node; thus, $f_{ce}^n = 0.85 \times 32.6 \times 0.8 = 22.17$ MPa

Struts C_b and C_c are prismatic struts; thus, $f_{ce}^s = 0.85 f_c' \beta_s = 0.85 \times 32.6 \times 1.0 = 27.71$ MPa

Strut C_d is a bottle-shaped stress field; thus, $f_{ce}^s = 0.85 f_c' \beta_s = 0.85 \times 32.6 \times 0.75 = 20.8$ MPa if transverse reinforcement to resist the lateral tension is provided; otherwise, $f_{ce}^s = 0.85 f_c' \beta_s = 0.85 \times 32.6 \times 0.6 = 16.63$ MPa

Node 1:
The nominal strength of strut C_c is $C_{cn} = 16.63 \times 108 \times 343 = 616\text{ kN} > C_c$

The width of strut C_d at node 1, $w_{C_d}^1 = 108 \sin\theta + 108 \cos\theta = 151.2$ mm

The nominal strength of strut C_d, $C_{dn} = 16.63 \times 151.2 \times 343 = 862.7\text{ kN} > C_d$, that is, the strut is safe without consideration for the transverse reinforcement.

Node 2:
The nominal strength of strut C_c is $C_{cn} = 22.17 \times 108 \times 343 = 821.3\text{ kN} > C_c$

The nominal strength of strut C_b is $C_{bn} = 22.17 \times 52.6 \times 343 = 400.0\text{ kN} > C_b$

The width of strut C_d at node 2, $w_{C_d}^2 = 108 \sin\theta + 52.6 \cos\theta = 106.7$ mm

The nominal strength of strut C_d, $C_{dn} = 16.63 \times 106.7 \times 343 = 608.6\text{ kN} > C_d$, that is, the strut is safe without consideration for the transverse reinforcement.

From the previous results, all the model components have nominal strength above the requirement from the forces acting on these elements. This finally means that the assembly can carry a moment applied to the beam equal to $M_{STM} = 105.0$ kN.m. This assembly failed in the test under a moment, $M_{EXP} = 119$ kN.m; that is, $M_{STM}/M_{EXP} = 88.2\%$

7.7 TEE BEAM–COLUMN CONNECTIONS

7.7.1 Joint Behavior

Under gravity loads, the tee beam–column joint transmits tension and compression forces due to the loads from the roof beams loads at the beam–column interface directly through the joint into the internal column without developing extensive cracking and diagonal tensile stresses in the joint.

In laterally loaded frames, the system of external action and the corresponding internal forces generated around the joint zone of the tee beam–column connection in cast-in-place frame construction is indicated in Figure 7.17. In this case, the joint behavior is similar to that of the exterior beam–column connection (MacGregor and Wight, 2005).

7.7.2 Role of Detailing

The most common detail, shown in Figure 7.18a, produces unacceptably low joint efficiency. Joints reinforced as shown in Figure 7.18b and c had a much better performance in experimental tests (Nilsson and Losberg, 1976). The hooks in those two reinforcement details act to restrain the opening of the inclined crack and to anchor the diagonal compressive strut in the joint. The reinforcement detail shown

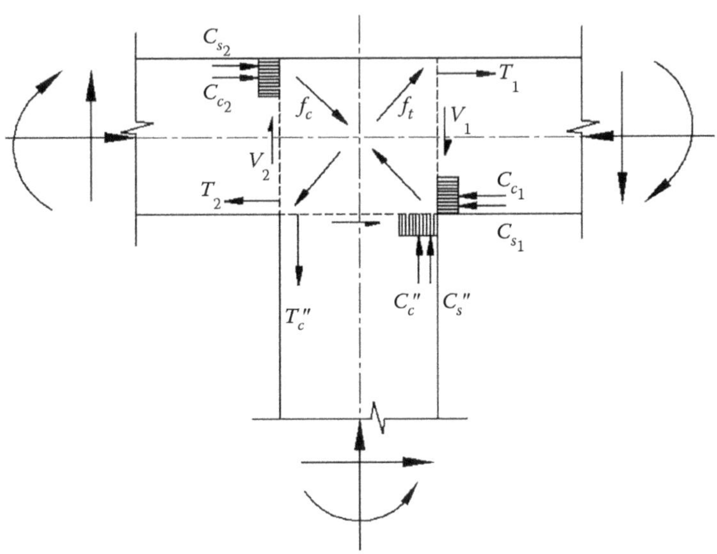

FIGURE 7.17 Internal forces of the tee beam–column connection.

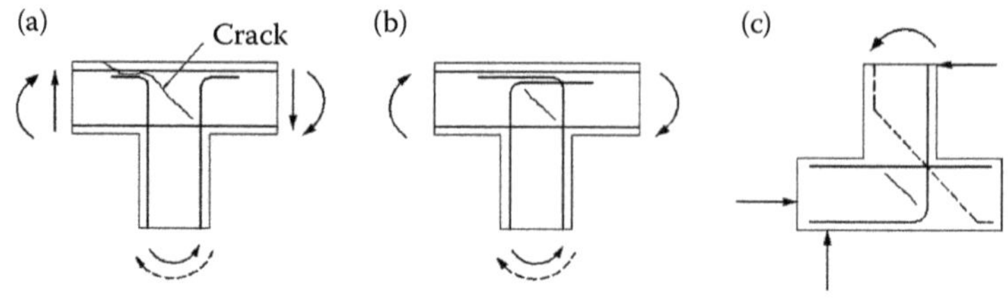

FIGURE 7.18 Different details of the tee beam–column connection: (a) unsatisfactory detail, (b) satisfactory detail, and (c) satisfactory detail in case of retaining walls.

in Figure 7.18c is satisfactory to develop the strength in the case of the retaining wall. The diagonal bar, shown in dashed lines, can be added if control of cracking at the base of the wall is desired.

7.7.3 Strut-and-Tie Modeling

In Figure 7.19a, the appropriate *STM* for the case of the gravity load is shown. In a laterally loaded frame, the forces acting on a T-joint can be idealized as shown by the simple *STM* in Figure 7.19b (or 7.19c). This model suggests that the internal forces resulting from the particular load pattern are largely transmitted by a diagonal strut across the joint.

Additional *STMs* of the tee beam–column connection for different geometry and detailing of the adjacent members connected with the joint are shown in Figure 7.19d and e.

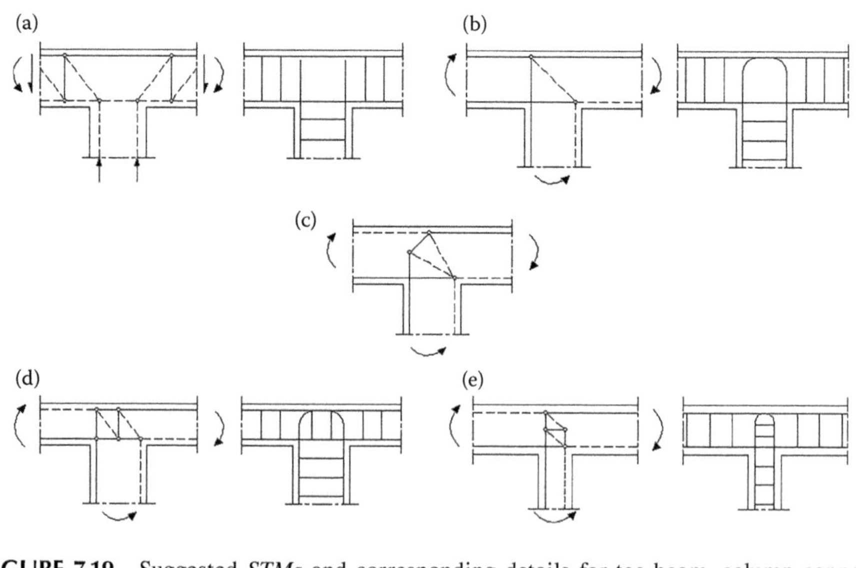

FIGURE 7.19 Suggested *STMs* and corresponding details for tee beam–column connections: (a) due to gravity loads and (b–e) due to lateral loads.

7.8 INTERIOR BEAM–COLUMN CONNECTIONS

7.8.1 JOINT BEHAVIOR

An interior joint under gravity loads transmits the tension and compression forces at the beam–column interface and the forces from the upper column, directly through the joint into the lower column without developing extensive cracking in the joint. In a laterally loaded frame, interior joints experience significant shear forces which develop extensive cracking in the joint core.

There are two mechanisms that are capable of transmitting the joint horizontal shear forces from one face of the joint to the other: the compression strut mechanism and the panel truss mechanism, as shown in Figure 7.20. In the first mechanism, the compression strut mechanism, a large strut is formed between the two opposite corners of the joint in compression. In the second mechanism, the panel truss mechanism, a truss is formed by the intermediate joint ties acting as tension members and smaller inclined struts acting in compression. Both of the two mechanisms represent the two extremes of the joint behavior. The compression strut in the first mechanism depends mainly on the compression blocks in the beams and columns, and on the crushing strength of the concrete in the strut. On the other hand, the panel truss in the second mechanism counts on the transfer by bond inside the joint and on the horizontal transverse steel and the vertical column reinforcement to sustain the truss action (Paulay, 1989).

In practice, the behavior of the joint probably falls somewhere between the two mechanisms. Since the joint is loaded to the first yield, both the strut and the truss mechanisms contribute to the joint shear resistance with the strut contributing more because large strains are required to activate the tension members in the panel truss. After the first yield had occurred, two paths are possible depending on the bond condition (Leon, 1990).

For design purposes, it is important to quantify the contribution of each of the aforementioned mechanisms. If a strut mechanism is accepted, compressive forces will be introduced into the joint by the bearing of the beam concrete blocks and the tensile forces by bond stresses where the joint is loaded to the first yield. In this case,

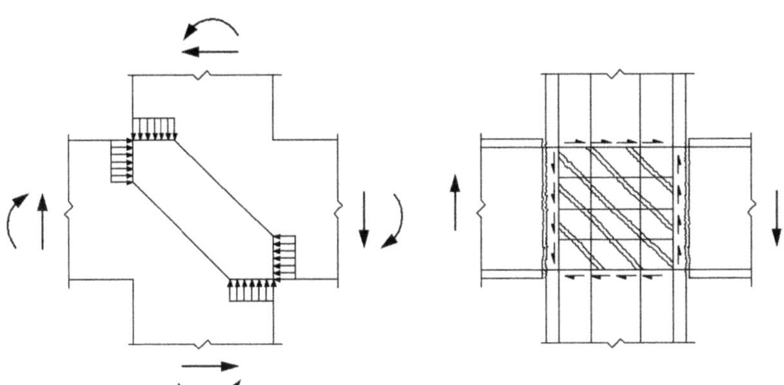

FIGURE 7.20 Shear transfer mechanisms in the interior beam column connection: (a) strut mechanism and (b) truss panel mechanism.

the required joint reinforcement reduces to that which is necessary to confine the concrete properly (Leon, 1990).

On the other hand, if a truss panel mechanism is accepted, then all forces will be introduced into the joint by bond from the bars because large cracks are presumed to exist at the joint beam interface where the beam steel has yielded. In this case, a large amount of transverse steel is required in the joint which is necessary to maintain the joint strength (Leon, 1990).

7.8.2 Bond Condition and Confinement

The anchorage length of the joint reinforcement is influenced by the bond condition which has a main role in joint behavior. After joint loading to the first yield has occurred, two paths are possible, depending on the bond conditions.

If a very good bond is present, and therefore all yielding takes place at the joint face with little or no yield penetration into the joint, the panel truss mechanism as shown in Figure 7.20b will prevail. Yielding of reinforcement at the column face results in a separation of the beams from the column, and the lack of a bar slip prevents the beam concrete compression blocks from confining the joint. Therefore, all the compressive forces are transferred to the joint by bond stress, and the efficiency of the strut mechanism is consequently reduced. To sustain this type of behavior through a severe load history, a large amount of transverse reinforcement as well as long anchorage lengths is required. Theoretically, there would be only a small amount of distributed cracking in the joint since shear stress levels are low, and therefore there is little or no loss of energy dissipation capacity with cycling.

If, on the other hand, the bond conditions are poor such that yield penetration and bar slip begin to occur, a completely different situation arises. The beam bars are then anchored in the beams at the opposite side of the joint, increasing the size of compressive blocks on the beams and thus increasing the effectiveness of the strut mechanism. The joint corner is therefore better confined and the strut mechanism is more effective. The increased use of the strut mechanism is associated with large compressive strains and therefore with more shear cracking of the joint. This results in larger shear deformations and losses of energy dissipation capacity (Leon, 1990).

7.8.3 Strut-and-Tie Modeling

The appropriate *STM* for the case of gravity load is shown in Figure 7.21a. For a laterally loaded connection, the simple *STM* as shown in Figure 7.21b suggests that the shearing and compression forces resulting from the particular load pattern are largely transmitted by a diagonal strut across the joint. The *STM* shown in Figure 7.21c considers both the strut and truss mechanism contribution in transferring shear.

7.8.4 Example 7.4: Strength Assessment of an Interior Joint

In this example, the strength capacity of a reinforced concrete interior beam–column connection tested by Abrams (1987) is assessed via the method of *STM*. The test specimen is a planner exterior beam–column assembly, Figure 7.22a. Nonlinear

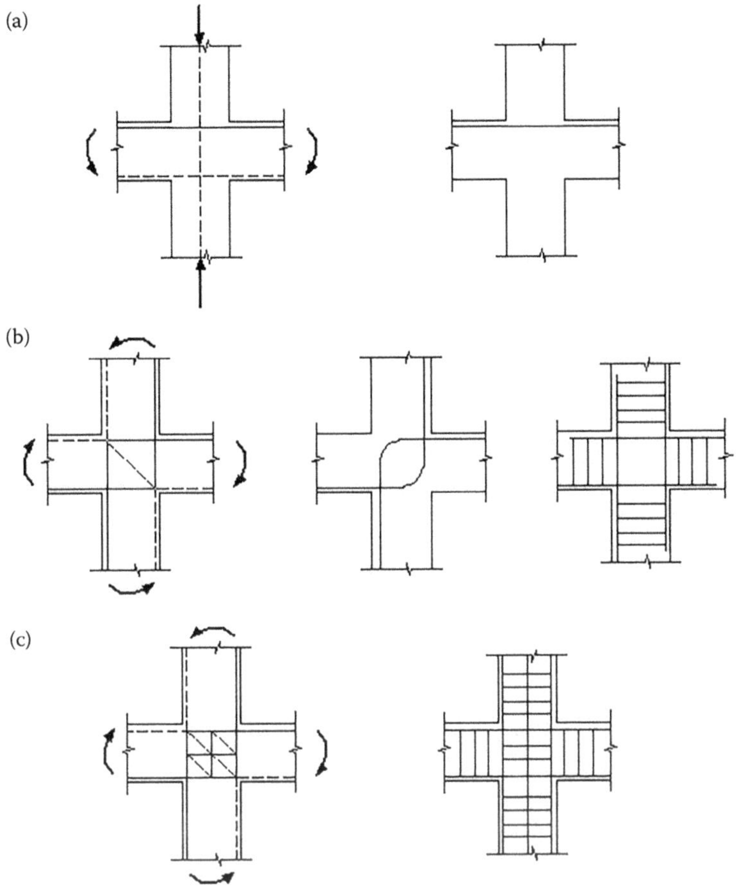

FIGURE 7.21 Suggested *STMs* and corresponding details for interior beam–column connections: (a) due to gravity loads and (b, c) due to lateral loads.

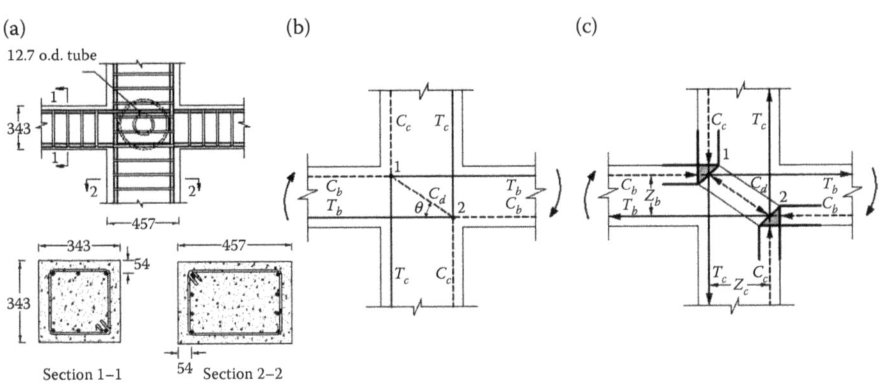

FIGURE 7.22 Example 7.4—an interior beam–column connection: (a) connection, (b) *STM*, and (c) *STM* details.

flexural behavior was expected to occur at the ends of the beams while the columns were expected to crack but not to yield. The connection upper and lower columns are 343 mm wide and 457 mm overall depth with 8ϕ16 reinforcing bars as shown in Figure 7.22a. The connection beams are 343 mm wide and 343 mm overall depth with 3ϕ19 top and bottom reinforcing bars. The beam depth $d_b = 289$ mm and the column depth $d_c = 403$ mm. The steel yield stress, $f_y = 470$ MPa and the concrete cylinder strength, $f_c' = 25.5\,\text{MPa}$.

It is obvious that the beams are weaker than the columns. Therefore, the solutions starts with investigating the beams. The appropriate *STM* is illustrated in Figure 7.22b.

Geometry of the STM and forces:
The tension steel of either beam is 3ϕ19 = 850.6 mm^2. Upon assuming that the beam tension reinforcement yields, the tension force of the beam, $T_b = 470 \times 850.6 = 399.8$ kN. The compression force of the beam, $C_b = 399.8\text{ kN} = (0.85 f_c' \beta_s) w_{C_b} b$. Though strut C_b is a prismatic strut, the strut is connected to node 1 (or 2), and therefore it should be influenced by the strength of this node. Node 1 is a $C - T - T$ node; however, the effectiveness factor of this node is taken as 0.8 since 50% of the load is transferred as a $C - C - C$ node and the other 50% is transferred as a $C - T - T$ node. Another way to look at the node is by dividing it diagonally, giving two equal $C - C - T$ nodes. Hence, the size of strut C_b is based on an efficiency factor 0.8, giving $w_{C_b} = 67.2\,\text{mm}$.

The beam lever arm, $z_b = 289 - 0.5 \times 67.2 = 255.4$ mm. Then the nominal moment of the beam, $M_{bn} = 399.8 \times 255.4 = 102.1$ kN.m. Thus, the column moment is $M_c = 102.1$ kN.m.

The column reinforcement on either side is located at 54 mm from the outer edge of the column. Therefore, the width of either C_c or T_c can be taken as $2 \times 54 = 108$ mm, giving a lever arm, $z_c = 403 - 54 = 349.0$ mm. The width of strut C_b can be taken as calculated before $w_{C_b} = 67.2\,\text{mm}$; however, for simplicity, it is taken to be the same as that of T_b. Then the width of either C_b or T_b is $2 \times (343 - 289) = 108$ mm. Hence, z_b is modified to $z_b = 289 - 0.5 \times 108 = 235.0$ mm. These geometrical relations are illustrated in Figure 7.22c. The force $C_c = T_c = M_c/z_c = 292.6$ kN. The angle $\theta = \tan^{-1} z_b/z_c = 33.95°$. The force in the diagonal strut is $C_d = (T_b + C_b)\cos\theta + (T_c + C_c)\sin\theta = 990.1$ kN.

Effective concrete strength of the nodes and the struts:
The strength of nodes 1 and 2, as explained before, is $f_{ce}^n = 0.85 f_c' \beta_n = 0.85 \times 25.5 \times 0.8 = 17.3\,\text{MPa}$

Struts C_b and C_c are prismatic and their strength is $f_{ce}^s = 0.85 f_c' \beta_n = 0.85 \times 25.5 \times 1.0 = 21.7\,\text{MPa}$

Strut C_d is a bottle-shaped stress field with sufficient transverse reinforcement, hence its strength is $f_{ce}^s = 0.85 f_c' \beta_n = 0.85 \times 25.5 \times 0.75 = 16.3\,\text{MPa}$.

The strength of either strut C_b or C_c is controlled by the strength of the node; whereas the strength of strut C_d is controlled by the strength of the strut itself.

Node 1 or 2:
The nominal strength of strut C_b is $C_{bn} = 17.3 \times 108 \times 343 = 640.9\text{ kN} > C_b$

The nominal strength of strut C_c is $C_{cn} = 17.3 \times 108 \times 343 = 640.9\text{ kN} > C_c$

The width of strut C_d at either node, $w_{C_d} = 108\sin\theta + 108\cos\theta = 149.9\,\text{mm}$

The nominal strength of C_d, $C_{dn} = 16.3 \times 149.9 \times 343 = 838.1$ kN < C_d.

Strut C_d is the only critical component, $C_{dn} = 0.85C_d$, and all other elements attain nominal strength exceeding their forces. Hence, the forces in all elements should be reduced to 85%, giving $C_b = 339.8$ kN and $M_b = 339.8 \times 235.0 = 79.9$ kN.m. The assembly failed due to a moment $M_{EXP} = 106.0$ kN.m; hence, $M_{STM}/M_{EXP} = 75.3\%$.

REFERENCES

Abdul-Wahab, H. M., and Ali, W. M., Strength and behavior of reinforced concrete obtuse corners under opening bending moments, *ACI Structural Journal*, 86(6), 1989, 679–685.

Abrams, D. P., Scale relations for reinforced concrete beam–column joints, *ACI Structural Journal*, 84(6), 1987, 502–512.

ACI 318-14, *Building Code Requirements for Structural Concrete and Commentary*, Detroit: American Concrete Institute, USA, 2014, 519 pp.

Balint, P. S. and Taylor, H. P. J., Reinforcement detailing of frame corner joints with particular performance to opening corners, Technical Report 42.462, Cement and Concrete Association, London, Feb., 1972, 16 pp.

El-Metwally, S. E., On the behavior and design of reinforced concrete beam-column connection, Engineering Research Bulletin, Matteria Faculty of Engineering and Technology, University of Helwan, Cairo, Egypt, Vol. 1, 2–6 January, 1992.

Hanson, N. W. and Conner, H. W., Seismic resistance of reinforced concrete beam-column joints, *ASCE, Journal of the Structural Division*, 93(ST5), 1967, 533–560.

Leon, R., Shear strength and hysteretic behavior of interior beam-column Joints, *ACI Structural Journal*, 87(1), 1990, 3–11.

MacGregor, J. G. and Wight, J. K., *Reinforced Concrete Mechanics and Design*, Pearson Prentice Hall, New Jersey, 2005.

Metawei, H., Behavior of beam-column joints in monolithic reinforced concrete structures, *Master's Thesis*, Ain Shams University, Egypt, 1990.

Nilsson, I. and Losberg, A., A discussion of the paper opportunities in bond research, by ACI Committee 408, *ACI Journal*, 68(5), 1971, 393–396.

Nilsson, I. and Losberg, A., Reinforced concrete corners and joints subjected to bending moments, *Journal of the Structural Division*, ASCE, 102(ST6), 1976, 1229–1254.

Park, R. and Paulay, T., *Reinforced Concrete Structures*, John Wiley & Sons, New York, 1975.

Paulay, T., Equilibrium criteria for reinforced concrete beam-column joints, *ACI Structural Journal*, 86(6), 1989, 635–643.

Schlaich, J. and Schäfer, K., Design and detailing of structural concrete using strut-and-,tthat is,models, *Journal of the Structural Engineer*, 69(6), 1991, 113–125.

Schlaich, J., Schäfer, K., and Jennewein, M., Toward a consistent design of structural concrete, *Journal of the Prestressed Concrete Institute*, 32(3), 1987, 74–150.

Scott, R. H., The effects of detailing on reinforced concrete beam-column connection behavior, *Journal of the Structural Engineer*, 70(18), 1992, 318–324.

Scott, R. H., Feltham, I., and Whittle, R. T., Reinforced concrete beam-column connections and BS 8110, *Journal of the Structural Engineer*, 72(4), 1994, 55–60.

Yuan, R. L., McLelland, G. R., and Chen, W. F., Experiments on closing reinforced concrete corners, *Journal of Structural Engineering*, ASCE, 108(4), 1982, 771–779.

8 Pile Caps

8.1 INTRODUCTION

If the state of stress is not predominantly plane, *three-dimensional strut-and-tie models* (3D *STMs*) should be used (Schlaich et al., 1987). With large 3D reinforced concrete blocks, load paths for the concentrated load may be considered in two directions, as illustrated in Figure 8.1, where the strut-and-tie method is applied in two perpendicular planes. Unlike *two-dimensional strut-and-tie models (2D STMs), 3D STMs* are required when the structure and loading are considerably spread over all the three dimensions, such as pile caps with two or more rows of piles.

Loads are essentially transferred in pile caps through 3D stress fields; therefore, *3D STMs* should be used, Figure 8.2 (Engström, 2011). Nevertheless, in some instances, 2D models are possible to employ. 3D models usually require less amount of reinforcement since the load path is direct and hence shorter. In 3D models, the geometry of struts and nodal zones may represent difficulty and therefore simplified geometry can be adopted as will be illustrated in a subsequent section. On the other hand, 2D models are easier to handle than 3D models.

As a result of the spatial nature of pile caps, Figure 8.2, their struts and nodal zones are distinct from those of 2D structure members. Struts in pile caps are surrounded by significant volumes of concrete with very low stress levels. These concrete volumes are called inactive concrete and they play a role in improving the strength of struts. Such effect will be discussed in a subsequent section.

Nodal zones in pile caps are bounded by 3D stress fields. Hence, such effect should be implemented in the estimate of their strength. This feature is discussed in a subsequent section.

In addition to the aforementioned features of pile caps, this chapter provides examples to illustrate the design procedure and strength assessment of pile caps. The design procedure and modeling of pile caps subjected to vertical loads and moments are covered. Nevertheless, the treatment of pile caps under vertical or vertical and lateral loads using 2D modeling is illustrated with two examples before proceeding to the 3D aspects. Since the solution of pile caps starts with the determination of pile loads, this point is discussed first in this chapter.

8.2 DISTRIBUTION OF PILE LOADS

The flexural rigidity of a pile cap significantly affects the distribution of pile loads. Design engineers very often assume that pile caps are rigid enough such that a uniform distribution of loads among piles is guaranteed. This assumption is acceptable if the piles are close to the column such that the angle of inclination of every strut deviating from underneath the column to a pile is larger than 45°. Of course, if the

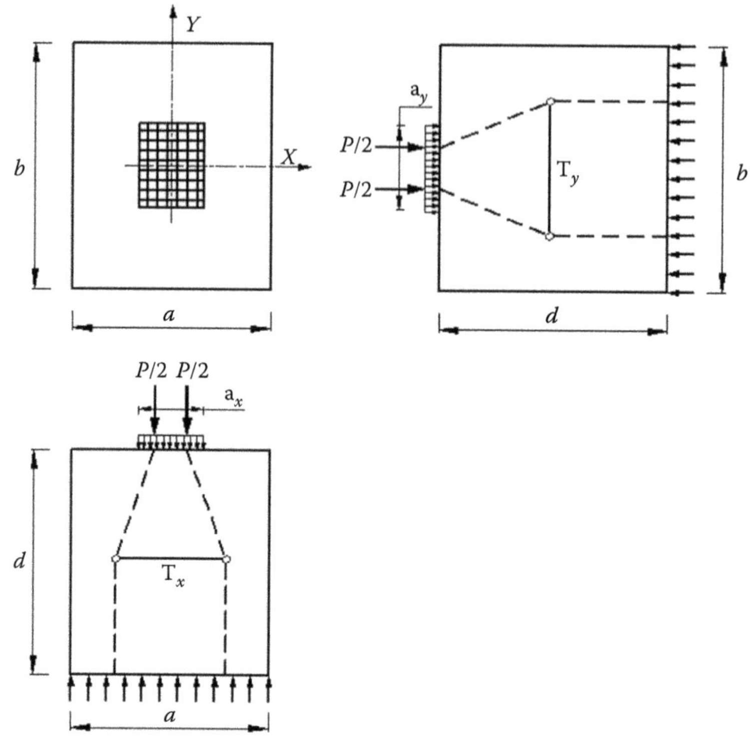

FIGURE 8.1 Applied *STM* in two different planes.

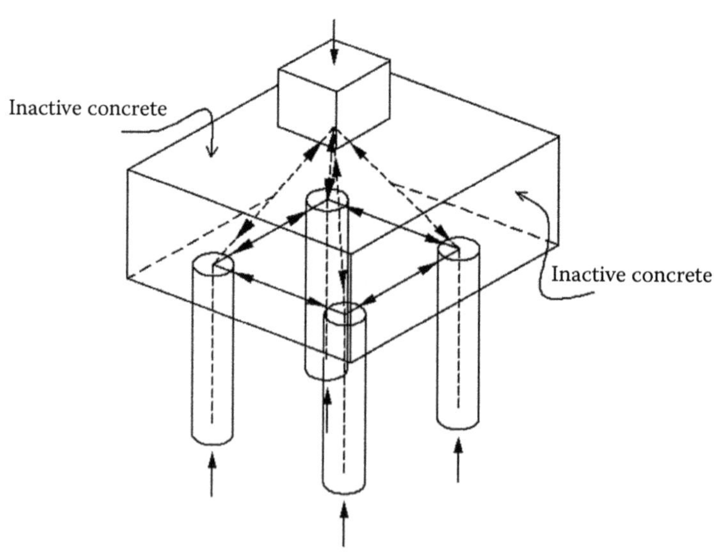

FIGURE 8.2 Example of a pile cap, showing a 3D *STM* and inactive concrete.

distance between the column and every supporting pile is the same, the pile loads will be equal.

On the basis of a linear elastic finite element analysis, Ghali (1999) stated that if the ratio ℓ/t exceeds 2.4, a flexible behavior is expected, where ℓ is the distance between the column centerline and the center of the furthermost corner pile and t is the thickness of the pile cap. Beredugo (1967) has shown that the load applied to a rigidity caped free-standing pile group is usually not uniformly distributed. Adebar et al. (1990) carried out tests on pile caps of different geometry and properties of concrete and reinforcement. They arrived at the same conclusion of nonuniform distribution of pile loads. *In such a case, a reliance on a linear elastic finite element analysis would be reasonable for pile load assessment.*

8.3 2D (INDIRECT) MODELING OF PILE CAPS

8.3.1 Design Example 8.1

A pile cap 4200 × 3600 × 1400 mm is supported by four piles of 600 mm diameter each and supports a 500 × 1500 mm column as shown in Figure 8.3. Using 2D *STM* design the reinforcement and check the nodes and struts of the pile cap for the following load combinations

Case 1: $N_u = 6000$ kN and $M_{uy} = 0.0$
Case 2: $N_u = 2700$ kN and $M_{uy} = \pm 4950$ kN.m

Assume that the concrete cylinder strength, $f_c' = 30.0$ MPa, and the yield stress of steel reinforcement, $f_y = 420.0$ MPa.

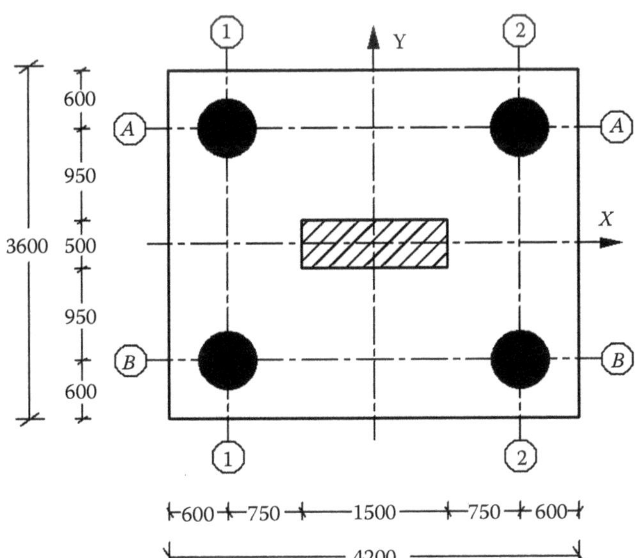

FIGURE 8.3 Example 8.1.

Column sectional design

For the given straining actions of the two cases of loading, the column reinforcement is 15ϕ28 on each short side. The bars are placed in three rows and their centroid is approximately located at 130 mm from the column short edges.

Load case 1

With reference to Figure 8.3, the column forces are first transferred in the x-direction to the lines connecting the piles on axes (1) and (2); then the forces are transferred to the piles. For this case, $N_u = 6000$ kN and $M_{uy} = 0.0$ are replaced by two stress resultants, 3000 kN each, located approximately at 330 mm from the column short edges, and transferred in the x-direction via the *STM* shown in Figure 8.4a. Of course, for design optimization, these resultant forces can be located closer to the column edges (at $\approx$160 mm). From the model illustrated in the figure, the tension force $T = 2700$ kN can be resisted by a reinforcement $= 8571.4$ mm^2 $= 14\phi28$ to be arranged in one layer within a width equal to twice the column breadth (1000 mm).

For either axis (1) or (2), Figure 8.3, the 3000 kN force is transferred to the piles on the axis via the *STM* shown in Figure 8.4b. From the model, the tension force

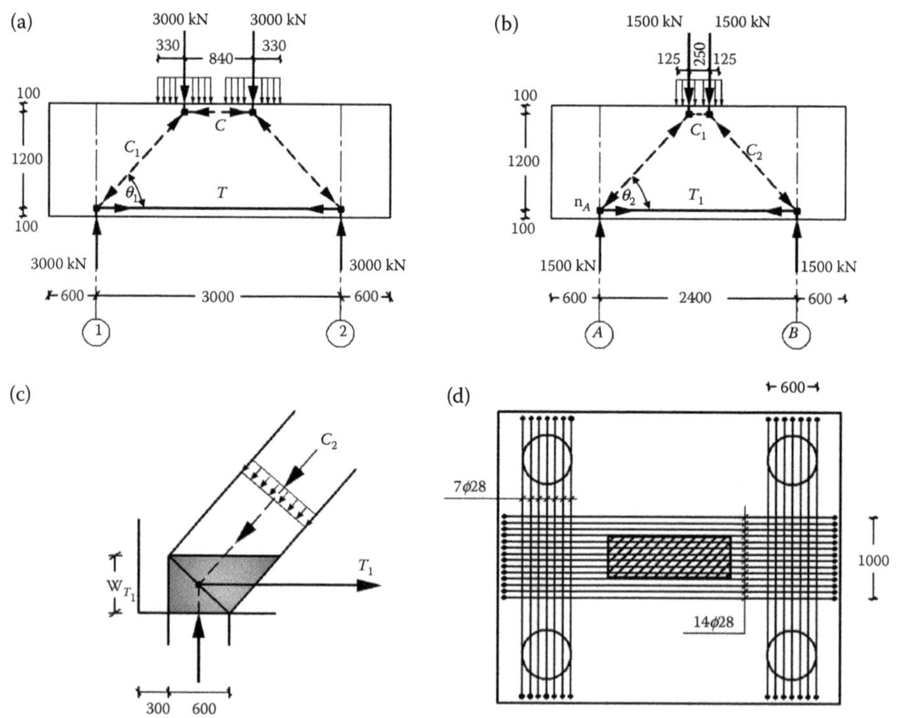

FIGURE 8.4 Example 8.1—*STM* details of loading case 1: (a) *STM* of the force transfer in the x-direction, (b) *STM* of the force transfer through axis (1) or (2), (c) nodal zone n_A of the model in Figure 8.4b, and (d) layout of bottom reinforcement from *STM*.

$T_1 = 1343.8$ kN, which requires a reinforcement = 4266 mm^2 = 7ϕ28 to be arranged within a width equal to the pile diameter (600 mm).

Design strength of nodes and struts

With reference to Figure 8.4b

The design strength of the upper nodes, $f_{cd}^n = 0.75^* \times 0.85 \times f_c' \times 1.0 = 19.13\,\text{MPa}$

*The strength reduction factor.

The design strength of the lower nodes, $f_{cd}^n = 0.75 \times 0.85 \times f_c' \times 0.8 = 15.3\,\text{MPa}$

The design strength of the upper strut, $f_{cd}^s = 0.75 \times 0.85 \times f_c' \times 1.0 = 19.13\,\text{MPa}$

The design strength of the diagonal struts, $f_{cd}^s = 0.75 \times 0.85 \times f_c' \times 0.6^\dagger = 11.48\,\text{MPa}$

†Unreinforced bottle-shaped strut.

Node n_A as an example of the lower nodes

Since the reinforcement will extend beyond the node a distance greater than twice the concrete cover and the reinforcement will be arranged in one layer, the node is detailed as shown in Figure 8.4c. Since the distance between the tension tie T_1 and the pile cap edge is assumed to be 100 mm, the height of the node $w_{T_1} = 200\,\text{mm}$.

The bearing stress = pile load/pile area = 1500 kN/[$\pi(600)^2/4$] = 5.31 MPa, which is less than the design strength, 15.3 MPa.

The upper side of the node is checked by the stress from the diagonal strut, $C_2 = 2013.9$ kN. The design strength of this side is the smaller of the node strength and the strut strength, which gives a value of 11.48 MPa. The strut width is 200 cos θ_A + 600 sin θ_A = 580.4 mm, and the breadth is the pile diameter, 600.0 mm. Thus, the node stress at the interface with the strut = 2013.9 kN/[600 × 580.4] = 5.78 MPa, which is safe.

Upon completing the check for the upper nodes, it is found that the nodes and struts are safe. The layout of the reinforcement from the *STM* is given in Figure 8.4d.

Load case 2

The column forces are first transferred in the x-direction to the lines connecting the piles on axes (1) and (2); then the forces are transferred to piles. Considering the case that the column moment causes compression to the piles on axis (2) and tension to the piles on axis (1), then $N_u = 2700$ kN and $M_{uy} = \pm 4950$ kN.m are replaced by equivalent two stress resultants on the column section, a downward force 5440 kN and an upward force 2740 kN, as shown in Figure 8.5a. Then the resultant force of both piles on axis (1) is ((2700/2) + (4950/3)) = 3000 kN (downward) and the resultant force of both piles on axis (2) is ((2700/2) – (4950/3)) = –300 kN (upward). The appropriate *STM* of this plane is shown in Figure 8.5b. From the model in Figure 8.5a, the tension force $T_3 = 2275$ kN can be resisted by a reinforcement = 7222.2 mm^2 = 12ϕ28, which is less than the corresponding reinforcement in case 1. The tension force $T_4 = 220$ kN requires reinforcement 4ϕ16 to be distributed within a width 500 mm, which is covered by the minimum reinforcement required by ACI 318-14. The tension force $T_5 = 2740$ kN requires a vertical reinforcement 8698.4 mm^2 = 15ϕ28, which is covered by the column reinforcement.

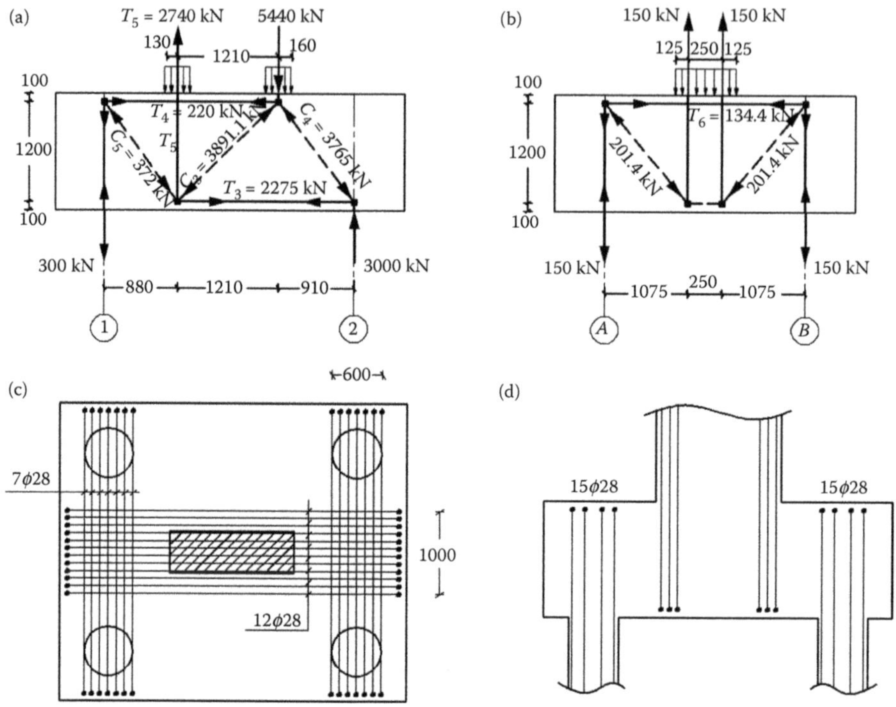

FIGURE 8.5 Example 8.1—*STM* details of loading case 2: (a) *STM* of the force transfer through the *x*-axis, (b) *STM* of the tension force transfer through axis (1) or (2), (c) layout of bottom reinforcement from *STM*, and (d) reinforcement of the model in (b).

The 3000 kN downward force on axis (2) is transferred to the piles on the axis via the *STM* shown in Figure 8.4b as for load case (1) and the reinforcement of the two cases is the same. The 300 kN upward force on axis (1) is transferred to the piles on the axis via the *STM* shown in Figure 8.5b. From the model, the tension force $T_6 = 134.4$ kN requires a reinforcement = 426.6 mm^2 = 3ϕ16 to be arranged within a width equal to the pile diameter (600 mm), which is covered by the minimum reinforcement required by ACI 318-14, and therefore it is not illustrated in the reinforcement layout.

The layout of the reinforcement from the *STM* is given in Figure 8.5c and d.

The final reinforcement layout from *STMs* is illustrated in Figure 8.6. It should be noted that a minimum reinforcement for temperature and shrinkage in the areas where there is no reinforcement from the model, equal to 0.2% of the concrete area, required by ACI 318-14, should be placed. This reinforcement is not illustrated in Figure 8.6.

8.3.2 Design Example 8.2

A pile cap 4200 × 8000 × 1400 mm is supported by eight piles of 600 mm diameter each and supports two columns of cross-section 500 × 1500 mm, as shown in

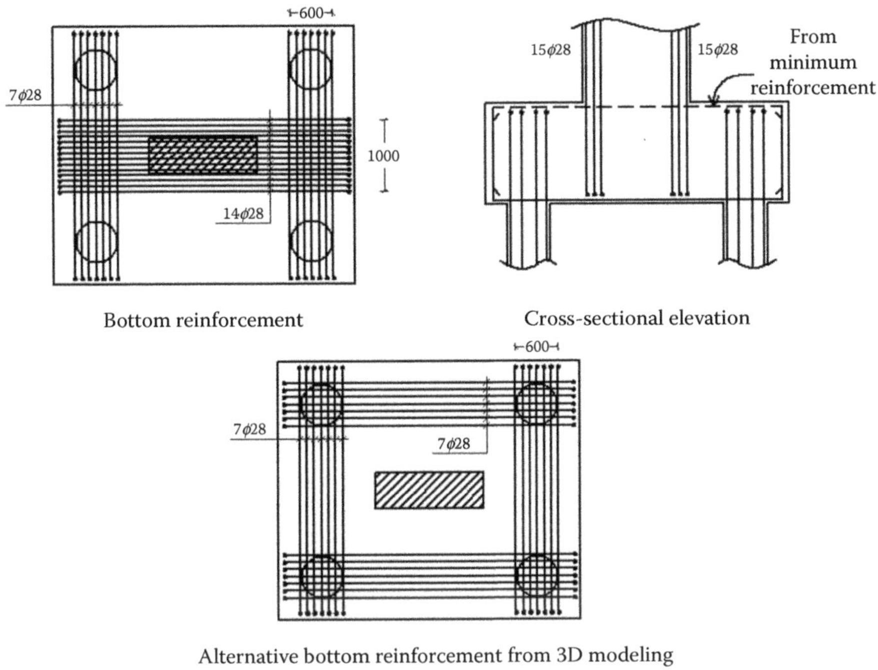

FIGURE 8.6 Example 8.1—final reinforcement layout from *STMs*.

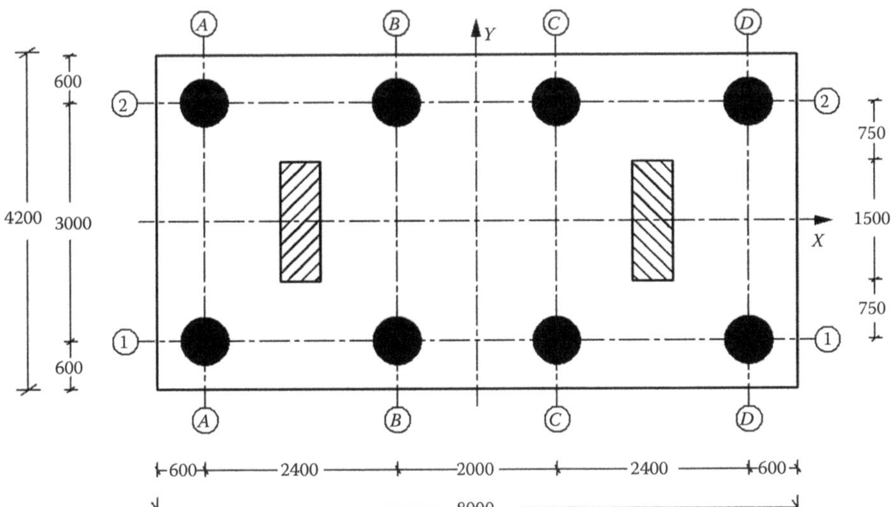

FIGURE 8.7 Example 8.2.

Figure 8.7. Using a 2D *STM* design, the reinforcement for the following load combinations if the concrete cylinder strength, $f_c' = 30.0\,\text{MPa}$, and the yield stress of steel reinforcement, $f_y = 420.0$ MPa.

Case 1: for each column $N_u = 6000$ kN and $M_{ux} = 0.0$
Case 2: for each column $N_u = 2700$ kN and $M_{ux} = \pm 4950$ kN.m

Column section and Reinforcement
Will be the same as in Example 8.1.

Load case 1
With reference to Figure 8.7, the forces of each column are first transferred in the *y*-direction to the lines connecting the piles on axes (1) and (2) as in Example 8.1 and as shown in Figure 8.8a. This will result in the load pattern illustrated in Figure 8.8b for either the piles on axis (1) or (2). The model in Figure 8.8b is twice statically indeterminate; however, upon utilization of symmetry, only one equation in one unknown can be formed leading to the indicated reactions. Then the pile forces will be as follows:

$$P_{1,A} = P_{1,D} = P_{2,A} = P_{2,D} = 1250\,\text{kN}$$

$$P_{1,B} = P_{1,C} = P_{2,B} = P_{2,C} = 1750\,\text{kN}$$

The appropriate *STM* for load transfer on either axis (1) or (2) is illustrated in Figure 8.8c, leading to the reinforcement layout shown in Figure 8.8d.

Load case 2
With reference to Figure 8.7 and considering the case that the column moment will cause compression to the piles on axis (2) and tension to the piles on axes (1), the forces of each column are first transferred in the *y*-direction to the lines connecting the piles on axes (1) and (2) as in Example 8.1 and shown in Figure 8.9a. This will result in the load pattern illustrated in Figure 8.8b for the piles on axis (2) and in Figure 8.9b for the piles on axis (1). For the load pattern of axis (2), the solution will proceed as in case (1) of this example. For the load pattern of axis (1), Figure 8.9b, the solution as explained in case (1) leads to the following pile forces.

$$P_{1,A} = P_{1,D} = -125\,\text{kN} \quad \text{(upward)}$$

$$P_{1,B} = P_{1,C} = -175\,\text{kN} \quad \text{(upward)}$$

The appropriate *STM* for load transfer on this axis is illustrated in Figure 8.9b. The reinforcement layout of this case will be as shown in Figure 8.9c.

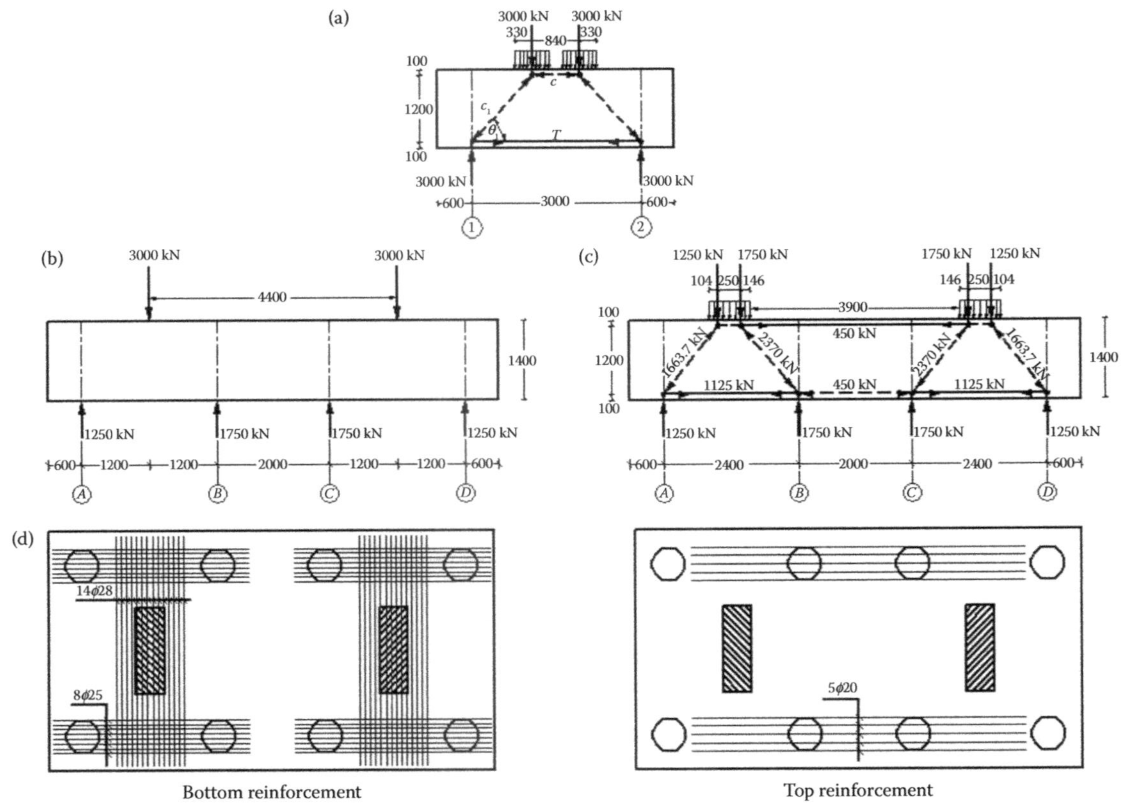

FIGURE 8.8 Example 8.2—*STM* details of loading case 1: (a) *STM* of the force transfer through the *y*-axis, (b, c) loads and corresponding *STM* of the force transfer through axis (1) or (2), and (d) layout of reinforcement from *STM*.

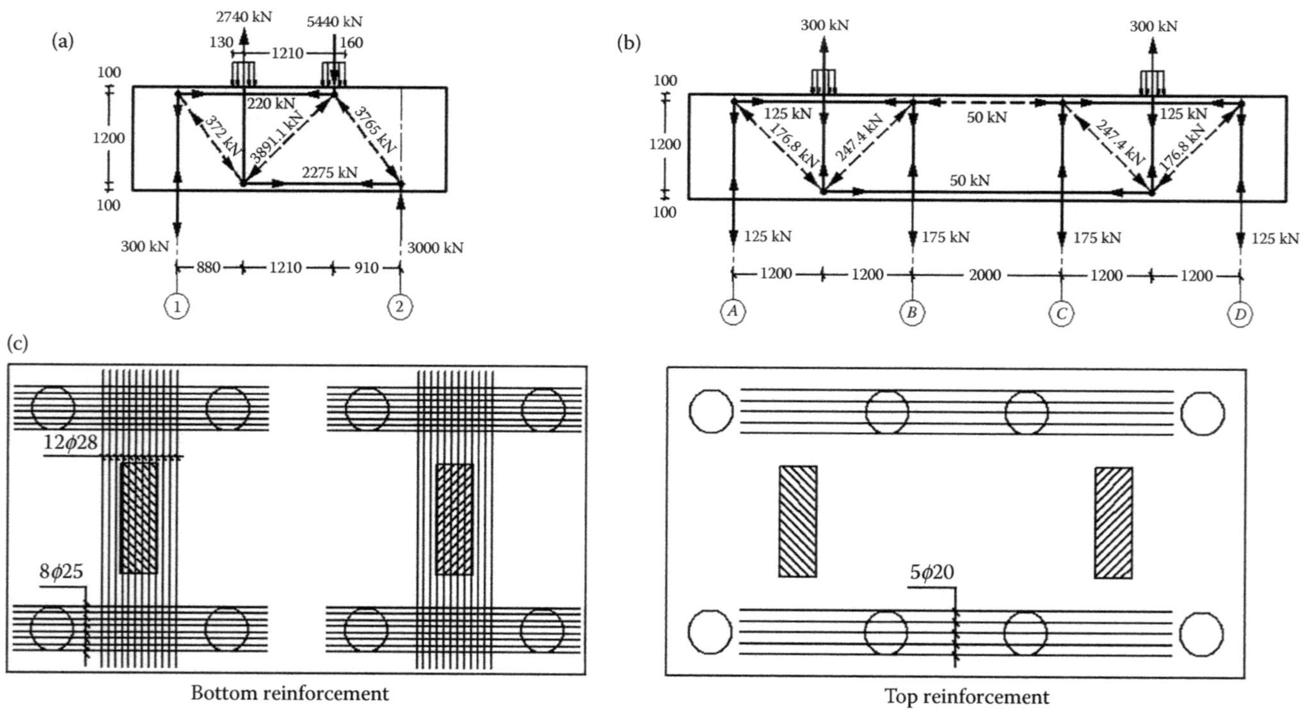

FIGURE 8.9 Example 8.2—*STM* details of loading case 2: (a) *STM* of the force transfer through the *y*-axis, and (b) *STM* of the tension force transfer through axis (1) or (2), and (c) layout of reinforcement from *STM*.

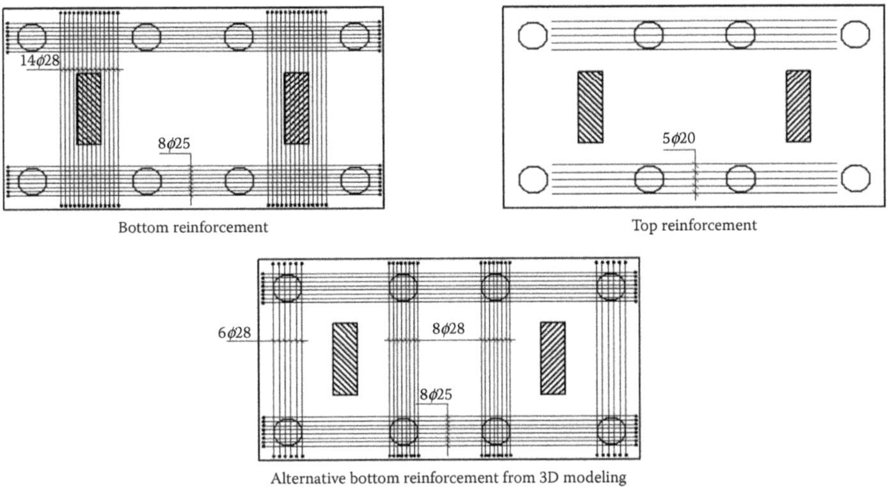

FIGURE 8.10 Example 8.2—final layout of reinforcement from *STMs*.

The final reinforcement layout which covers both cases (1) and (2) is shown in Figure 8.10.

8.4 GEOMETRY OF 3D *STMs*

8.4.1 Challenges

As explained in the introduction section, 3D models mean direct transfer of the column loads to the piles, which is associated with shorter load path and hence less tension reinforcement. However, it is more difficult to deal with the geometry of struts and nodal zones than in the case of 2D models. To facilitate dealing with 3D *STMs*, revisions were made to the ACI Building Code where Section 23.9.5 was added to simplify the detailing of nodal zones in 3D, by not requiring exact geometry compatibility between the struts and the faces of the nodal zone. ACI 318-14 recommended the following: "*23.9.5 In a three-dimensional strut-and-tie model, the area of each face of a nodal zone shall be at least that given in 23.9.4, and the shape of each face of the nodal zone shall be similar to the shape of the projection of the end of the strut onto the corresponding face of the nodal zone.*"

8.4.2 Simplification of Nodal Zone Geometry

In order to simplify the calculation procedure of 3D models, the geometry of nodal zones, and subsequently the geometry of struts, can be simplified to the convenience of the designer but keeping the axes of struts and locations of nodes unchanged. As an example, in order to simplify the nodes geometry of the pile cap in Figure 8.11a underneath the column if subjected to vertical loads, the column cross-section can be divided into five equal areas as shown in Figure 8.11d. In this division, the central square area corresponds to the central pile, and the other four trapezoidal areas

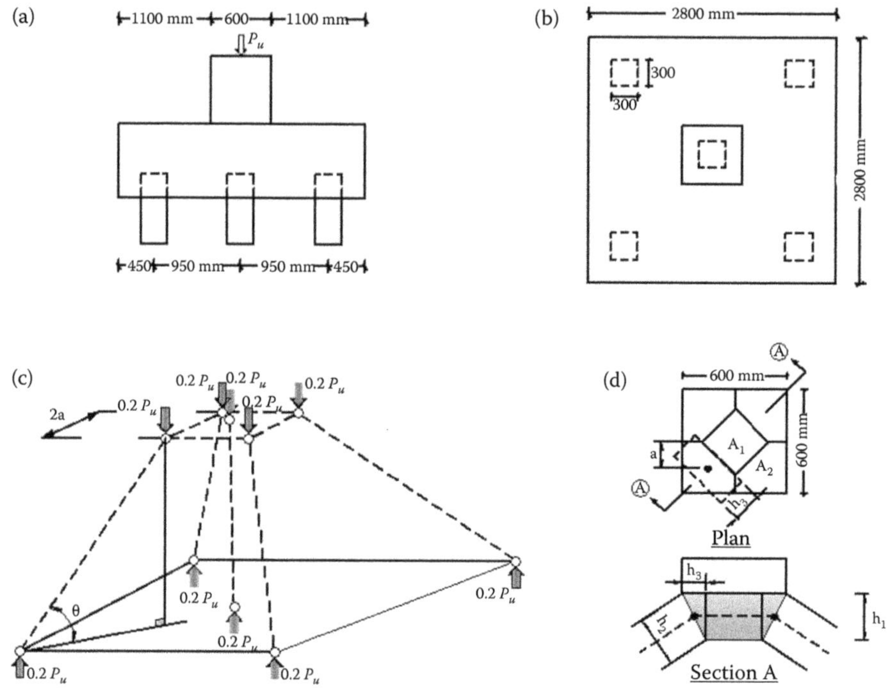

FIGURE 8.11 Example 8.1 of geometry simplification of 3D nodes.

correspond to the outer four piles. The centroid of the trapezoidal area can be determined and then this area can be replaced with an equivalent rectangular area that has the same centroid as the trapezoid, as shown in Figure 8.11d.

For the pile cap example in Figure 8.12a, under vertical loads, the column cross-sectional area is divided into four equal triangular areas, each corresponding to a

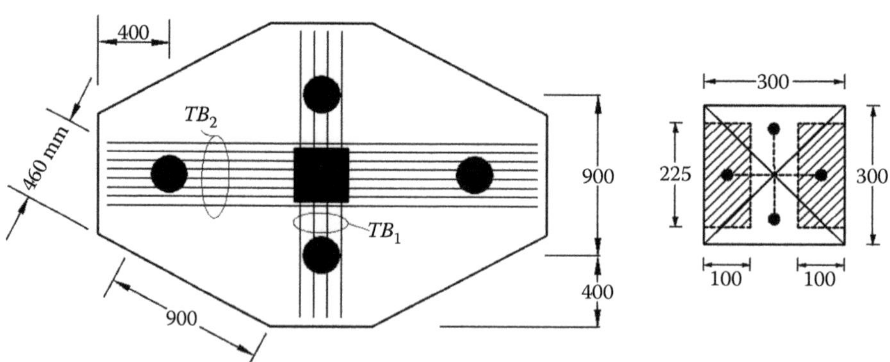

FIGURE 8.12 Example 8.2—geometry simplification of 3D nodes.

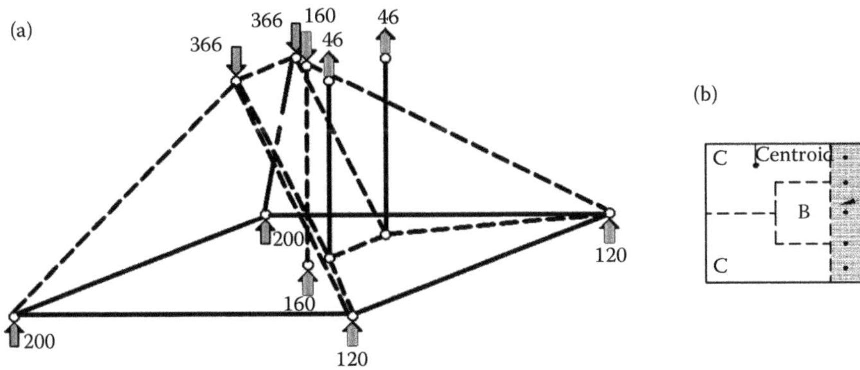

FIGURE 8.13 Example 8.3—geometry simplification of 3D nodes.

pile of the four supporting piles. The triangular area is replaced with a rectangular area of the same centroid as shown in Figure 8.12b for the two farther piles.

For the pile cap in Figure 8.11a, if the column is subjected to vertical force and bending moment such that one side of the column reinforcement is under tension, the 3D model is shown in Figure 8.13a. For the column section, the shaded area (area A) surrounding the tension reinforcement, Figure 8.13b, has the same centroid as the tension reinforcement. The rest of the column area is divided as shown in the figure such that equilibrium is satisfied. Both the area B corresponding to the central load and the area A corresponding to the tension reinforcement, are easy to deal with. The areas C, L-shaped, can be replaced with equivalent rectangular areas having the same centroids as the original L-shaped areas.

8.4.3 Limits of Strut Angle

As in 2-D *STMs*, the limitation of the 3D strut angle has to be respected due to the need for strain compatibility and the limited ductility of concrete. In 3D, the limitations should apply to the real angle between the tie and the strut, which is different from the angle between the tie and the projection of the strut in the vertical plane of the tie. As in 2-D, this angle should be within the range from 26.56 degrees ($\tan^{-1} 1/2$) to 60 degrees.

8.5 STRENGTH OF STRUTS IN PILE CAPS

8.5.1 Effect of Inactive Concrete on Bearing Struts

As a result of the spatial nature of pile caps, very significant volumes of concrete, far from any strut or tie or confinement, are subjected to low stresses, and therefore are called "inactive concrete," Figure 8.2. These volumes surround the struts deviating from underneath the column and heading to the piles. When the column is loaded, these important volumes of inactive concrete give rise to significant internal restraint. *Confinement by inactive concrete greatly reduces the tendency of compressive struts to develop transverse tensile stresses within the strut, hence increasing the compressive capacity of these struts.*

8.5.2 Bearing Strength of Struts Confined by Inactive Concrete

In deep members such as pile caps, Adebar and Zhou (1993) stated that compressive struts do not fail by crushing of concrete and the failure may occur due to longitudinal splitting of these struts as a result of the transverse tension caused by the stress deviation and spreading of compressive stresses in order to maintain compatibility. Such deviation will cause biaxial or triaxial compression in the nodal zone, but transverse tension near the mid-length of the strut, Figure 8.14.

In the case of plane stress, when the tension is resisted only in one direction field as in walls or deep beams, the influence of the "amount of spreading" on the bearing stress to cause transverse splitting is shown in Figure 8.15 (Schlaich et al., 1987). In 3D structures such as pile caps, the compressive stresses spread in two directions, reducing the transverse tension in any one direction. Hence, in such a case, the risk of splitting is less decisive than for the 2D case because the tension in the bottle shape is reduced due to the confinement provided by the surrounding concrete, and the tensile stresses are reduced in each single direction.

For a better understanding of the transverse splitting in the struts of pile caps, Adebar and Zhou (1993) carried out linear elastic finite element analysis to determine the triaxial stresses at first cracking within cylinders of various diameters, D, and heights, H, subjected to concentric axial compression over a constant-size circular bearing area of diameter d, Figure 8.16a. The geometry of the problem can be summarized in terms of two parameters, namely, the ratio of the cylinder diameter to the load diameter D/d and the ratio of the cylinder height to the load diameter H/d.

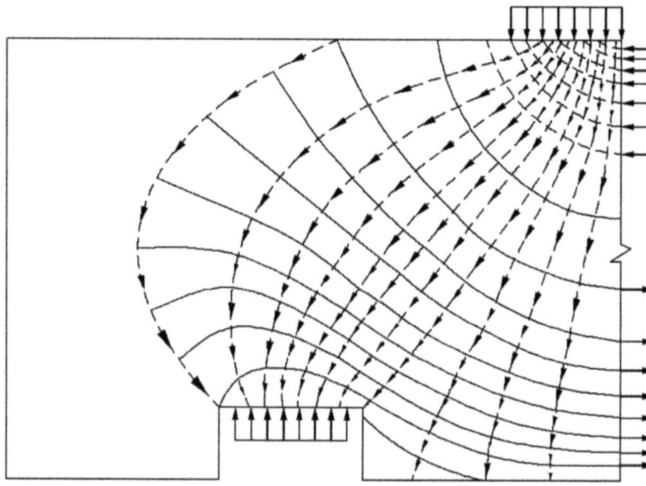

FIGURE 8.14 Linear elastic stress trajectories with transverse tension due to spreading of compression stress trajectories (Adebar and Zhou, 1993).

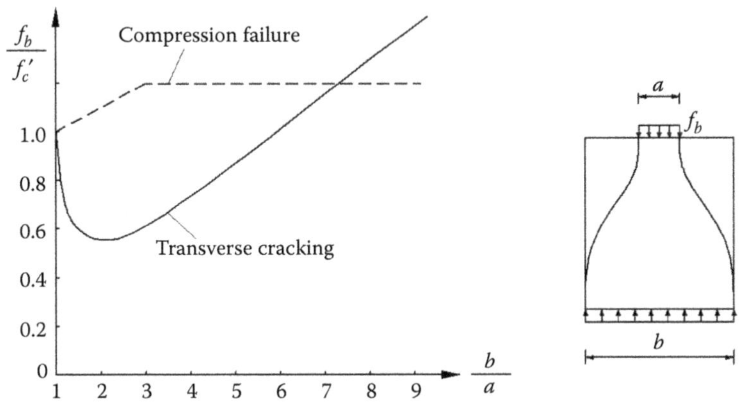

FIGURE 8.15 Maximum bearing stress to cause transverse splitting in a biaxial stress field.

Figure 8.17 summarizes the influence of *D*/*d* and *H*/*d* on the bearing stress at first cracking, with the assumption that the ratio of compressive strength to tensile strength $f_c'/f_t' = 15$. When $D/d = 1.0$, the cylinder is subjected to cylinder uniaxial compression, while $D/d \gg 1.0$, the compression stresses are spread out, creating a state of triaxial compression close to the loaded surface and biaxial tension near the mid-height of the cylinder, Figure 8.16b. Increasing the values of *D*/*d* and *H*/*d* beyond a certain limit does not noticeably affect the internal stress flow and hence the bearing stress which causes first cracking, Figure 8.17.

Upon carrying out experimental tests to verify the finite element results of Figure 8.17, Adebar and Zhou (1993, 1996) proposed that when designing deep member

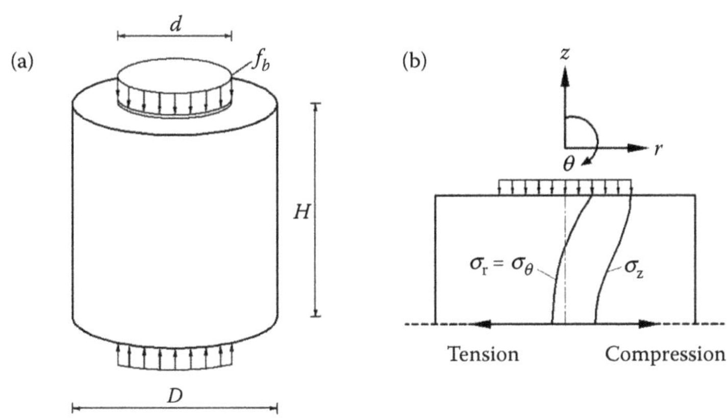

FIGURE 8.16 Analytical study of transverse tension in a triaxial stress field: (a) geometrical parameters and (b) typical stress distribution under a bearing plate (Adebar and Zhou, 1993).

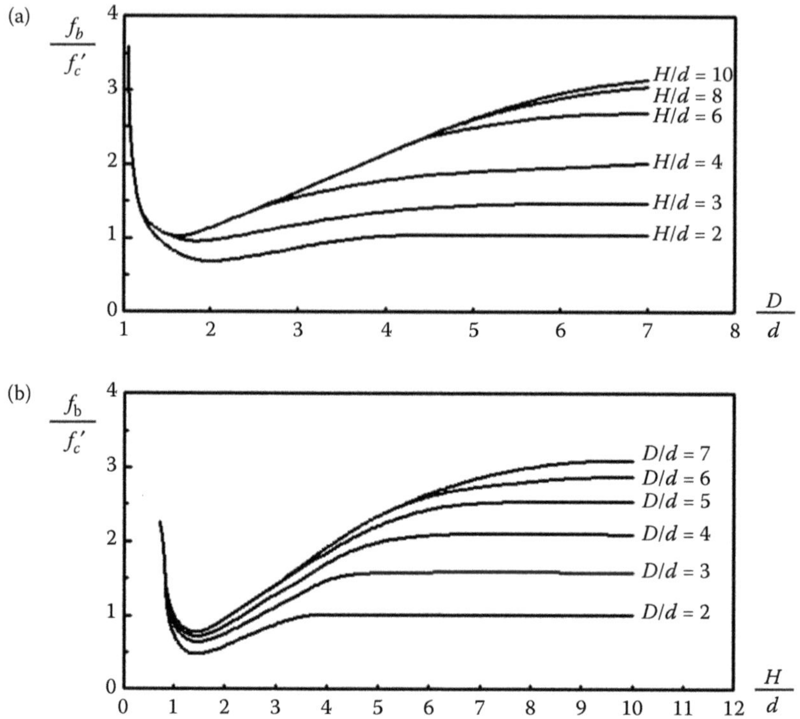

FIGURE 8.17 Analytical study of the ratio between stresses at cracking, f_b, to concrete strength, f'_c (Adebar and Zhou, 1993): (a) influence of confinement and (b) influence of height.

(disturbed region) without sufficient reinforcement, to ensure redistribution after cracking, the maximum bearing stress should be limited to

$$f_b \leq 0.6 f'_c(1+2\alpha\beta), \quad f'_c \leq 34.5\ \text{MPa} \tag{8.1}$$

$$\alpha = 0.33\left(\sqrt{(A_2/A_1)} - 1\right) \leq 1.0, \tag{8.2}$$

$$\beta = 0.33((h/b) - 1) \leq 1.0 \tag{8.3}$$

where the ratio representing the aspect ratio (height/width) of the compression strut, h/b, should not be taken to be less than 1.0 (i.e., $h/b \geq 1.0$). The parameter α accounts for the amount of confinement, while the parameter β accounts for the geometry of the compression stress field. In the calculation of the strut aspect ratio, h/b, some simplification can be introduced as will be illustrated in Examples 8.3 and 8.4.

The formulas for confinement and geometry (aspect ratio), α and β, were chosen to give reasonably simple expressions and yet correspond well with the finite element predictions and the experimental results. The lower bearing stress limit of $0.6f_c'$ is appropriate if there is no confinement (i.e., $A_2/A_1 \approx 1.0$), regardless of the height of the compression strut, as well as when the compression strut is relatively short, $h/b \approx 1.0$, regardless of the amount of confinement. The upper limits on α and β were set by Adebar and Zhou (1993) to guarantee an upper limit on f_b which corresponds approximately to the upper limit of bearing strength given in ACI-318-14.

For actual structural members, the size effect has to be implemented, leading to the following design value,

$$f_b \leq 0.6(0.85f_c')(1+2\alpha\beta),\ f_c' \leq 34.5\ \text{MPa} \tag{8.4}$$

Since the concrete bearing strength is actually proportional to the concrete tensile strength, Equation 8.1 may give unsafe values if the concrete compressive strength is significantly greater than 34.5 MPa. Therefore, the following equation was given by Adebar and Zhou (1993) in this case.

$$f_b \leq 0.6\ f_c' + 6\sqrt{f_c'}(\alpha\beta),\ f_c' > 34.5\ \text{MPa} \tag{8.5}$$

Equation 8.5 expresses the bearing strength enhancement in a form similar to what was proposed by Hawkins (1968).

Thus, the design bearing strength of a strut in a pile cap, if the concrete compressive strength is significantly greater than 34.5 MPa, should be

$$f_b \leq 0.6(0.85f_c') + 6\sqrt{(0.85f_c')}(\alpha\beta),\ f_c' > 34.5\ \text{MPa} \tag{8.6}$$

where f_c' is in MPa.

8.6 STRENGTH OF NODAL ZONES IN PILE CAPS

8.6.1 Strength of the Nodal Zone underneath the Column

The nodes underneath the column are essentially subjected to a state of triaxial compression, which increases the node strength. Wang et al. (1987) conducted some experimental tests on cubes loaded by triaxial compression, and the results indicated that rather low transversal stresses produce an important increase in the bearing strength. For instance, if the two transversal stresses are equal to the uniaxial cube strength, the strength in the third direction is raised to approximately five times the uniaxial strength. Also, if the two transversal stresses are equal to 20% of the uniaxial cube strength, the strength in the third direction is of the order of two times the uniaxial strength.

The strength of concrete under a triaxial state of compressive stress, f_{cc}', can be expressed as Park and Paulay (1975),

$$f_{cc}' = f_c' + 4.1f_l \tag{8.7}$$

where f_l is the lateral confining pressure (i.e., the other two perpendicular stresses). If f_l is assumed any small value such that $f_l \geq 0.24f_c'$, the corresponding value of $f_{cc}' \geq 2.0f_c'$, which reduces to $f_{cc}' \geq 1.7f_c'$ if the size effect is accounted for.

Several standards give recommendations for the triaxial compressive strength. Eurocode 2 gives the following upper limit for the concrete design strength for triaxial compression, f_{cd4}, which may be used if the transverse stresses are known and larger than $0.75f_{ck}$,

$$f_{cd4} = k_4[1-(f_{ck}/250)]f_{cd} \tag{8.8}$$

With $k_4 = 3.0$ as recommended by Eurocode 2, f_{ck} is the characteristic value of concrete compressive strength at 28 days and f_{cd} is the design value of concrete compressive strength.

In the Recommendations of FIP (1999), the following value is recommended.

$$f_{cd4} = 3.88f_{cd} \tag{8.9}$$

In ACI 318-14, the design bearing strength of concrete, f_{ce} shall not exceed $0.85f_c'$, except when the supporting surface area, A_2, is wider on all sides than the loaded area, A_1, Figure 8.18; then the following value is recommended:

$$f_{ce} = 0.85f_c'\sqrt{A_2/A_1}, \quad \sqrt{A_2/A_1} \leq 2.0 \tag{8.10}$$

The ACI 318-14 upper limit on $\sqrt{A_2/A_1}$ limits the maximum strength of a node under a triaxial state of stress, f_{ce}, to $1.7f_c'$.

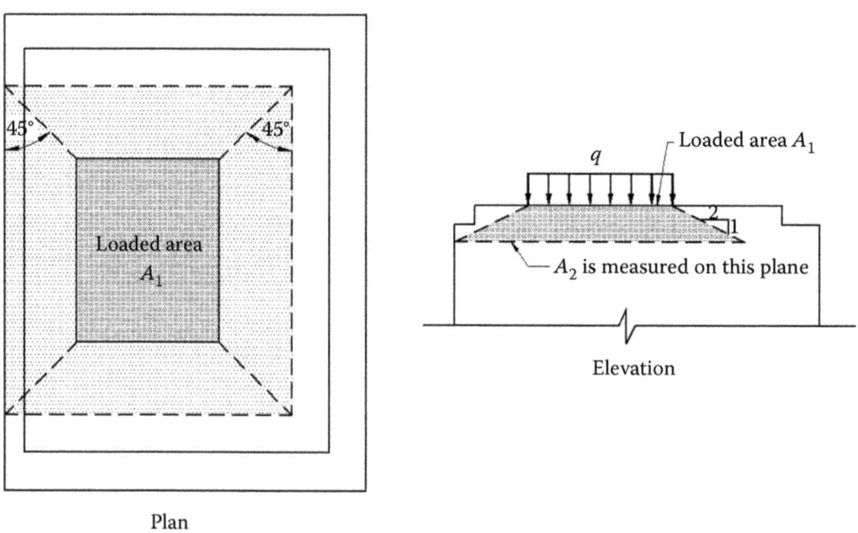

FIGURE 8.18 Application to find A_2 in stepped or sloped supports (ACI 318-14).

8.6.2 Strength of the Nodal Zone above the Piles

The pile cap lower nodal zones, above the piles, are bounded by the inclined strut coming from underneath the column, the pile force, and the reinforcement tension. Any of these nodes represents one end of the inclined strut. The strength of the strut at this node is derived from either Equations 8.4 or 8.6, based on the concrete strength, with the assumption that there is no tension anchored at this node. In order to account for the tension anchored at the node, the obtained strength should be reduced by 20% or 40% based on the number of tension ties connected to the node; that is, the node strength is $0.8f_b$, obtained from the strut strength equation at this node, if one tension tie is connected to the node, and $0.6f_b$ in case of two or more ties.

8.7 EXAMPLE 8.3: STRENGTH ASSESSMENT OF PILE CAP SUPPORTED BY 4 PILES VIA 3D MODELING

In this example, the strength of pile cap B, Figure 8.19, tested by Adebar et al. (1990), is assessed using 3D strut-and-tie modeling. The pile cap size: overall depth $h = 600$ mm, number of piles = 4, square column of size 300 mm, circular piles of diameter 200 mm, with 100 mm embedded into the underside of the pile cap. The cylinder compressive strength of concrete, $f_c' = 24.8$ MPa, the yield stress of longitudinal steel, $f_y = 479$ MPa, the area of main reinforcement, $A_s(TB1) = 12$ No. 10 (1200 mm²), with depth to steel centroid = 390 mm in the short direction and $A_s(TB2) = 22$ No. 10 (2200 mm²), with depth to steel centroid = 400 mm in the long direction. The solution procedure is given next.

Strut-and-Tie modeling

Figure 8.20 illustrates a simple 3D *STM* of pile cap B of this example. The concentrated column load is transmitted directly to the four supports (piles) by four inclined

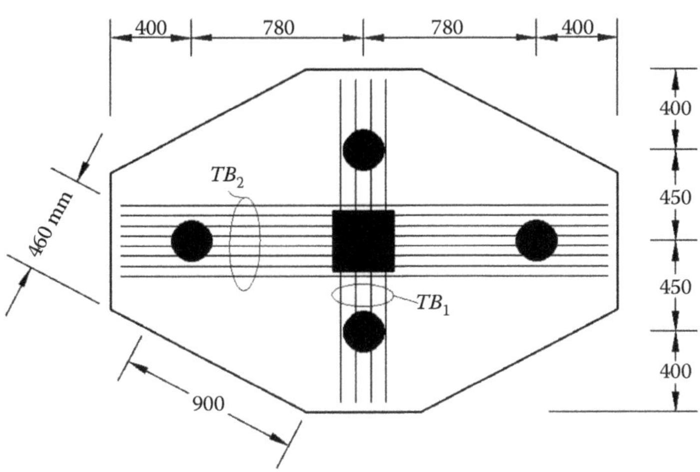

FIGURE 8.19 Example 8.3—geometry of pile cap B.

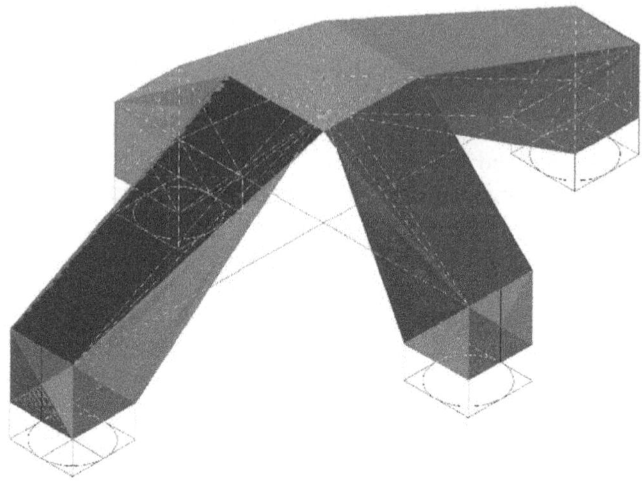

FIGURE 8.20 Example 8.3—3D *STM* of pile cap B.

compression struts, while the horizontal tension ties (longitudinal reinforcement) are required to prevent the piles from being spread apart. The analysis will be performed in both the short and long directions in the following sections.

Analysis of the short direction

1. The internal lever arm, h_{s_1}:
   As shown in Figure 8.21, the height of the lower node is termed as w_{T_1} and can be computed from:

$$w_{T_1} = n\phi_{bar} + 2c + (n-1)s$$

where n is the number of steel layers, ϕ_{bar} is the longitudinal steel diameter, c is the clear concrete cover, and s is the clear distance between bars. This can be alternatively expressed as $w_{T_1} = 2 \times$ (overall thickness—depth to steel centroid—embedded part of pile)

$$w_{T_1} = 2\times(600-390-100) = 220\,\text{mm}$$

The column cross-sectional area, 9.0×10^4 mm^2, is divided to four equal triangles of an area 2.25×10^4 mm^2 each and each triangle corresponds to a strut. This triangular bearing area can be simplified to a rectangle of the same centroid as the triangle as shown in Figure 8.21.
The width of the horizontal strut $C_{hor,1}$, a_1, can be computed as follows.

$$T_1 = A_s(TB1)\times f_y = C_{hor,1} = 1.7 f_c' b a_1$$

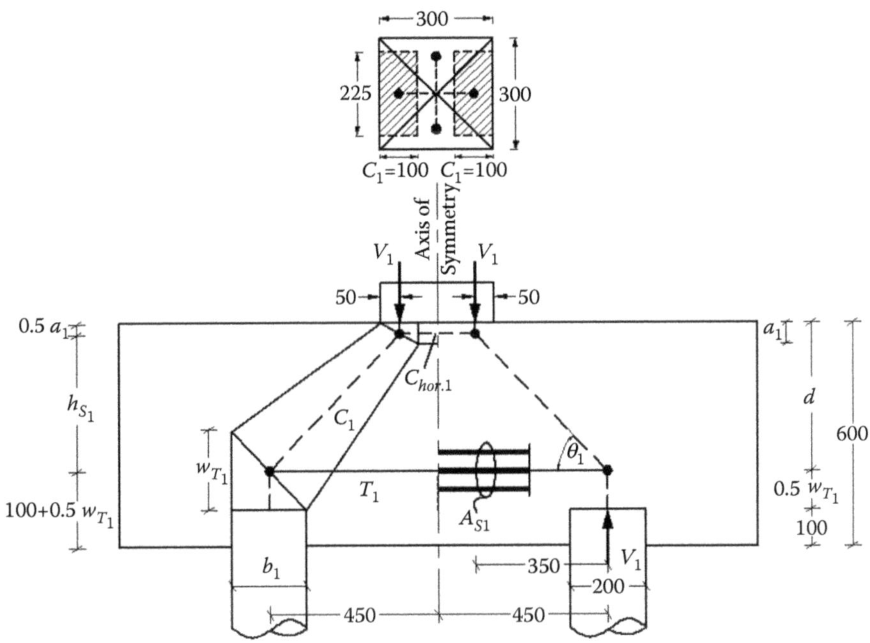

FIGURE 8.21 Example 8.3—*STM* for pile cap B in the short direction.

where $1.7f_c'$ is the assumed strength of the upper nodal zone $C-C-C$ node and b is the equivalent breadth of the upper node (i.e., $b = 225$ mm); then

$$T_1 = 1200 \times 479 = 574.8\,\text{kN} = C_{hor,1} = 1.7 \times 24.8 \times 225 \times a_1$$

$$a_1 = 60.6\,\text{mm}$$

Thus, h_{s_1} = depth to steel centroid—$0.5a_1$

$$h_{s_1} = 390 - 0.5 \times 60.6 = 359.7\,\text{mm}$$

2. Angle of inclined strut, θ_1:
   From pile cap geometry,

$$\theta_1 = \tan^{-1}\frac{359.7}{350} = 45.8^{\circ}$$

3. Widths of struts:
   The upper width of strut C_1, $w_{C_1}^u$, and the lower width, $w_{C_1}^l$, can be calculated as shown in Figure 8.21 based on the dimension of column and piles as follows:
   - With reference to Figure 8.22a, the width of strut C_1 at the upper node, $w_{C_1}^u$, is

$$w_{C_1}^u = a_1 \cos\theta_1 + c_1 \sin\theta_1 = 60.6 \times \cos 45.8 + 100 \times \sin 45.8 = 113.9\,\text{mm}$$

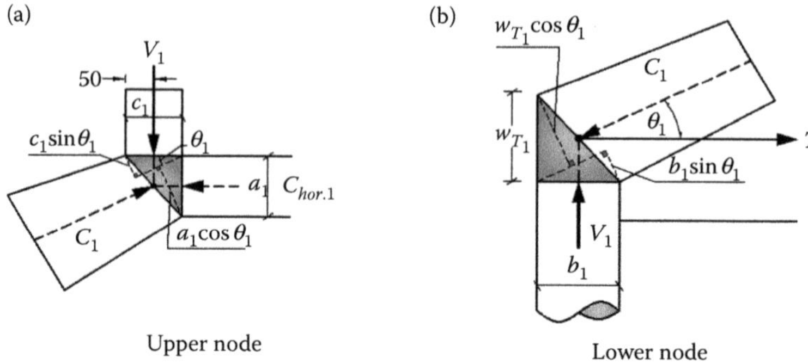

FIGURE 8.22 Example 8.3—nodes of 3D *STM* of pile cap B in the short direction.

- With reference to Figure 8.22b, the width of strut C_1 at the lower node, $w^l_{C_1}$, is

$$w^l_{C_1} = b_1 \sin\theta_1 + w_{T_1}\cos\theta_1 = 200 \times \sin 45.8 + 220 \times \cos 45.8 = 296.8 \text{ mm}$$

4. *STM* forces:

$$V_1 = \frac{T_1}{350} \times h_{s_1} = \frac{574.8}{350} \times 359.7 = 590.7kN$$

$$C_1 = \frac{T_1}{\cos\theta_1} = \frac{574.8}{\cos 45.8} = 824.5kN$$

5. Checking stress limits:

Struts:

Horizontal strut, $C_{hor,1}$
The width of strut $C_{hor,1}$ was determined based on its effective strength; hence, there is no need to carry out any further checks for this strut.

Diagonal strut, C_1
The width of the diagonal strut C_1 changes linearly between the upper and lower nodes (uniformly tapered). The effective compression strength of this strut f^s_{ce} differs at its two ends, due either to different nodal zone strength at the two ends (upper node and lower node), or to different bearing areas.

The compressive strength of concrete in this strut is significantly enhanced with the confinement from the surrounding inactive concrete. This strength can be computed from Equation 8.4, $f_b \le 0.6(0.85f'_c)(1+2\alpha\beta)$. The parameter α accounts for the amount of confinement (Figure 8.23), $\alpha = 0.33\left(\sqrt{(A_2/A_1)} - 1\right) \le 1.0$,

At the column nodal zone (upper node):
$\sqrt{(A_2/A_1)} = \sqrt{\dfrac{1244 \times 1244}{300 \times 300}} = 4.15$, this gives $\alpha = 0.33(4.15 - 1) > 1.0$, take $\alpha = 1.0$.

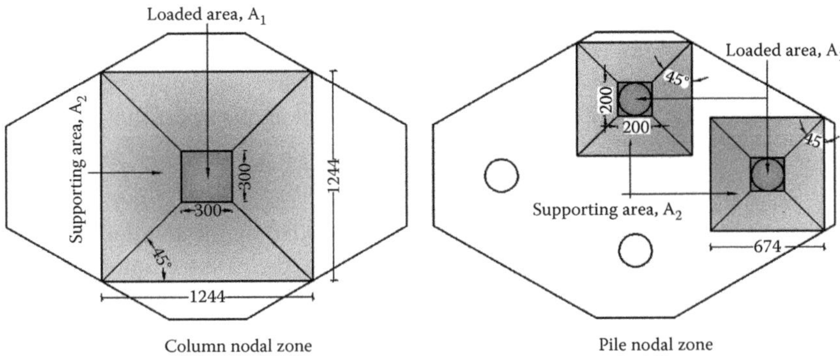

FIGURE 8.23 Amount of confinement for the pile cap example.

- *At the pile nodal zone (lower node):*

$\sqrt{(A_2/A_1)} = \sqrt{\dfrac{674\times 674}{200\times 200}} = 3.37$, which gives $\alpha = 0.33(3.37 - 1) = 0.78$.

The parameter β accounts for the geometry of the compression stress field, $\beta = 0.33((h/b) - 1) \leq 1.0$, where (h/b) signifies the geometry of the compression strut (i.e., h is the strut length and b is the strut width at either the upper or lower node). For easiness in the calculations, the ratio (h/b) can be expressed, based on trigonometry, as (h_s/b_b), where h_s is the strut vertical height and b_b is the equivalent diameter of the bearing area of the nodal zone connected with the strut.

For strut C_1, $h_s = h_{s_1} = 359.7\,\text{mm}$. For the upper node of the strut, the bearing area is the column cross-sectional area divided by the number of piles; thus, $b_b = \sqrt{(300^2/4)(4/\pi)} = \sqrt{22500\times(4/\pi)} = 169\,\text{mm}$. For the lower node of the strut, the bearing area is the pile area; hence, b_b is the pile diameter, $b_b = 200$ mm. Thus,

- *At the column nodal zone (upper node):*

$h/b = h_s/b_b = 359.7/169 = 2.13$, which gives $\beta = 0.33(2.13\text{-}1) = 0.37$.

- *At the pile nodal zone (lower node):*

$h/b = h_s/b_b = 359.7/200 = 1.8$, which gives $\beta = 0.33(1.8\text{-}1) = 0.264$.

Thus, the bearing strength of strut C_1 *at the upper node* is

$$f_b = 0.6(0.85f_c')(1 + 2\times 1.0\times 0.37) = 0.89f_c'$$

The bearing strength of strut C_1 *at the lower node* is

$$f_b = 0.6(0.85f_c')(1 + 2\times 0.78\times 0.264) = 0.72f_c'$$

For the strength of strut C_1, take the average strength at the upper and lower nodes;

$$f_{ce}^{C_1} = 0.5(0.89f_c' + 0.72f_c') = 0.81f_c'$$

The strength of strut C_1 at the upper node is controlled by the smaller of the strut strength, $0.81f_c'$, and the node strength, $1.7f_c'$. Hence, the strut strength at the upper node, $f_{ce}^{C_1} = 0.81 f_c'$. Upon substitution in $C_{1n} = f_{ce}^{C_1} b w_{C_1}^u$, where b is the equivalent breadth of strut at the upper nodal zone, $b = 225$ mm, $C_{1n} = 514.8$ kN.

The lower node is a $C - C - T$ node, $\beta_n = 0.8$. Thus, the limiting compressive strength of this node is 80% of the strength of the strut at this node; that is,

$$f_{ce}^n = 0.8(0.72 f_c') = 0.58 f_c'$$

The strength of strut C_1 at the lower node is controlled by the smaller of the strut strength, $0.81 f_c'$, and the node strength, $0.58 f_c'$. Hence, the strut strength at the lower node, $f_{ce}^{C_1} = 0.58 f_c'$. Upon substitution in $C_{1n} = f_{ce}^{C_1} b w_{C_1}^l$, where b is the pile diameter, $b = 200$ mm for the lower node, $C_{1n} = 853.8$ kN.

It is noted that there is a significant difference between the strength of strut C_1 at the upper and lower nodes, and this can be attributed to the significant difference between the strength of the upper nodal zone, $1.7 f_c'$, and the strut strength, $0.81 f_c'$. In order to reduce the gap between the two values of strut strength, the height of the upper node should be re-estimated based on a strength lower than $1.7 f_c'$, $\leq (514.8/853.8) \times 1.7 f_c' \approx 1.0 f_c'$.

Upon using a strength of the upper node of $0.8 f_c'$, and redoing the calculations, the following has been obtained.

The width of the horizontal strut $C_{hor,1}$, $a_1 = 128.8$ mm, $h_{s1} = 325.6$ mm, and the angle $\theta_1 = 42.93°$. The width of strut C_1 at the upper node, $w_{C_1}^u = 162.4$ mm and at the lower node, $w_{C_1}^l = 297.3$ mm. The pile force $V_1 = 534.7$ kN, the force in the diagonal strut, $C_1 = 785.0$ kN.

- *At the column nodal zone (upper node):*
  $h/b = h_s/b_b = 325.6/169 = 1.93$, which gives $\beta = 0.31$.
- *At the pile nodal zone (lower node):*
  $h/b = h_s/b_b = 325.6/200 = 1.63$, which gives $\beta = 0.21$.

Thus, the bearing strength of strut C_1 *at the upper node* is

$$f_b = 0.6(0.85 f_c')(1 + 2 \times 1.0 \times 0.31) = 0.83 f_c'$$

The bearing strength of strut C_1 *at the lower node* is

$$f_b = 0.6(0.85 f_c')(1 + 2 \times 0.78 \times 0.21) = 0.68 f_c'$$

For the strength of strut C_1, take the average strength at the upper and lower nodes;

$$f_{ce}^{C_1} = 0.76 f_c'$$

The strength of strut C_1 at the upper node is thus $f_{ce}^{C_1} = 0.76 f_c'$. Upon substitution in $C_{1n} = f_{ce}^{C_1} b w_{C_1}^u$, $C_{1n} = 688.8$ kN.

The limiting compressive strength of the lower node is

$$f_{ce}^n = 0.8(0.68 f_c') = 0.54 f_c'$$

The strength of strut C_1 at the lower node is $0.54 f_c'$, giving $C_{1n} = 796.3$ kN. *Take the smaller value of C_{1n}* at the two ends of the strut, *upper and lower nodes*. Then $C_{1n} = 688.8$ kN.

Nodes:

There is no need to carry out any checks for the upper node since the strength of the node is greater than or equal to the strength of the surrounding struts. The critical node is the lower node, and as estimated before, $f_{ce}^{n} = 0.54 f_c'$. Based on this, the nominal value of the pile reaction is

$$V_{1n} = 0.54 f_c' \times 200 \times 200 = 535.7\,\text{kN}$$

Since the reinforcement of the tie, T_1, is arranged in multiple layers, the assumed height of the node w_{T_1} should not exceed a maximum limiting value, $w_{T_1,\max}$. This is checked next.

$$w_{T_1,\max} = \frac{T_1}{f_{ce}^{n} b} = \frac{T_1}{0.54 f_c' \times 200} = 214.6\,\text{mm}$$

Since a value of $w_{T_1} = 220\,\text{mm}$ ($> w_{T_1,\max} = 214.6\,\text{mm}$) was used in estimating the width of strut C_1 at the lower node, a new value of the strut width should be calculated. This value is $w_{C_1}^{l} = 293.4\,\text{mm}$, leading to a new nominal value of the strut strength at this node $C_{1n} = 785.7$ kN. This last value of the nominal strength of the strut is still greater than the value at the upper node, and therefore, the solution should proceed with no change for this strut.

In order to determine the predicted failure load in this direction, Table 8.1 gives the final results of the short direction.

From the results in the table, the size of the upper nodal zone and hence the forces should be reduced to 88%. This finally gives a value of $V_{1n} = 470.5$ kN. It should be noted that the reinforcing steel does not yield at this obtained pile load.

Analysis of the long direction

As performed for the short direction

1. The internal lever arm, h_{s_2}:
   As shown in Figure 8.24, the term w_{T_2} (height of the lower node)

$$w_{T_2} = 2 \times (600 - 400 - 100) = 200\,\text{mm}$$

TABLE 8.1
Example 8.3—Summary of Calculation Results of the Short Direction of Pile Cap B

Model Label	Member	Actual Force, kN	Maximum Capacity, kN	Satisfaction
Struts	$C_{hor,1}$	574.8	574.8	yes
	C_1	785.0	688.8	No, 88%
Lower node, $C-C-T$	V_1	534.7	535.7	yes

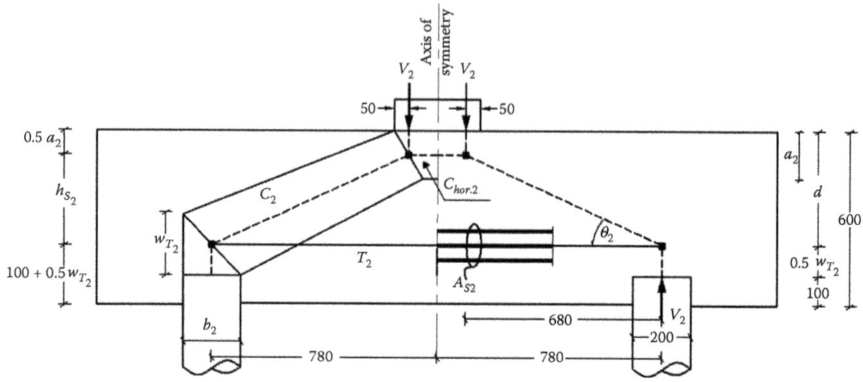

FIGURE 8.24 Example 8.3—*STM* of pile cap B in the long direction.

The width of the horizontal strut $C_{hor,2}$, a_2, can be computed as follows.

$$T_2 = A_s(TB2) \times f_y = 2200 \times 479 = 1053.8\,\text{kN}$$
$$= C_{hor,2} = 1.7 f_c' b a_2 = 1.7 \times 24.8 \times 225 \times a_2$$
$$a_2 = 111.1\ \text{mm}$$
$$h_{s_2} = 400 - 0.5 \times 111.1 = 344.5\,\text{mm}$$

2. Angle of inclined strut, θ_2:

$$\theta_2 = \tan^{-1} \frac{344.5}{680} = 26.9°$$

3. Widths of struts:
   - With reference to Figure 8.25a, the width of strut C_2 at the upper node, w^u_{C2}, is

$$w^u_{C_2} = a_2 \cos\theta_2 + c_2 \sin\theta_2 = 111.1 \times \cos 26.9 + 100 \times \sin 26.9$$
$$= 144.3\,\text{mm}$$

   - With reference to Figure 8.25b, the width of strut C_2 at the lower node, w^l_{C2}, is

$$w^l_{C_2} = b_2 \sin\theta_2 + w_{T_2} \cos\theta_2 = 200 \times \sin 26.9 + 200 \times \cos 26.9$$
$$= 268.8\,\text{mm}$$

4. *STM* forces:

$$V_2 = \frac{T_2}{680} \times h_{s2} = \frac{1053.8}{680} \times 344.5 = 533.9\,\text{kN}$$

$$C_2 = \frac{T_2}{\cos\theta_2} = \frac{1053.8}{\cos 26.9} = 1181.7\,\text{kN}$$

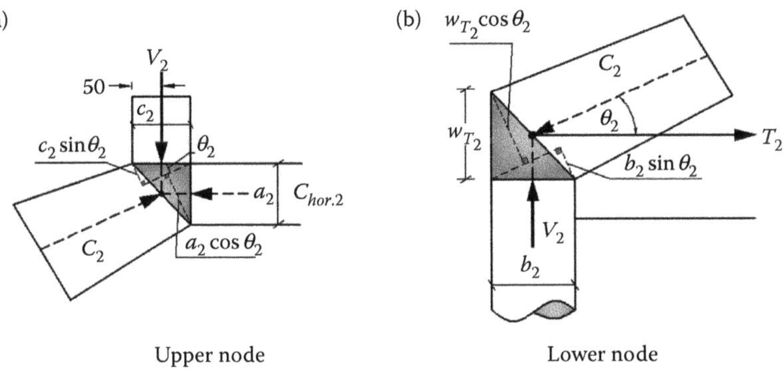

FIGURE 8.25 Example 8.3—nodes of 3D *STM* of pile cap B in the long direction.

5. Checking stress limits:

Struts:

Horizontal strut, $C_{hor,2}$
The width of strut $C_{hor,2}$ was determined based on its strength; hence, there is no need to carry out any further checks for this strut.

Diagonal strut, C_2
With reference to Figure 8.23, and as explained before, $\alpha = 1.0$ for the upper node and $\alpha = 0.78$ for the lower node. In order to obtain the parameter β, the ratio (h/b) is calculated in the same manner as for the short direction. The strut vertical height $h_s = h_{s2} = 344.5$ mm. For the upper node, the equivalent diameter of the bearing area is $b_b = 169$ mm and for the lower node, $b_b = 200$ mm. Thus,

- *At the column nodal zone (upper node):*
  $h/b = h_s/b_b = 344.5/169 = 2.04$, which gives $\beta = 0.33(2.04\text{-}1) = 0.34$.
- *At the pile nodal zone (lower node):*
  $h/b = h_s/b_b = 344.5/200 = 1.72$, which gives $\beta = 0.33(1.72\text{-}1) = 0.24$.

Thus, the bearing strength of strut C_2 *at the upper node* is

$$f_b = 0.6(0.85f_c')(1+2\times1.0\times0.34) = 0.86f_c'$$

The bearing strength of strut C_2 *at the lower node* is

$$f_b = 0.6(0.85f_c')(1+2\times0.78\times0.24) = 0.7f_c'$$

For the strength of strut C_2*, take the average strength at the upper and lower nodes;*

$$f_{ce}^{C2} = 0.5(0.86f_c' + 0.7f_c') = 0.78f_c'$$

The strength of strut C_2 at the upper node is controlled by the smaller of the strut strength, $0.78f_c'$, *and the node strength,* $1.7f_c'$. *Hence, the strut strength at the upper node,* $f_{ce}^{C2} = 0.78f_c'$. Upon substitution in $C_{2n} = f_{ce}^{C2} \times b \times w_{C2u}$, where b is the equivalent breadth at the upper node width, $b = 225$ mm, $C_{2n} = 628.1$ kN.

The lower node is a $C - C - T$ node, $\beta_n = 0.8$. Thus, the limiting compressive strength of this node is 80% of the strength of the strut at this node, that is,

$$f_{ce}^n = 0.8(0.7f_c') = 0.56f_c'$$

The strength of strut C_2 at the lower node is controlled by the smaller of the strut strength, $0.78f_c'$, *and the node strength,* $0.56f_c'$. *Hence, the strut strength at the lower node,* $f_{ce}^{C2} = 0.56f_c'$. Upon substitution in $C_{2n} = f_{ce}^{C2} b w_{C_2}^l$, where b is the pile diameter, $b = 200$ mm for the lower node, $C_{2n} = 746.6$ kN.

Upon taking the smaller value of C_{2n} at the two ends of strut nodes, *upper and lower nodes,* $C_{2n} = 628.1$ kN.

Nodes:

There is no need to carry out any checks for the upper node since the strength of the node is greater than or equal to the strength of the surrounding struts. The critical node is the lower node, and as estimated before, $f_{ce}^n = 0.56f_c'$. Based on this, the nominal value of the pile reaction is

$$V_{2n} = 0.56f_c' \times 200 \times 200 = 555.5\,\text{kN}$$

Since the reinforcement of the tie, T_2, is arranged in multiple layers, the assumed height of the node w_{T_2} should not exceed a maximum limiting value $w_{T_2,\max}$, as illustrated next.

$$w_{T_2,\max} = \frac{T_2}{f_{ce}^n b} = \frac{T_2}{0.56f_c' \times 200} = 379.4\,\text{mm}$$

Since the assumed value of $w_{T_2} = 200\,\text{mm}$, employed in the estimate of the width of C_2 at the lower node, is less that its maximum limit, $w_{T_2,\max}$, the assessment of the nominal strength of C_2 is correct and the solution should proceed.

The final results of the long direction are given in Table 8.2.

From the results in the table, the size of the upper nodal zone and hence the forces should be reduced to 53.2%. This finally gives a value of $V_{2n} = 284.0$ kN. It should be noted that the reinforcing steel does not yield at this obtained pile load.

If an iterative process is followed with the objective to have the nominal strength of strut C_2 at its two ends close, the strength prediction will improve.

Finally, the nominal strength of pile cap B from *STM*,

$$P_{STM} = 2(V_1 + V_2) = 2(470.5 + 284.0) = 1509.0\,\text{kN}$$

$$P_{STM}/P_{EXP} = \frac{1509.0}{2189} = 69\%$$

TABLE 8.2
Example 8.3—Summary of Calculation Results of the Long Direction of Pile Cap B

Model Label	Member	Actual Force, kN	Maximum Capacity, kN	Satisfaction
Struts	$C_{hor,2}$	1053.8	1053.8	yes
	C_2	1181.7	628.1	No, 53.2%
Lower node, $C-C-T$	V_2	533.9	555.5	yes

8.8 EXAMPLE 8.4: STRENGTH ASSESSMENT OF PILE CAP SUPPORTED BY 6 PILES VIA 3D MODELING

In this example the strength of pile cap C, Figure 8.26, tested by Adebar et al. (1990), is assessed using the strut-and-tie method. The pile cap, with an overall depth, $h = 600$ mm, is supporting a square column of size 300 mm, and is supported by six circular piles of diameter 200 mm, with 100 mm embedded into the underside of the pile cap. The cylinder compressive strength of concrete, $f_c' = 27.1$MPa, and the yield stress of the reinforcing steel, $f_y = 479$ MPa. The area of the main reinforcement in the short direction, $A_s(TB1) = 12$ No. 10 (1200 mm^2), with depth to steel centroid = 390 mm and $A_s(TC2) = 11$ No. 10 (1100 mm^2) with depth to steel centroid = 390 mm, and in the long direction and $A_s(TC3) = 21$ No. 10 (2100 mm^2), with depth to steel centroid = 400 mm. The solution procedure is given as follows.

Strut-and-Tie modeling:

Figure 8.27 illustrates a simple 3D *STM* of pile cap C. The concentrated column load is transmitted directly to the six supports (piles) by six inclined compression struts while

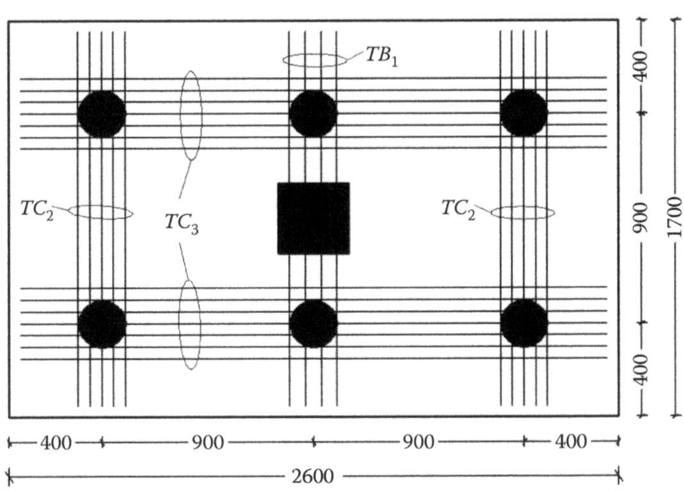

FIGURE 8.26 Example 8.4—geometry of pile cap C.

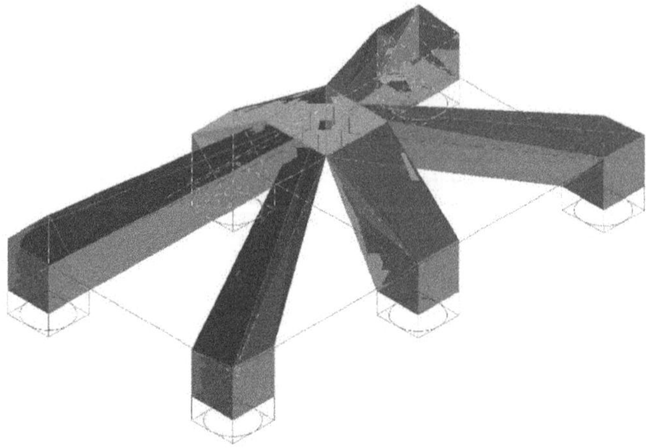

FIGURE 8.27 Example 8.4—3D-*STM* of pile cap C.

the horizontal tension ties (longitudinal reinforcement) are required to prevent the piles from being spread apart. Figure 8.28 illustrates a top view of the proposed *STM*.

The analysis will be performed in both the short direction, strut C_1, and diagonal direction, strut C_2, as follows.

Analysis of the short direction

1. The internal lever arm, h_{s_1}:
   As shown in Figure 8.29, the height of the lower node is termed as w_{T_1} and can be computed from:

$$w_{T_1} = n\phi_{bar} + 2c + (n-1)s$$

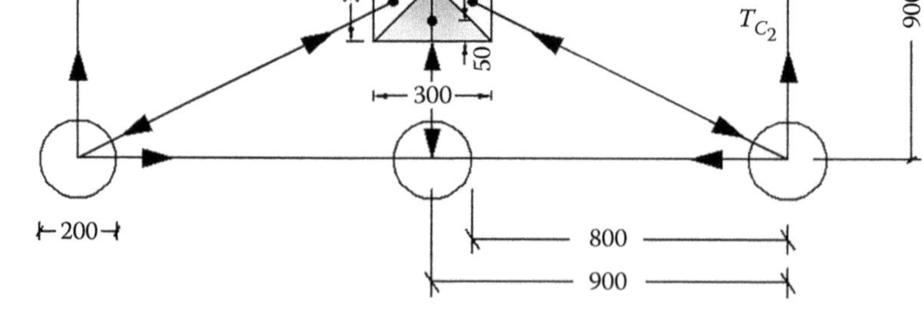

FIGURE 8.28 Example 8.4—top view of *STM* of pile cap C.

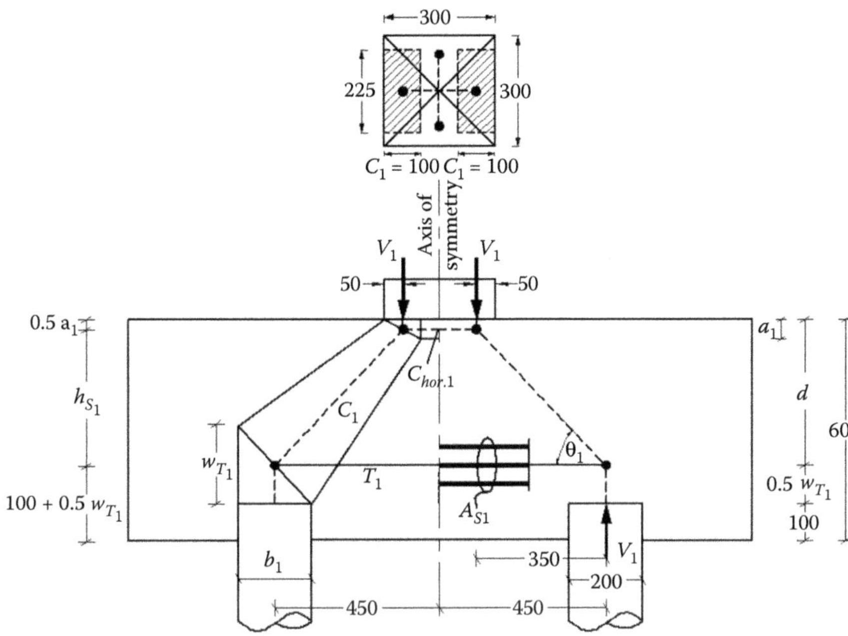

FIGURE 8.29 Example 8.4—*STM* for pile cap C in the short direction.

where n is the number of steel layers, ϕ_{bar} is the longitudinal steel diameter, c is the clear concrete cover, s is the clear distance between bars. This can be alternatively expressed as
$w_{T_1} = 2 \times$ (overall thickness—depth to steel centroid—embedded part of pile),

$$w_{T_1} = 2 \times (600 - 390 - 100) = 220\,\text{mm}$$

The width of the horizontal strut $C_{hor,1}$, a_1, can be computed as follows:

$$T_1 = A_s(TB1) \times f_y = C_{hor,1} = 1.7 f_c' b a_1$$

where $1.7f_c'$ is the strength of the upper nodal zone, $C - C - C$ node, and b is the equivalent breadth of the upper node (i.e., $b = 225$ mm); then, $T_1 = 1200 \times 479 = 574.8$ kN $= C_{hor,1} = 1.7 \times 27.1 \times 225 \times a_1$, from which $a_1 = 55.5$ mm
Thus, h_{s_1} = depth to steel centroid—$0.5a_1$

$$h_{s_1} = 390 - 0.5 \times 55.5 = 362.3\text{ mm}$$

2. Angle of inclined strut, θ_1:
   From pile cap geometry

$$\theta_1 = \tan^{-1}\frac{362.3}{350} = 46.0°$$

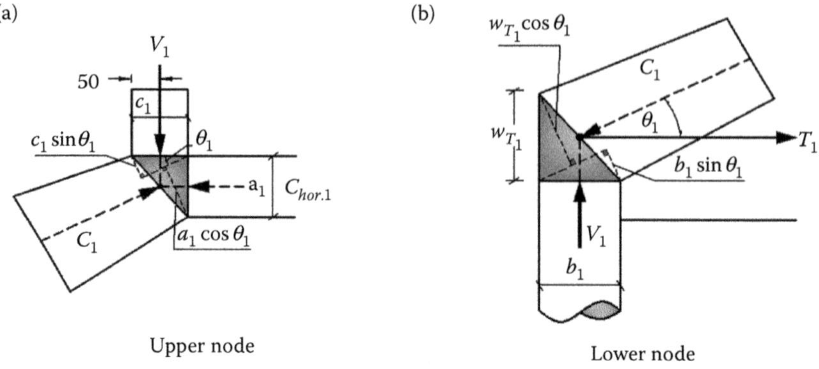

FIGURE 8.30 Example 8.4—nodes of *STM* of pile cap C in the short direction.

3. Widths of struts:
   The upper width of strut C_1, $w^u_{C_1}$, and the lower width, $w^l_{C_1}$, can be calculated as shown in Figure 8.30 based on the dimensions of column and piles as follows:
   - With reference to Figure 8.30a, the width of strut C_1 at the upper node, $w^u_{C_1}$, is

     $$w^u_{C_1} = a_1 \cos\theta_1 + c_1 \sin\theta_1 = 55.5 \times \cos 46.0 + 100 \times \sin 46.0 = 110.5\,\text{mm}$$

   - With reference to Figure 8.30b, the width of strut C_1 at the lower node, $w^l_{C_1}$, is

     $$w^l_{C_1} = b_1 \sin\theta_1 + w_{T_1} \cos\theta_1 = 200 \times \sin 46.0 + 220 \times \cos 46.0 = 296.7\,\text{mm}$$

4. *STM* forces:

$$V_1 = \frac{T_1}{350} \times h_{s_1} = \frac{574.8}{350} \times 362.3 = 595.0\,\text{kN}$$

$$C_1 = \frac{T_1}{\cos\theta_1} = \frac{574.8}{\cos 46.0} = 827.5\,\text{kN}$$

5. Checking stress limits:

   Struts:

   Horizontal strut, $C_{hor,1}$
   The width of strut $C_{hor,1}$ was determined based on its design strength; hence, there is no need to carry out any further checks for this strut.

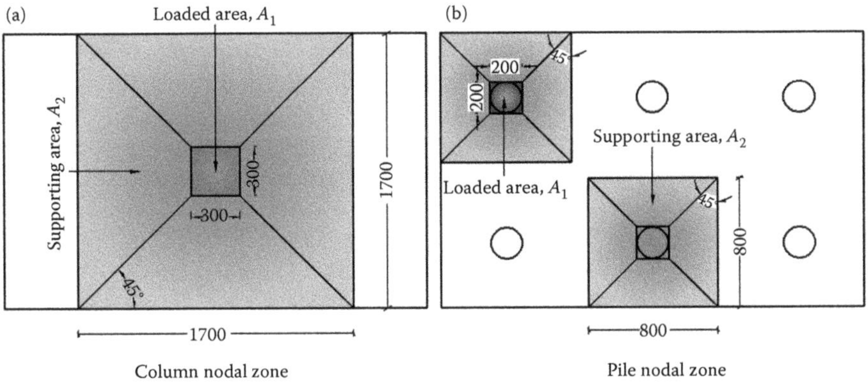

FIGURE 8.31 Example 8.4—amount of confinement for pile cap C.

Diagonal strut, C_1

With reference to Figure 8.31, the parameter α is calculated as follows:

- *At the column nodal zone (breadth at the upper node);*

$\sqrt{(A_2/A_1)} = \sqrt{\dfrac{1700\times1700}{300\times300}} = 5.7$, which gives $\alpha = 1.0$.

- *At the pile nodal zone (lower node);*

$\sqrt{(A_2/A_1)} = \sqrt{\dfrac{800\times800}{200\times200}} = 4.0$, this gives $\alpha = 0.99$.

In order to calculate the parameter β, the same procedure followed in Example 8.3 is followed here. The vertical height of strut C_1, $h_s = h_{s_1} = 362.3\,\text{mm}$. The equivalent diameter of the bearing area of the upper nodal zone connected with strut C_1, $b_b = \sqrt{(22500)(4/\pi)} = 169\,\text{mm}$. For the lower node, the bearing area is the pile area; hence, b_b is the pile diameter, $b_b = 200$ mm. Thus,

- *At the column nodal zone (upper node):*

$h/b = h_s/b_b = 362.3/169 = 2.14$, which gives $\beta = 0.33(2.14 - 1) = 0.38$.

- *At the pile nodal zone (lower node):*

$h/b = h_s/b_b = 362.3/200 = 1.81$, which gives $\beta = 0.33(1.81 - 1) = 0.27$.

Thus, the bearing strength of strut C_1 *at the upper node* is

$$f_b = 0.6(0.85f_c')(1+2\times1.0\times0.38) = 0.9f_c'$$

The bearing strength of strut C_1 *at the lower node* is

$$f_b = 0.6(0.85f_c')(1+2\times0.99\times0.27) = 0.78f_c'$$

For the design strength of strut C_1, take the average strength at the upper and lower nodes:

$$f_{ce}^{C_1} = 0.5(0.9f_c' + 0.78f_c') = 0.84f_c'$$

The strength of strut C_1 at the upper node is controlled by the smaller of the strut strength, $0.84f_c'$, and the node strength, $1.7f_c'$. Hence, the strut strength at the upper node, $f_{ce}^{C_1} = 0.84f_c'$. Upon substitution in $C_{1n} = f_{ce}^{C_1} b w_{C_1}^u$, where b is the equivalent upper node width, $b = 225$ mm, $C_{1n} = 566.0$ kN.

The lower node is a $C - C - T$ node, $\beta_n = 0.8$. Thus, the limiting compressive strength of this node is 80% of the strength of the strut at this node; that is,

$$f_{ce}^n = 0.8(0.78f_c') = 0.62f_c'$$

The strength of strut C_1 at the lower node is controlled by the smaller of the strut strength, $0.84f_c'$, and the node strength, $0.62f_c'$. Hence, the strut strength at the lower node, $f_{ce}^{C_1} = 0.62f_c'$. Upon substitution in $C_{1n} = f_{ce}^{C_1} b w_{C_1}^l$, where b is the pile diameter, $b = 200$ mm for the lower node, $C_{1n} = 997.4$ kN.

Take the smaller value of C_{1n} at the two end nodes of the strut, upper and lower nodes. Then $C_{1n} = 566.0$ kN.

The significant difference between the strength of strut C_1 at the upper and lower node is a result of the significant difference between the strength of the upper nodal zone, $1.7f_c'$, and the strut strength, $0.84f_c'$. In order to reduce the gap between the two values of strut strength, the height of the upper node is re-estimated based on a strength lower than $1.7f_c'$, $\leq (566.0/997.4) \times 1.7f_c' \approx 0.96f_c'$. Upon using a strength of the upper node $= 0.8f_c'$, and redoing the calculations, the following has been obtained.

The width of the horizontal strut $C_{hor,1}$, $a_1 = 117.9$ mm, $h_{s_1} = 331.0\,\text{mm}$, and the angle $\theta_1 = 43.4°$. The width of strut C_1 at the upper node, $w_{C_1}^u = 154.4\,\text{mm}$ and at the lower node, $w_{C_1}^l = 297.3\,\text{mm}$. The pile force $V_1 = 543.6$ kN, the force in the diagonal strut, $C_1 = 791.1$ kN.

- *At the column nodal zone (upper node):*

$h/b = h_s/b_b = 331.0/169 = 1.96$, which gives $\beta = 0.32$.

- *At the pile nodal zone (lower node):*

$h/b = h_s/b_b = 331.0/200 = 1.66$, which gives $\beta = 0.22$.

Thus, the bearing strength of strut C_1 *at the upper node* is

$$f_b = 0.6(0.85f_c')(1 + 2 \times 1.0 \times 0.32) = 0.84f_c'$$

Thus, the bearing strength of strut C_1 *at the lower node* is

$$f_b = 0.6(0.85f_c')(1 + 2 \times 0.78 \times 0.22) = 0.69f_c'$$

For the design strength of strut C_1, take the average strength at the upper and lower nodes;

$$f_{ce}^{C_1} = 0.77f_c'$$

The strength of strut C_1 at the upper node is thus $f_{ce}^{C_1} = 0.77f_c'$. Upon substitution in $C_{1n} = f_{ce}^{C_1} b w_{C_1}^u$, $C_{1n} = 724.9$ kN.

The limiting compressive strength of the lower node is

$$f_{ce}^{n} = 0.8(0.69f_c') = 0.55f_c'$$

The strength of strut C_1 at the lower node is $0.55f_c'$, giving $C_{1n} = 886.3$ kN. *Take the smaller value of C_{1n} at the two end nodes of the strut, upper and lower nodes.* Then $C_{1n} = 724.9$ kN.

Nodes:
Since in Example 8.3 the critical node is the lower node, and as estimated before, $f_{ce}^{n} = 0.55f_c'$, based on this, the nominal value of the pile reaction is

$$V_{1n} = 0.55f_c' \times 200 \times 200 = 596.2\,\text{kN}$$

Since the reinforcement of the tie, T_1, is arranged in multiple layers, the height of the node w_{T_1} should be checked against a maximum value as illustrated next.

$$w_{T_1,\max} = \frac{T_1}{f_{ce}^{n} b} = \frac{T_1}{0.55f_c' \times 200} = 192.8\,\text{mm}$$

Since a value of $w_{T_1} = 220\,\text{mm}$ ($> w_{T_1,\max} = 192.8\,\text{mm}$) was used in estimating the width of strut C_1 at the lower node, a new value of the strut width should be calculated based on $w_{T_1,\max}$. This value is $w_{C_1}^{l} = 277.5\,\text{mm}$, leading to a new nominal value of the strut strength at this node $C_{1n} = 885.7$ kN. This last value of the nominal strength of the strut is still greater than the value at the upper node, and therefore the solution should proceed with no change for this strut.

The final results of the short direction are given in Table 8.3.

From the results in the table, the size of the upper nodal zone and hence the forces should be reduced to 92%. This finally gives a value of $V_{1n} = 500.1$ kN.

Analysis of the diagonal direction

As performed for the short direction

1. The internal lever arm, h_{s_2}:

   As shown in Figure 8.32, the term w_{T_2} (height of the lower node)

$$w_{T_2} = 2 \times (600 - 400 - 100) = 200\,\text{mm}$$

TABLE 8.3
Example 8.4—Summary of Calculation Results of Short Direction of Pile Cap C

Model Label	Member	Actual Force, kN	Maximum Capacity, kN	Satisfaction
Struts	$C_{hor,1}$	574.8	574.8	yes
	C_1	791.1	724.9	No, 92%
Lower node, $C-C-T$	V_1	543.6	596.2	yes

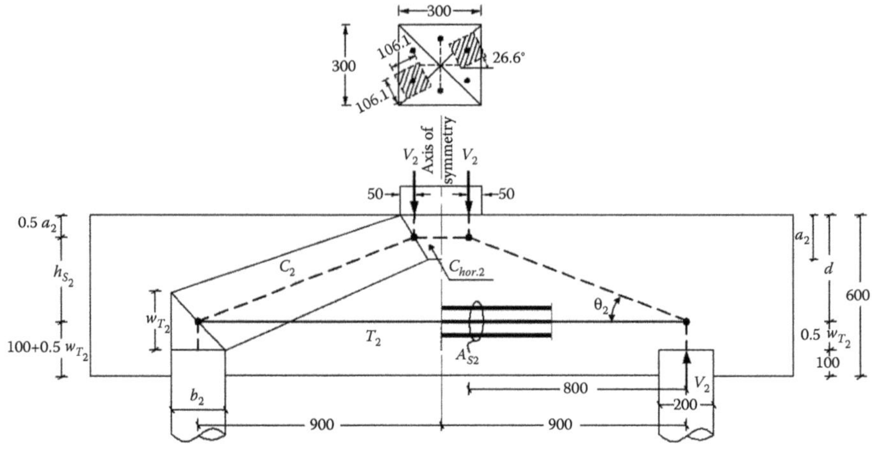

FIGURE 8.32 Example 8.4—*STM* for pile cap C in the long direction—elevation.

From the *STM* top view of pile cap C, Figure 8.28,

$$T_2 = T_{C2}\sin 26.6 + T_{C3}\cos 26.6 = 1100\times 479\times \sin 26.6 + 2100\times 479$$
$$\times \cos 26.6 = 1135.4\,\text{kN}$$

With reference to Figure 8.32, the breadth of the horizontal strut $C_{hor,2}$, b, is 106.1 mm. Then the width of the strut, a_2, can be computed as follows.

$T_2 = 1135.4kN = C_{hor,2} = 1.7 f_c' b a_2 = 1.7\times 27.1\times 106.1\times a_2$, which leads to a_2 = 232.3 mm.

$$h_{s2} = 400 - 0.5\times 232.3 = 283.9\,\text{mm}$$

2. Angle of inclined strut, θ_2
   With reference to Figures 8.28 and 8.32,

$$\theta_2 = \tan^{-1} 283.9/\sqrt{800^2 + 400^2} = 17.61°$$

3. Widths of struts
   - With reference to Figure 8.33a, the width of strut C_2 at the upper node, w_{C2}^u, is

$$w_{C_2}^u = a_2\cos\theta_2 + c_2\sin\theta_2 = 232.3\times\cos 17.61 + 106.1\times\sin 17.61$$
$$= 253.5\,\text{mm}$$

   - With reference to Figure 8.33b, the width of strut C_2 at the lower node, w_{C2}^l, is

$$w_{C_2}^l = b_2\sin\theta_2 + w_{T_2}\cos\theta_2 = 200\times\sin 17.61 + 200\times\cos 17.61$$
$$= 251.1\,\text{mm}$$

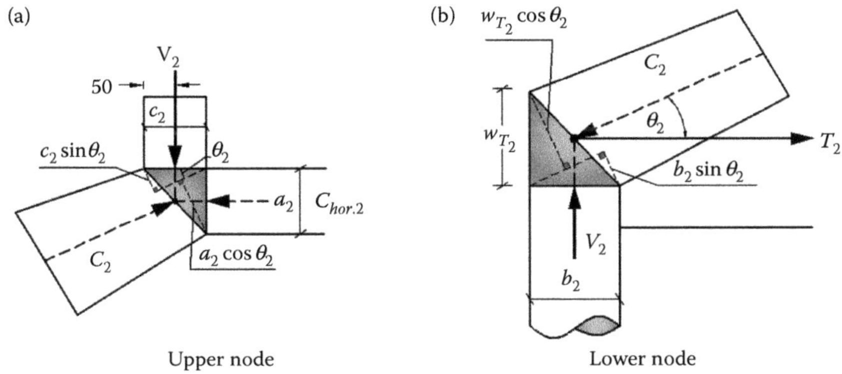

FIGURE 8.33 Example 8.4—nodes of 3D *STM* of pile cap C in the long direction.

4. *STM* forces:

$$V_2 = T_2 \times \frac{h_{s2}}{\sqrt{800^2 + 400^2}} = 1135.4 \times \frac{283.9}{894.43} = 360.4\,\text{kN}$$

$$C_2 = \frac{T_2}{\cos\theta_2} = \frac{1135.4}{\cos 17.61} = 1191.2\,\text{kN}$$

5. Checking stress limits:

Struts:

Horizontal strut, $C_{hor,2}$
The width of strut $C_{hor,2}$ was determined based on its designed strength; hence, there is no need to carry out any further checks for this strut.

Diagonal strut, C_2
With reference to Figure 8.31 and as explained for the case of short direction, $\alpha = 1.0$ for the upper node and $\alpha = 0.99$ for the lower node. In order to calculate the parameter β, the vertical height of strut C_2, $h_s = h_{s_2} = 283.9\,\text{mm}$. The equivalent diameter of the bearing area of the upper nodal zone connected with strut C_2, $b_b = \sqrt{(11250)(4/\pi)} = 119.7\,\text{mm}$. For the lower node, the bearing area is the pile area; hence, b_b is the pile diameter, $b_b = 200$ mm. Thus,

- *At the column nodal zone (upper node):*
  $h/b = h_s/b_b = 283.9/119.7 = 2.37$, this gives $\beta = 0.33(2.37 - 1) = 0.45$.
- *At the pile nodal zone (lower node):*
  $h/b = h_s/b_b = 283.9/200 = 1.42$, this gives $\beta = 0.33(1.42 - 1) = 0.14$.

Thus, the bearing strength of strut C_2 *at the upper node* is

$$f_b = 0.6(0.85 f_c')(1 + 2 \times 1.0 \times 0.45) = 0.97 f_c'$$

The bearing strength of strut C_2 *at the lower node* is

$$f_b = 0.6(0.85f_c')(1+2\times0.99\times0.14) = 0.65f_c'$$

For the design strength of strut C_2, *take the average strength at the upper and lower nodes*:

$$f_{ce}^{C_2} = 0.5(0.97f_c' + 0.65f_c') = 0.81f_c'$$

The strength of strut C_2 *at the upper node is controlled by the smaller of the strut strength,* $0.81f_c'$, *and the node strength,* $1.7f_c'$. *Hence, the strut strength at the upper node,* $f_{ce}^{C_2} = 0.81f_c'$. Upon substitution in $C_{2n} = f_{ce}^{C_2}bw_{C_2}^{u}$, where b is the equivalent upper node width, $b = 106.1$ mm, $C_{2n} = 590.4$ kN.

The lower node is a $C-T-T$ node, $\beta_n = 0.6$. Thus, the limiting compressive strength of this node is 60% of the strength of the strut at this node; that is,

$$f_{ce}^{n} = 0.6(0.65f_c') = 0.39f_c'$$

The strength of strut C_2 *at the lower node is controlled by the smaller of the strut strength,* $0.39f_c'$, *and the node strength,* $0.81f_c'$. *Hence, the strut strength at the lower node,* $f_{ce}^{C_2} = 0.39f_c'$. Upon substitution in $C_{2n} = f_{ce}^{C_2}bw_{C_2l}$, where b is the pile diameter, $b = 200$ mm for the lower node, $C_{2n} = 530.8$ kN.

Take the smaller value of C_{2n} *at the two end nodes of the strut, upper and lower nodes*. Then $C_{2n} = 530.8$ kN.

Nodes:

There is no need to carry out any checks for the upper node since the strength of the node is greater than or equal to the strength of the surrounding struts. The critical node is the lower node, and as estimated before, $f_{ce}^{n} = 0.39f_c'$. Based on this, the nominal value of the pile reaction is

$$V_{2n} = 0.39f_c'\times200\times200 = 422.8\,\text{kN}$$

Since the reinforcement of the tie, T_2, is arranged in multiple layers, the height of the node w_{T_2} should be checked against a maximum value as illustrated next.

$$w_{T_2,\max} = \frac{T_2}{f_{ce}^{n}b} = \frac{T_2}{0.39f_c'\times200} = 537.1\,\text{mm}$$

Since the assumed value of $w_{T_2} = 200\,\text{mm}$, employed in the estimate of the width of C_2 at the lower node, is less that its maximum limit, $w_{T_2,\max}$, the assessment of the nominal strength of C_2 is correct and the solution should proceed.

The final results of the long direction are given in Table 8.4.

From the results in the table, the size of the upper nodal zone and hence the forces should be reduced to 45%. This finally gives a value of $V_{2n} = 162.2$ kN.

Upon comparing the strength of strut C_2 at the two ends, there is a noticeable difference between the two values. This difference can be reduced by another

TABLE 8.4
Example 8.4—Summary of Calculation Results of Long Direction of Pile Cap C

Model Label	Member	Actual Force, kN	Max. Capacity, kN	Satisfaction
Struts	$C_{hor,2}$	1135.4	1135.4	yes
	C_2	1191.2	530.8	No, 45%
Lower node, $C-T-T$	V_2	360.4	422.8	yes

cycle of calculations, which starts with reducing the force in tie T_2. Upon assuming $T_2 = C_{hor,2} = 1000.0$ kN, then

$a_2 = 204.6$ mm, $h_{s_2} = 297.7$ mm; $\theta_2 = 18.41°$; $w^u_{C_2} = 227.6$ mm; $w^l_{C_2} = 252.9$ mm; $V_2 = 332.8$ kN; $C_2 = 1053.9$ kN; $C_{2n} = 548.3$ kN $= 0.52C_2$ (the weakest link).

Hence, $V_{2n} = 173.1$ kN

Finally, the nominal strength of pile cap C from *STM*,

$$P_{STM} = 2V_1 + 4V_2 = 2 \times 500.1 + 4 \times 173.1 = 1692.8\,\text{kN}$$

$$P_{STM}/P_{EXP} = 1692.8/2892 = 59\%$$

REFERENCES

Adebar, P., Kuchma, D., and Collins, M. P., Strut-and-tie models for the design of pile caps: An experimental study, *ACI Structural Journal*, 87(1), 1990, 81–92.

Adebar, P., and Zhou, Z., Bearing strength of compressive struts confined by plain concrete, *ACI Structural Journal*, 90(5), 1993, 534–541.

Adebar, P., and Zhou, Z., Design of deep pile caps by strut-and-,tthat is,models, *ACI Structural Journal*, 93(4), 1996, 1–12.

ACI 318-14, *Building Code Requirements for Structural Concrete and Commentary*, Detroit: American Concrete Institute, USA, 2014, 519 pp.

Beredugo, Y. O., An experimental study of the load distribution in pile groups in sand, *Canadian Geotechnical Journal*, 3(3), 1967, 145–166.

EN 1992-1-1:2004, Eurocode 2: Design of Concrete Structures, Part 1-1: General rules and rules for buildings, December 2004.

Engström, B., *Design and Analysis of Deep Beams, Plates and Other Discontinuity Regions*, Department of Structural Engineering, Chalmers University of Technology, Göteborg, Sweden, 2011.

Fédération Internationale Du BétonFIB; Fip, Practical Design of Structural Concrete, fédération internationale de la précontrainte, September 1999, 113 pp.

Fédération Internationale Du BétonFIB, Design examples for the 1996 FIP recommendations - Practical design of structural concrete, Technical Rep., International Federation for Structural Concrete fib, London, 2002.

Ghali, M. K., *Effect of Pile Cap Flexural Rigidity on the Behaviour of Bridge Foundations*, IABSE Colloquium: "Foundations for Major Bridges: Design and Construction", New Delhi, 1999, Vol. 80, 241–246 pp.

Hawkins, N. M., Bearing strength of concrete loaded through rigid plates, *Magazine of Concrete Research (London)*, 20(62), 1968, 31–40.

Park, R., and Paulay, T., *Reinforced Concrete Structures*, John Wiley & Sons, New York, 1975.

Schlaich, J., Schäfer, K., and Jennewein, M., Toward a consistent design of structural concrete, *Journal of the Prestressed Concrete Institute*, 32(3), 1987, 74–150.

Wang, C. Z., Guo, Z. H., and Zhang, X. Q., Experimental investigation of biaxial and triaxial compressive concrete strength, *ACI Materials Journal*, March-April, 1987, 92–100.

Index

D

E

T

U

V

W

Y

Z

For Product Safety Concerns and Information please contact our EU
representative GPSR@taylorandfrancis.com
Taylor & Francis Verlag GmbH, Kaufingerstraße 24, 80331 München, Germany

www.ingramcontent.com/pod-product-compliance
Ingram Content Group UK Ltd.
Pitfield, Milton Keynes, MK11 3LW, UK
UKHW020158200926
475702UK00004B/519
* 9 7 8 1 4 9 8 7 8 3 8 4 2 *